ACCEPTABLE EVIDENCE

ENVIRONMENTAL ETHICS AND SCIENCE POLICY SERIES

Kristin Shrader-Frechette, GENERAL EDITOR

Acceptable Evidence:
Science and Values in Risk Management
Edited by Deborah G. Mayo and Rachelle D. Hollander

Experts in Uncertainty:
Expert Opinion and Subjective Probability in Science
Roger M. Cooke

ACCEPTABLE EVIDENCE

Science and Values in Risk Management

Edited by
Deborah G. Mayo
Rachelle D. Hollander

New York   Oxford
OXFORD UNIVERSITY PRESS

Oxford University Press

Oxford   New York   Toronto
Delhi   Bombay   Calcutta   Madras   Karachi
Kuala Lumpur   Singapore   Hong Kong   Tokyo
Nairobi   Dar es Salaam   Cape Town
Melbourne   Auckland   Madrid

and associated companies in
Berlin   Ibadan

Published by Oxford University Press, Inc.
200 Madison Avenue, New York, New York 10016

First issued as an Oxford University Press paperback, 1994.

Oxford is a registered trademark of Oxford University Press

Library of Congress Cataloging-In-Publication Data
Acceptable evidence: science and values in risk management / edited
by Deborah G. Mayo, Rachelle D. Hollander.
p. cm. (Environmental ethics and science policy)
Includes bibliographical references.
ISBN 0-19-506372-4
ISBN 0-19-508929-4 (pbk)
1. Technology—Risk assessment.   2. Risk management.
I. Mayo, Deborah G.   II. Hollander, Rachelle D.   III. Series.
T174.5.A23   1991   363.1—dc20   90-43855

9 8 7 6 5 4 3 2 1

Printed in the United States of America
on acid-free paper

In memory of Remy A. Johnston (1966–1989),
a philosophy student and friend

                         R. HOLLANDER

To George W. Chatfield
For all his support and encouragement

                         D. MAYO

Acknowledgments

This volume grows out of a symposium entitled "Ethics, Evidence, and the Management of Technological Hazards," which the editors organized for the May 1986 annual meeting of the American Association for the Advancement of Science. Five contributors to this volume—E. W. Colglazier, R. Kasperson, D. Mayo, V. Miké, and E. Silbergeld—were participants in that session. Other participants were Jack Campbell, June Fessenden MacDonald, Lesley Russell, Robert Moolenaar, and Sheldon Samuels. We benefited a great deal from the ideas that came out of that forum: they gave us the impetus to begin discussing these issues with one another and with other people over the next few years, which led to this book.

The AAAS symposium—the first time the editors talked together about these ideas—proved to be enlightening. It became clear that participants were becoming aware of some of the problems, oversimplifications, and misconceptions in the ways the relationships between risk assessment and values had been construed in earlier interdisciplinary work. Enough threads were available to begin weaving a new approach to a whole variety of risk assessment issues: one that is not victimized by the view that scientific rationality requires value-free methods or the view that no choice about risk management is better than another.

The editors owe special thanks to Kristin Shrader-Frechette, general editor of the Environmental Ethics and Science Policy Series, for encouraging us throughout this project and for giving many extremely helpful suggestions at numerous stages. To Oxford editors, Cynthia Read and Irene Pavitt, we are grateful for cheerful editorial assistance, flexibility, and excellent advice. The careful comments of Oxford's anonymous reviewer contributed to the clarity and coherence of this volume. For all the work that went into photographing our cover, we are thankful to Mayo Studios and to photographer Karineh Gurjian.

Finally, we wish to express our appreciation to the contributors, for their patience and for working with us to synthesize research from a wide variety of fields into a unified work on acceptable evidence in risk assessment. Our goals for this volume required contributions from a multiplicity of fields, making it all the more important to ensure that authors used terms consistently and took special care to consider the questions that may arise about their implications. Many disputes in interdisciplinary risk research stem from equivocal uses of key terms, resulting in people talking past one another. Without the special willingness of the contributors to work along with us, it would have been impossible to overcome this obstacle. Whether we have done so adequately or well, will be up to the reader to decide.

Contents

Introduction

Scientific and technical experts play a continually growing role in societal processes of hazard management, and controversies over scientific evidence and method are increasingly visible in these processes. However, the manner in which issues of evidence interrelate with issues of risk policy often goes unattended. Discussions on science and values in risk management have largely focused on the entry of values in judging the importance of risks and protection from risks, that is, issues of acceptable risk. This volume, instead, concentrates on the entry of values in collecting, interpreting, communicating, and evaluating the evidence of risks, that is, issues of the acceptability of evidence of risk. Questions about the acceptability of evidence of risk are logically more fundamental than are questions about acceptable risk, in the sense that knowing how to answer the latter questions depends on knowing how to answer the former.

By focusing on acceptable evidence, this volume attempts to avoid two extremes that we believe have been barriers to progress in this area. The first extreme supposes that issues of evidence of risk can and ought to be separated from the individual and social values that necessarily enter in reaching policies about risk (i.e., risk management). This view assumes that evidence of risk is largely a matter of objective scientific data, which may be captured in standard quantitative measures of risk. The second extreme is found in varying degrees in much of the interdisciplinary work relating science and values in risk assessment. Here, the recognition that values (methodological, political, and others) may enter into every stage of risk management—even at the level of establishing evidence of risk—has often been taken to imply that there is little objective or empirical basis on which to criticize risk assessments.

This volume denies both these extremes but argues instead for a more constructive conclusion: that understanding the interrelations of scientific with value issues enables a critical scrutiny of risk assessments. It allows the contributors to begin to address such questions as: How much disagreement about risk assessments is due to different assumptions about the causes and nature of harm, to the application of different methods or models, or to interests or biases—conscious or unconscious? To help disentangle sources of uncertainty and conflicts about risk assessments, the chapters in this book provide important tools for distinguishing issues of scientific knowledge and evidence from issues of personal or social values.

Although we can discuss only a few of these issues, we intend this book as the start of work in that direction and in the direction of developing some agreement on the criteria for judging the acceptability of evidence on which policy decisions are based. Understanding the values that infuse both the development of evidence and the uses to which evidence is put in individual, group, and societal decision making can help us act more responsibly and make better decisions.

Our goals require that both empirical and normative material be included. The empirical work elucidates current social and political mechanisms in risk management, how judgments of value and evidence actually enter, and the problems that result from mistakes in relating values and evidence. In analyzing environmental, medical, and occupational cases, the contributors describe scientific and statistical tools that have been used, misused, or overlooked. The normative work includes theoretical and philosophical analyses of scientific and statistical models that are used in risk management. It also includes analyses of assumptions underlying various views about risk assessment, like that of the "fact–value" distinction. Both empirical and normative work are required for more effective frameworks for evaluating the merit of these models and views and of the risk assessments based on them.

This book is divided into three parts. Part I considers the entry of values in deciding how to respond to evidence of harms and benefits associated with risks. Because the evidence does not come divorced from its acceptability, the only way to understand perceptions and evaluations of evidence is to take these values into account. In contrast with those who would regard the value ladenness of evidence as leading to relativism or irrationality, the chapters in Part I show that individual and social judgments about the acceptability of risks can be analyzed as rigorously as can quantitative judgments of morbidity or mortality. The thesis of Part II is that sophistication about uncertainty requires acknowledging that values color its presence in both risk assessment and risk management. Although resolving questions of uncertainty is thus intertwined with policy values, the contributors show that it is a mistake to suppose it is therefore a question to be answered by policymakers alone. Without a clear understanding of the uncertainties, scientists, engineers, policymakers, and citizens will not be able to behave responsibly. Part III demonstrates the central role that the philosophy of science plays in understanding and evaluating the merit of risk evidence. The contributions in this section point us toward better models for decision making, causal inquiry, and statistical reasoning. They lay the groundwork for an approach to risk assessment that is not victimized by the view that scientific rationality requires value-free methods or the view that all choices concerning risk management are equally good.

Overcoming false dichotomizations of facts and values allows us to return values to the realm of reasoned deliberation. It enables us to consider how values enter into making evidential claims and whether these are the values we wish to use. This is important just now, when human choices involving science and technology have such great influence for the present, the immediate future, and the far future.

I

PERCEIVING
AND COMMUNICATING
RISK EVIDENCE

In Part I, the contributors consider how values and perceptions, both cultural and individual, color the identification, interpretation, evaluation, and management of risks. Ostensibly, rational perceptions of risks and their importance or weight should and would be functions of the best evidence of the harms associated with the technologies, substances, and practices in question. For the past twenty-five or so years, people thought that the best evidence of such harms would be captured by such measures as mortality and morbidity rates. The relatively new field of quantitative risk assessment then adapted these measures for various technologies. The scientific culture in which risk assessment evolved, in turn, supported this approach, seeking objectivity and rationality in quantitative measurement. However, despite the increased sophistication of quantitative analyses of risk, they fail to match the public perceptions of risk and the weights placed on risk and do not foster adequate communication about risk. Reports of various risk numbers also fail to predict public reactions, giving rise to serious difficulties in the ability to manage such risks.

The questions now are these: If perceptions and the weighing of risks do not correspond simply to the quantitative evidence of these kinds of harm, does this show that the harmfulness of risks has little to do with the way that people perceive and judge risks? Does the whole subject of perceiving, communicating, and weighing risks lack logic or rationality? Many have suggested as much. Mary Douglas and Aaron Wildavsky, cited by several of the contributors to this volume, claim that our risk judgments have little, if anything, to do with actual incidences of harms. If this is so, there is little hope for rationally resolving problems in risk perception and communication: Appealing to risk evidence will have little force in a rational adjudication among competing views.

But our contributors deny this conclusion. Their research questions the premise that the risk quantities typically used capture the best evidence regarding risks. These quantities admit no parameters in which to factor the various individual and cultural values that color our evaluation of evidence about risks. The very existence of certain risks thus will be hidden to professional assessors who rely on unsophisticated techniques that do not include ideological and psychological influences. Relying on them has created many of the obstacles we now face in understanding and communicating risk evidence.

The contributions to this part are in an area of research about values, perceptions, and risks that can perhaps be said to have begun in response to the Rasmussen report, a report from the U.S. Atomic Energy and the U.S. Nuclear Regulatory commissions, entitled *An Assessment of Accident Risks in U.S. Commercial Nuclear Power Plants,* resulting from a study directed by Norman Rasmussen and published in 1974. It attempted to quantify those risks associated with nuclear power reactor operations. Rasmussen measured risks by injury, morbidity, and mortality rates and compared them with those from natural hazards and nonnuclear technological accidents. William Lowrance's book *Of Acceptable Risk,* published in 1976, was more sophisticated, stating that such measures could never tell us what makes risks acceptable. Determining the acceptability of risks (e.g., against what risks should the public be protected?) was a social judgment. This and other works increased the recognition that social values enter into judgments of the acceptability of risk.

What has remained far less well understood is the manner in which value judgments also influence the characterization of the evidence of risk. Many still suppose that quantitative measures, free of "extrascientific" values, provide adequate evidence of risk. It is here that the message of this part of our book enters.

Where Lowrance's *Of Acceptable Risk* stressed the entry of values in deciding how to respond to risky substances and practices, this part of our book stresses the use of values in deciding how to respond to the evidence of harm and benefit associated with these substances and practices. It concerns the ways in which we value evidence about risks; that is, it concerns the issue of acceptable evidence. But because evidence does not come divorced from acceptability of evidence, the only way to understand perceptions and evaluation of evidence is to take these values into account.

In contrast with those who regard value-laden evidence as leading to relativism or irrationality, the conclusion of this part is that questions about the acceptability of evidence have answers. The chapters by Roger and Jeanne Kasperson and Sheila Sen Jasanoff review the recent research and thinking about the ways in which characteristics of societies or cultures help explain the process by which some risks are acknowledged and others remain unnoticed. Paul Slovic describes the psychological and other factors that influence perceptions of risks, in order to show how these factors may be introduced into more sophisticated risk calculations. Vincent Covello, Peter Sandman, and Paul Slovic provide guidelines for communicating risk evidence that do not preempt personal or social decisions as to their acceptability.

In effect, what these researchers have discovered is that individual and social judgments about the acceptability of risks can be analyzed as rigorously as can quantitative judgments of morbidity or mortality. Indeed, they must be analyzed in this way if risk is to be adequately understood and evaluated. These individual and social judgments obey a certain "logic," albeit not the logic of early risk assessors at the time of the Rasmussen report. The contributions in this part help us understand this logic.

SOCIETAL VALUES IN RISK EVALUATION

Chapters 1 and 2 focus on the roles of social values in the identification, interpretation, and evaluation of risk. In "Hidden Hazards," Roger and Jeanne Kasperson show why certain hazards take center stage and others, which may cause no less injury, disease, or death, take a back seat; why the media are replete with daily accounts of some hazards from nature and technology, while others pass unnoticed. If only that evidence that is addressed in formal risk assessments of the incidence or prevalence of mortality or morbidity is considered, this differential awareness cannot be explained.

Many hazards go unnoticed, however, not because they are associated with low estimates of risk to health or environment but, rather, because societal values have led to the acceptance of the injuries and diseases that they are believed likely to cause. Elusive hazards, like ozone depletion or global warming, go unattended in part because of the difficulty of detecting them and in part because of the political challenge their management requires. Ideological priorities may account for the acceptance of other hazards; for instance, people who believe strongly in a right to own handguns state: Guns don't kill people; people do. Hazards to marginal groups like AIDS (acquired immune deficiency syndrome) victims may be downgraded. Other hazards, perhaps from nuclear power, are amplified because they are strange, invisible, and not subject to personal control. Experts assessing risk overlook this phenomenon, and the hazard of amplification goes unremarked. Some technologies, like genetic engineering and communication technologies, may have subtle and cumulative effects that our social structures and political processes are not equipped to identify and discuss, and so their impact remains hidden. Finally, many technologies fall into more than one of these categories, thereby making it even harder to secure a place for their consideration on the public agenda. Given these varieties of hidden hazards and cultural responses, it is likely that their assessment and managment will continue to perplex us, but attending to these different dimensions may mitigate these efforts.

Sheila Sen Jasanoff compares making decisions about risks in two societies, the United States and Great Britain, in "Acceptable Evidence in a Pluralistic Society." She attempts to answer the question whether the quality or character of scientific evidence determines policy outcomes, or is scientific evidence itself shaped by the dynamics of regulation? Her answer, at least in some cases, is the latter.

Jasanoff addresses this question by examining and comparing criteria used in British and U.S. regulatory processes for judging the acceptability of evidence for regulating lead. In Britain, uncertainty was granted and a decision was based on the perception that for certain groups, the current standard was unacceptably close to the margin of safety. The authority and legitimacy of the Royal Commission's decision went unchallenged. In the United States, the focus is on making decisions in which science is expected to provide political authority and legitimacy; hence the concentration on three techniques designed to improve the objectivity and reliability of evidence used in policy decisions. These techniques are *cost–benefit analysis, risk assessment* (especially in regulating carcinogens), and *expert elicitation.* In each case, Jasanoff shows how the pressure to adopt a formal analytical technique grew out of a desire to move policy decisions beyond the reach of political manipulation. She indicated how these attempts nevertheless failed to ensure the political neutrality of scientific and technical assessments and how each analytical technique eventually fell prey to social criticisms. For instance, in the risk assessment of formaldehyde, the shape of the dose response curve led to disputes about whether the data should be interpreted using the standard extrapolation model: Some experts interpreted its use as a political, proregulatory decision. The dynamic of science ensures its accessibility to political challenge.

PSYCHOLOGICAL VALUES IN RISK PERCEPTION AND COMMUNICATION

Chapters 3 and 4 are concerned with psychological rather than sociological phenomena. Chapter 3 characterizes our risk perceptions—those elements that influence individual judgments about the severity of risks—and Chapter 4 accepts the legitimacy of individual and cultural concerns and public participation in decisions about the acceptability of risks. Its task is to reveal how values enter into presenting risk information and to provide advice and guidance for persons wishing to discuss risks responsibly and effectively.

Paul Slovic's chapter, "Beyond Numbers: A Broader Perspective on Risk Perception and Risk Communication," examines psychometric studies that show that the concept of risk means different things to different people. Experts' judgments of risk correlate closely with technical estimates of annual fatalities, whereas laypersons' relate more to other characteristics of risks. Risk perceptions and risk-taking behaviors appear to be determined not only by accident probabilities, annual mortality rates, or mean losses of life expectancy but also by numerous other characteristics of risks such as uncertainty, controllability, catastrophic potential, equity, and threat to future generations. Disagreement between experts and others may often be due not to irrationality or even lack of information on the public's part but, rather, to the public's operating with a conception of risk that reflects legitimate social values typically omitted from standard risk statistics.

Slovic recognizes the value judgments involved but does not take this to indicate that risk eludes systematic treatment. Rather, he shows these values as amenable to a quantitative, if more complex, analysis. Using factor analysis, many values can be condensed into a small set of factors. Factor 1, *dread risk* is defined as perceived lack of control, dread, catastrophic potential, fatal consequences, and the inequitable distribution of risks and benefits. Factor 2, *unknown risk* is defined as those hazards judged unobservable, unknown, new, and delayed in manifesting harm. This yields a new model of risk, which, by quantifying key values, affords a better way to predict and understand public responses and attitudes and to communicate them to experts and citizens. It may help policymakers and elites make policies that do not rely on how they wish citizens would behave but on how they are likely to behave.

Citizens' demands to know about the risks to which they are exposed, and to hold officials and institutions accountable for these risks, are not likely to diminish. Because people evaluate risks in the ways described in preceding chapters, attempts to withhold information or present it in a light favorable to certain interests over others are not likely to be persuasive.

In this part's final chapter, "Guidelines for Communicating Information About Chemical Risks Effectively and Responsibly," Vincent T. Covello, Peter M. Sandman, and Paul Slovic offer advice to people who need to communicate information about risk to individuals and public groups. They identify ways to communicate about risk without presuming values that the audience may not share. The chapter lists seven rules for discussing risk in a way that will be perceived as credible and trustworthy, noting that an important part of these communications is information about actions being taken to ameliorate such risks. The authors state that these rules are often violated and ask readers to consider why. The second part of the chapter explains risk-related numbers and statistics in ways that people will understand and consider fair and relevant. The third section provides guidance for the use of risk comparisons. Risk comparisons can put risks into perspective when they are part of a sound long-term program to help people evaluate the acceptability of risks. They will not work, however, when they are part of a campaign to preempt people's judgments, that is, to present information in a way that prejudges the acceptability of risk. This section describes comparisons that are helpful and others that are likely to create problems of credibility and community resentment. The final section identifies those objections likely to be made to any risk-related information, because of data uncertainties, the desire to withhold information, and the desire for zero risk. It also outlines some of the ways to acknowledge and overcome these difficulties.

Each of these chapters provides information about and insights into those values and perceptions that are part of the notion of risk. Each of them shows that an analysis of individual and social judgments about risks provides a basis for understanding the acceptability of risk evidence. All of them form the basis for ways in which risks can be better evaluated and managed.

1

Hidden Hazards

ROGER E. KASPERSON AND JEANNE X. KASPERSON

In this last decade of the twentieth century, hazards have become a part of everyday life as they have never been before. It is not that life, at least in advanced industrial societies, is more dangerous. Indeed, by any measure, the average person is safer and is likely to live longer and with greater leisure and well-being than at earlier times. Nevertheless, the external world seems replete with toxic wastes, building collapses, industrial accidents, groundwater contamination, and airplane crashes and near collisions. The newspapers and television news daily depict specific hazard events, and a parade of newly discovered or newly assessed threats—the "hazard-of-the-week" syndrome—occupies the attention of a host of congressional committees, federal regulatory agencies, and state and local governments. Seemingly any potential threat, however esoteric or remote, has its day in the sun.

How is it, then, that certain hazards pass unnoticed or unattended, growing in size until they have taken a serious toll? How is it that asbestos pervaded the American workplace and schools when its respiratory dangers had been known for decades? How is it that after years of worry about nuclear war, the threat of a "nuclear winter" did not become apparent until the 1980s? How is it that the Sahel famine of 1983 to 1984 passed unnoticed in the hazard-filled newspapers of the world press, until we could no longer ignore the specter of millions starving? How is it that America "rediscovered" poverty only with Michael Harrington's vivid account of the "other Americans" and acknowledged the accumulating hazards of chemical pesticides only with Rachel Carson's *Silent Spring*? How is it that during this century a society with a Delaney amendment and a $10 billion Superfund program has allowed smoking to become the killer of millions of Americans? And why is it that the potential long-term ecological catastrophes associated with burning coal command so much less concern than do the hazards of nuclear power?

These oversights or neglects, it might be argued, are simply the random hazards or events that elude our alerting and monitoring systems. After all, each society has its "worry beads," particular hazards that we choose to rub and polish assiduously (Kates 1985). Because our assessment and management resources are finite, certain hazards inevitably slip through and surface as surprises or outbreaks. Or are hazards simply part of the overall allocation of goods and

Table 1.1. Dimensions of Hidden Hazards

	Global Elusive	Ideological	Marginal	Amplified	Value Threatening
Abortion					X
AIDS (early stages)			X		
Asbestos in manufacturing		X	X		
Assembly lines		X	X		X
Automation/unemployment		X	X		X
Botulism				X	
Climatic change	X				
Computers					X
Deforestation	X				
Desertification	X				
Export of hazardous technologies			X		
Famines			X		
Gene-pool reduction	X				
Genetic engineering				X	X
Global poverty		X	X		
Handguns		X			
Hazardous wastes				X	
Indoor air pollution			X		
Information banks					X
Nuclear power				X	X
Nuclear winter	X				
Occupational hazards		X			
Oral contraceptives			X		X
Ozone depletion	X				
Pesticides	X			X	
Sickle-cell anemia			X		
Soil erosion	X				
Televised violence		X			X
Terrorism				X	X
Toxic material				X	
VDTs			X		X

bads in a global political economy, so that the epidemiology of hazard events is only one of many expressions of underlying social and economic forces? Alternatively, are the "hidden hazards" simply those that occur in distant times, distant places, or distant (i.e., not proximate) social groups?

The following discussion explores the phenomenon of hidden hazards, arguing that neither randomness nor any single structural hypothesis adequately explains the diversity of their causes and manifestations. Like the world of hazards of which they are part, hidden hazards have to do with both the nature of the hazards themselves and the nature of the societies and cultures in which they occur. The "hiding" of hazards is at once purposeful and unintentional, life threatening and institution sustaining, systematic and incidental. We see five distinct aspects of hidden hazards—global elusive, marginal, ideological, amplified, and value threatening—associated with differing causal agents and processes. As Table 1.1 indicates, some multiple hazards defy pigeonholing and span one or more categories.

GLOBAL ELUSIVE HAZARDS

Some hazards remain hidden from society's close scrutiny and management efforts because of the nature of the hazards themselves. Writing in 1985, Kates noted a series of significant changes in society's management agenda. Governmental concern has traditionally focused on the visible issues of automobile smog and raw sewage and only recently on the less visible problems posed by low concentrations of toxic pollutants. Corporations have historically allocated most of their hazard-management budgets to ensuring industrial safety; only now are the more subtle long-term effects to health receiving much attention. Meanwhile, a shift has begun in the temporal perspective, from the commonplace accidents of everyday existence, for which experience and actuarial data exist, to the possibility of rare but catastrophic accidents that may occur in the future, for which simulation and scenarios must substitute for experience.

Similarly, most hazards commanding our attention have been those that could be pinpointed at places and localities, where cause and effect could often be closely linked. The newer, more elusive hazards involve problems (e.g., acid rain) associated with complex regional interactions and, most troublesome, global sources, distribution, and effects (stratospheric ozone depletion, climatic change, global warming). Lengthy lag times between activities and alterations in global fluxes of energy and materials, geographical separation in source regions and the regions where the effects are manifested, complex interactions of human and physical systems, and the slowly developing accumulation of materials in environmental sinks characterize the class of elusive hazards.

It is these global elusive hazards that promise to be particularly troublesome in the years ahead. With the exception of long-term climatic change, it is only recently that such hazards have attracted substantial scientific and political attention. Soil degradation and erosion, global deforestation, desertification, and the accumulation of trace pollutants in the oceans and the stratosphere all have received only limited scientific scrutiny and yet have the capacity to undermine the long-term sustainability of the earth. For the first time in history, the unintended consequences of human actions and social processes approximate the scale of the processes of nature in terms of their potential effects on the earth as a life-support system (International Council of Scientific Unions 1986, p. 2). Despite the belated and limited attention, the warning signs of what the Brundtland report (World Commission on Environment and Development 1987) has characterized as "interlocking environmental crises" show up in (1) fossil-fuel burning linked to widespread acid deposition in North America, Europe, and China; (2) the recently discovered hole in the stratospheric ozone layer over Antarctica; (3) the extensive tropical deforestation implicated in species extinction and the flooding of lowland farms; and (4) large-scale air and water pollution threatening human health in the emerging megacities of developing countries.

The potential severity of global elusive hazards tends to be hidden not only because of their diffuseness and complexity but also because their management challenges a politically compartmentalized world, one in which there is a significant North–South conflict. These hazards typify the archetypal global-commons problems in which individual nations have incentives to allow others to undertake ameliorative initiatives and expend scarce resources. Lacking a global environmental ethic or a coherent international legal or regulatory regime, it is—despite the recent success of the Montreal Protocol (1987) concerning ozone depletion—unclear whether international initiatives to deal with such hazards will be successful.

Encouraging in this regard is the newly formed International Geosphere– Biosphere Program, a massive undertaking of the International Council of Scientific Unions, whose aim is "to describe and understand the interactive physical, chemical, and biological processes that regulate the total earth system, the unique environment that it provides for life, the changes that are occurring in that system, and the manner by which these changes are influenced by human actions" (Malone 1986, p. 8). Even here, however, although the program salutes the importance of human actions in global environmental transformations, it currently includes only a modest effort to forge the links between human causes and environmental transformations. (New initiatives, we should note, are currently under way by the Human Dimensions of Global Change Programme (Toronto) and the International Social Sciences Council, as well as by the National Science Foundation, the National Research Council, and the Social Science Research Council in the United States to begin to fill this void). Similarly, there is no provision for mobilizing the results into international policies or management programs.

The chance of these elusive hazards' becoming visible and high-priority concerns depends on whether (1) the potential changes can be expressed in terms that indicate vividly the long-term risks associated with particular human activities, (2) an international constituency of public concern over these hazards can be built and sustained, and (3) the serious equity problems between developed and developing countries can be overcome. Global elusive hazards could become the major environmental challenge of the 1990s and might well revise long-held notions of national and international security (Mathews 1989).

IDEOLOGICAL HAZARDS

Some hazards remain hidden or unattended because they lie embedded in a societal web of values and assumptions that either denigrates the consequences or deems them acceptable, elevates associated benefits, and idealizes certain notions or beliefs. Since the advent of television, violence has been an intrinsic part of news and program content, including Saturday morning cartoons aimed at children. The effort of several decades to regulate televised violence has run aground on the twin shores of the political power of the networks and the belief that violence is a part of American reality and that the protection of free speech should override the need to prevent antisocial behavior.

Handguns are a similar matter. Despite an extraordinary annual national toll from handgun-related violence and the assassination or attempted assassination of a succession of the nation's political leaders, control efforts have failed to overcome the credo that the right to bear arms is one of the most inalienable of American rights. Or to take a different case, the notion that unemployment is primarily a failure of individuals rather than economic systems accords this social hazard a status very different from that prevailing in socialist societies. In the latter, social programs are enacted to correct the structural imperfections in the economy and to ensure that the victims of these imperfections can provide for basic needs.

Ideological hazards are present no less, of course, in Marxist and socialist societies. There the growth of bureaucracy and extensive controls over the economy have produced widespread concerns over the lack of incentives for personal initiative and means for career fulfillment. The emphasis on economic growth over other societal goals has often led to widespread air and water pollution.

The resulting environmental damage has played a significant role in the independence movements in Eastern Europe and the autonomous Soviet republics, as the centralized control of information has eroded the recognition of hazards by those who bear the risks.

A striking illustration of ideological hazards at work in the American context is the little-noticed differential in prevailing standards for the protection of workers and publics. This discrepancy in protection is rooted in the idealization of private enterprise and the assumption that it should enjoy wide latitude of freedom from government interference. In the U.S. hazard-management system, the Occupational Safety and Health Administration (OSHA) is the primary regulator of occupational hazards, whereas the Environmental Protection Agency (EPA) is the primary regulator of hazards to the general public. Other agencies' regulations govern special groups of workers (for example, miners), particular classes of hazards (consumer products), or special types of hazards (such as radiation), but OSHA and EPA have the principal responsibility for most occupational and general environmental hazards.

Typically, OSHA's standards regulating the exposure of workers to hazardous substances involve limitations on the concentration of a substance in the air (typically averaged over an eight-hour shift), as respiration is the most likely pathway for the hazard and as, at least in principle, it is not too difficult to monitor exposure levels. The EPA's regulations for most substances, on the other hand, limit emissions (the rate at which the substance may be discharged into the air or water). Hazard management through emission control is natural, as the population at risk and the environment in which exposure to the substance occurs are likely to be difficult to characterize. Thus direct comparison with OSHA's regulations is no easy task.

To overcome this problem, Derr et al. (1981, 1983) examined a small but important class of substances that are emitted in sufficiently large volume and by sufficiently dispersed sources that the EPA has found it practical to establish standards for ambient concentrations. Even for these substances, the comparison with OSHA's standards has its problems: The time periods over which the concentrations are to be averaged are frequently different, and uncertainties in dose-response information make it difficult to determine whether a short exposure to high concentrations is better or worse than is a long exposure to lower concentrations.

Despite these difficulties, a comparison is possible. Figure 1.1 shows, for ten substances for which it is possible to compare limits on concentration, the current

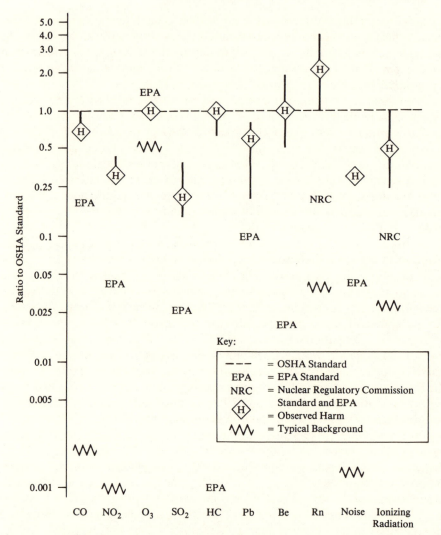

Figure 1.1 Public (EPA) and worker (OSHA) standards, levels of observed harm, and normal background levels for ten hazards. (Derr et al. 1981, p. 14)

EPA and OSHA standards information on medical effects and background level, and the ratio of the environmental standard to the occupational standard. Except for ozone and carbon monoxide, for which there are only short-term standards and for which the primary health concerns are short-term stresses, the comparison is based on the limits that the standard imposes on the cumulative annual dose. The figure also shows two quantities useful in characterizing each hazard: the natural background level for the hazard and the lowest level of concentration at which adverse human health effects have been observed.

It is apparent that regulators afford a greater measure of protection to members of the general public than to workers. Although one might reasonably expect that the standards for both groups would be set at or below the point at which measured harm occurs, this is not the case. General environmental standards are set below levels of measured harm, but workplace standards tend to be set above that level. In fact, the OSHA's standards are set at about the level of observed harm in four cases, at two times higher in two cases, and at four times higher in three cases. In only one case, radon, is the standard below the level of observed harm, and given recent advisories by the surgeon general and the Environmental Protection Agency (Hilts 1988), that is a wise decision. The standards generally reflect a belief that U.S. workers may be appropriately exposed to hazards between ten and one hundred times the level considered to be unsafe for the U.S. public.

The double standard, of course, reflects a long-term societal reluctance to intervene in the operation of private industry in the United States. It is not surprising that OSHA was an early target of the Reagan administration's efforts to weaken regulatory efforts and enforcement capabilities (Simon 1983). Not infrequently, workplace hazards have grown into serious occupational health problems as their presence has remained obscure because of corporate concerns over the release of "proprietary information."

Despite its affluence and leadership in health research, the United States does not possess the most stringent standards for the protection of workers. Table 1.2 shows comparative occupational standards for some twelve airborne toxic substances in thirteen industrialized countries. There is a general East–West dichotomy, with the Soviet Union and the East European countries exhibiting the greatest stringency in standards (ICPS n.d.) and the United States the greatest permissiveness. Although most of the differences are within an order of magnitude, factors of 20 or more appear for heptachlor, lead, and parathion. A transnational analysis of some 169 workplace standards found that only 19 were most stringent in the United States, whereas some 147 (87 percent) were most stringent in the Soviet Union (Kates 1978; Winell 1975). The reasons for the differences between U.S. and U.S.S.R. standards are historical but also clearly ideological. Equally notable, it should be added, is the tendency in Marxist societies to set rigorous workplace standards (which are often not met in practice) while permitting widespread degradation of the general environment (Jancar 1987).

Ideological mechanisms exist, of course, to explain and justify the tolerance of high workplace risks. Workers, it is argued, are accustomed to dealing with hazards, are generally well informed and able to protect themselves, and are, in any case, compensated in workers' wages. A lengthy refutation of these myths is beyond the scope of this discussion. Suffice it to note that the available evidence shows that

- Few workers are given information about risks at the time of employment and only unevenly after employment (Melville 1984; Nelkin and Brown 1984), although OSHA's (OSHA 1983) Hazard Communication standard has upgraded past practices.
- Few workers are genuinely free to accept or reject a job, or to move to a different job, on the basis of risk information (Melville 1984).
- Some workers in the major unionized manufacturing sectors receive a

Table 1.2. International Comparison of Occupational Standards (all standards in milligrams per cubic meter mg/m^3)

Countries	Benzene	Cadmium Oxides	Carbon Monoxide	DDT	Heptachlor	Lead	Beryllium	Mercury	Nickel	NO$_x$	Ozone	Parathion
Belgium	30	0.05	55	1	0.5	0.15	0.002	0.05	0.1	9	0.2	0.1
Finland	32	0.01	55	1	0.5	0.15	0.002	0.005	—	9	0.2	0.1
German Democratic Republic	50	—	35	1	—	0.15	0.002	0.1	—	10	0.2	—
Federal Republic of Germany	50	0.1	55	1	0.5	0.2	—	0.1	—	9	0.2	0.1
Italy	20	0.01	55	1	—	0.15	0.002	—	1	5	0.2	0.1
Japan	80	0.1	55	1	—	0.15	0.002	0.05	1	9	0.2	0.1
Netherlands	30	0.05	55	1	0.5	0.15	0.002	0.05	0.1	9	0.2	0.1
Romania	50 (max)	0.2	30	0.7	0.3	0.1	0.001 (max)	0.05	—	10 (max)	0.1	0.05
Sweden	30	0.02	40	1	—	0.1	0.002	0.05	0.01	9	0.2	—
Switzerland	32	0.1	55	1	0.5	0.15	0.002	0.05	—	9	0.2	0.1
USSR	5	0.1	20	0.1	0.01	0.01	0.001	0.01	0.5	5	0.1	0.05
Poland	30	0.1	30	0.1	—	0.05	0.001	0.01	—	5	0.1	—
United States	30	0.1	55	1	0.5	0.2	0.002	0.05	1	9	0.2	0.11

Source: ILO (1977).

small risk premium in wages, but most workers in secondary labor sectors receive no risk premium, despite frequent high risks (Graham and Shakow 1981).

Despite the evidence, claims are nonetheless commonplace in U.S. political discourse that the American workplace is safe, that workers are knowledgeable about hazards and well equipped to protect themselves, and that they receive compensation for any damage to their health and safety (Viscusi 1983).

MARGINAL HAZARDS

Those who occupy the margins of human populations, cultures, societies, and economies often are exposed to hazards hidden from or concealed by those at the center or in the mainstream. Marginal existence in itself heightens vulnerability because many of those who live in this way are already weakened from malnutrition and access to societal resources and alternative means of coping are few (Kates et al. 1988, 1989). Then, too, the effects themselves are often quite literally invisible and unrecognized, such as the chronic hunger of many rural populations throughout the Third World or the malnutrition of many elderly in the United States. In some cases, the imposition and concealment of the hazard can be quite purposeful, as in the periodic actions in Ethiopia and Sudan to withhold food relief from political opponents and to hide the results from the world's view (Harrison and Palmer 1986). Although the social sciences have devoted much attention to marginality as a social and political phenomenon and have made some effort to assess its role in the experience of hazards and disasters, it is a concept that remains imperfectly understood and lacking an adequate theory.

Marginality can arise from position in age or gender. Young children, who are highly susceptible to neurological damage from lead exposure, have also often been exposed to particular sources of lead (e.g., eating lead paint) at higher levels than adults have (Chisolm and O'Hara 1982; Kane 1985). Although the hazards of lead have been long known and documented (Nriagu 1983), the hazards for children were allowed to exist and even grow for decades before effective action was taken. Both children and women have traditionally spent much more time in the home than have adult men, but only recently has indoor pollution been recognized as a serious source of contaminants. The household, not officially recognized as a workplace in the United States, does not come under OSHA's jurisdiction. It has received from the EPA little attention as an element of a public environment to be protected, although recent warnings on radon may signal a change. Because women tend to receive the highest exposures to a wide variety of contaminants (including benzyopyrene, suspended particles, and nitrogen oxides) in working over the stove, they are the major group at risk. Recent evidence has laid bare the extraordinary level of this as a global hazard, with a conservatively estimated 300 million to 400 million people (mostly women in developing countries) at risk (United Nations Environment Programme 1986, p. 33). Despite these figures, Bowonder (1981) specifically describes how this large hazard has remained hidden in developing countries.

Managing marginal hazards poses its own dangers because societal interventions may actually exacerbate existing social problems or even spawn new ones. It is already worrisome, for example, that so much attention to the management of high susceptibility to reproductive and occupational hazards has been devoted to women when so few data exist indicating significant sex differences in susceptibility (Bingham 1985, p. 80). Much of the current concern, for example, is about the possible effects of a variety of chemicals in the workplace on female reproductive ability or on the fetus (OTA 1985). The temptation in such cases is to identify all those who are capable of becoming pregnant (whether or not they plan to do so) so as to deny them employment (thereby reducing both adverse health effects and potential liability). Such actions pose the danger that science will be used to secure safety by altering the work force rather than by reducing exposure in the workplace. Meanwhile, there may be abuses centered on the equality of opportunity in the workplace.

Marginality may also have its genesis in social class and political economy, as particular groups are pushed out to the edge. A body of radical social theory has emerged that seeks to explain this phenomenon. Wisner (1976) sees marginality as the social allocation of space, shaped by the forces of colonialism and capitalism. Dominant classes gain control over more fertile lands and resources, forcing others into more marginal areas. Blaikie and Brookfield (1987) regard what ensues as a series of changes leading to marginalization. The more marginal lands have a high sensitivity and a low resilience to environmental change (e.g., droughts) or management (e.g., overuse). Such environmental degradation may be irreversible as areas lose their capacity to sustain life (Texler 1986). Not infrequently, a vicious circle sets in of increasing impoverishment and further marginalization of land and land managers. Thus, land degradation becomes both a result and cause of social marginalization (Blaikie and Brookfield 1987, p. 23), a situation that, in O'Keefe and Wisner's (1975) terms, requires only "trigger" events (e.g., drought, smallpox, locusts, or civil war) to produce a disaster. Through marginalization, therefore, famine becomes a "normal" event.

This concept of marginalization has also been used to explain the occurrence of environmental and social disasters. Susman, O'Keefe, and Wisner (1983) see an international retrogressive trend from a state of *undevelopment* to *underdevelopment,* in which marginalization is a central ingredient (Figure 1.2). Underdevelopment, in their view, causes the peasants to make their livelihood in more hazardous environments or to change their uses of resources in ways that exacerbate their vulnerability (Sewell and Foster 1976). Witness, for example, the plight of Bangladesh, as three-fourths of the nation lay under floodwaters in the summer of 1988. Ironically, provision of relief usually reinforces the *status quo ante,* that is, the process of underdevelopment that produced the vulnerability in the first place.

Much remains to be done before we can have a comprehensive and accepted theory of marginalization and its role in hazard experience. But whatever the contribution of marginalization to explaining the occurrence of hazards and disasters, there can be little doubt that hazards concentrated in the margin are

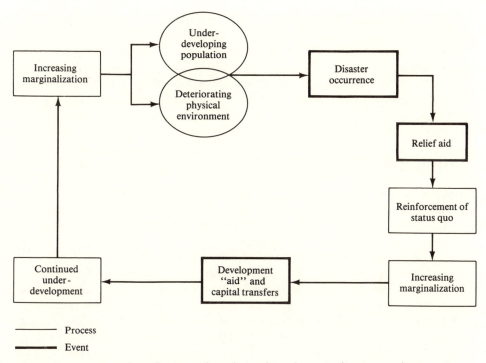

Figure 1.2   Diagram hypothesizing the relationship of marginalization to disaster. (Susman, O'Keefe, and Wisner 1983, p. 279)

often hidden and neglected. Such was the case with the Sahel drought of 1983–84 which, despite its eventual toll as one of the great environmental disasters of the twentieth century, passed largely unnoticed by the world press, international organizations, and national development agencies until the famine reached its height during 1984 (Harrison and Palmer 1986). Yet experts had predicted the potential for continuing famine in the Sahel for some time. As early as 1982, for example, the United Nation's Food and Agriculture Organization (FAO) had issued alarming reports (partly contradicted by other sources) on the situation in Ethiopia (Harrison and Palmer 1986; Downing et al. 1987). The Reagan administration was clearly reluctant to deal with Marxist–Leninist regimes with whom its diplomatic relations were strained. The instability of governments, the diplomatic strains, and the remoteness of the affected areas also made accurate food-related information difficult to obtain. Within the U.S. government, policymakers debated whether the appropriate response should be humanitarian or political (Downing et al. 1987). Not until the NBC evening news aired a BBC special about Ethiopia in October 1984 did the specter of emaciated, fly-ridden bodies dying of starvation illuminate the scale of the tragedy (Li 1987, p. 415).

Although this was a dramatic case of a disaster's remaining invisible and growing in scale and severity until the world belatedly responded, it is not surprising that marginal hazards are often hidden. Logically, the same forces that produced the marginality may be expected to shroud the associated consequences. Then, too, the margins are not a high priority with the center; information flow and interaction are characteristically weak; ideological and political differences often cause separations; and the margins lack the power and resources to project the hazard event into the world view. So it should be expected that hazards occurring among native peoples, ethnic and tribal minorities, isolated regions and locales, and secondary labor forces and peasants will often pass unheeded or remain obscure to those in the mainstream of power and society.

AMPLIFIED HAZARDS

Hidden hazards arise not only from social opaqueness but also because of limitations in the practice of science and the place of science in politics. The passage of the 1970 National Environmental Policy Act (NEPA) in the United States established a precedent—that the assessment of the major environmental and social effects on projects must precede their implementation—that has been emulated worldwide. Indeed, it is instructive to see China coping with this innovation in the late 1980s (Ross 1987). A host of procedural requirements has emerged regarding the environmental impact statement, and a variety of assessment methodologies have flourished. More recently the field of risk analysis has arisen to supplement and extend earlier modes of analysis. These new techniques, concepts, and methods are currently germinating in corporate and governmental programs designed to protect human health, the environment, and individual well-being.

Innovations in assessment practices have demonstrated an extraordinary power to evaluate and compare many of the environmental prices of new and existing technologies and major projects. But ironically, as these innovative techniques illuminate some hazards, they obscure others. One reason for this is that the technical approach to hazards (usually labeled risk analysis) often focuses narrowly on the probability of certain events and the magnitude of specific consequences (often fatalities, morbidity, or environmental damage). Yet studies of risk perception and cultural-group response reveal that most people experience hazards much more broadly; they are concerned, to be sure, about health and environmental effects but also how the risk came about in the first place, whether it is imposed or voluntary, how "fair" the distribution of benefits and risks associated with a particular technology is, and whether catastrophes are possible (Rayner and Cantor 1987; Slovic 1987). It is increasingly clear that technology conveys a variety of impacts, of great concern to the public, that cannot be gauged, or sometimes not even identified, by conventional assessment methodologies. Thus, some hazards are "hidden" to the professional assessors of technology impacts. These hazards interact with social structures and social groups, individuals, society, and the economy in ways unanticipated by technical conceptions of risk, thereby creating *amplified hazards*.

Recently, we have conceptualized the structure and processes that compose the social amplification of risk (Figure 1.3) (Kasperson et al. 1988). People experience risks through personal experience or the depiction of the event in communication sources. Risk-related information is processed by social and

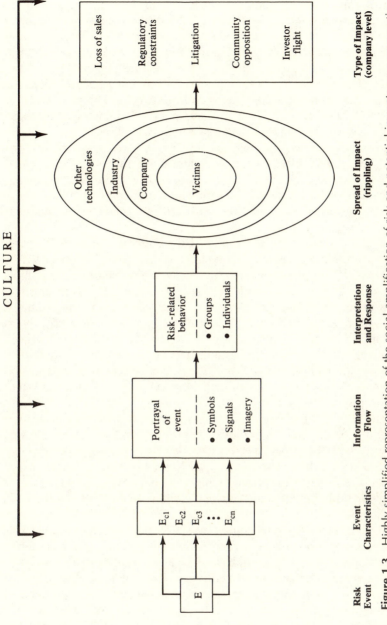

Figure 1.3 Highly simplified representation of the social amplification of risk and potential impacts on a corporation.

individual amplification "stations," including the scientist who communicates the risk assessment, the news media, activist social organizations, informal networks of friends and neighbors, and public agencies. These information systems amplify risk events in two ways, by intensifying or weakening signals of the risk and by filtering the signals with respect to the attributes of risk and their importance. The informational amplification of risk both generates and reacts to behavioral responses in social groups and individuals. These, in turn, result in such secondary impacts as mental imagery and antitechnology attitudes, impacts on business sales and property values, and political and social pressures. Through such secondary effects, the impacts of the risk event may "ripple" to other parties, distant locations, or future generations. Such was the case with the Three Mile Island accident of 1979 which, although probably resulting in no fatalities, shut down nuclear plants worldwide, cost billions of dollars, and eroded public confidence in nuclear power and (perhaps) other high technologies, industry, and regulatory institutions.

Not all technologies or projects, of course, involve amplified hazards. For these, conventional assessment techniques suffice to make visible the hazards of concern. But the most controversial technologies and projects are those with amplified hazards and for which the existing assessment and management approaches miss the mark. An example is hazardous facility siting in the United States. After decades of siting industrial facilities in rural areas, the intense public opposition to a variety of such facilities—prisons, refineries, hazardous waste incinerators, power plants, and genetic-engineering laboratories—has perplexed decision makers and confounded existing institutional processes (Greenberg and Anderson 1984; Lindell and Earle 1983; OTA 1987; von Winterfeldt and Edwards 1984).

Social scientists have abetted the concern among public officials over the "irrational" public and "hysterical" media, by resorting to overly simplistic characterizations of public responses as so-called LULU (Locally Unwanted Land Uses) or NIMBY (Not in My Back Yard) syndromes (Freudenberg 1984; Popper 1983). What is in fact occurring is a set of complex but rational public reactions to both the risk and the social circumstances involving the social amplification of risk.

Instead of laying bare and addressing the hidden hazards, policymakers have fashioned siting processes addressed to misconceptions. When the public has sought assurance of the safety of hazardous facilities, developers and siters have offered compensation (not infrequently construed by the public as bribery) for risk. When fairness in site selection has been demanded, poor communities with high unemployment levels have been the siting targets. When institutional trust has been lacking, developers have engaged in public relations to improve their image, rather than in sharing power. Thus, the national effort to site such facilities has stumbled on the rocks of ungauged hazards amplified through public concern, community conflict, and institutional distrust. The effects have been quite dramatic. By 1987, twenty-two states in the United States still, despite various federal and state efforts, had no commercial hazardous waste facilities, and thirty-five states had no commercial land disposal facilities. Of eighty-one siting applications over the fifty states for hazardous waste disposal facilities between 1980 and 1987, thirty-one have been withdrawn or denied permits. Meanwhile, only six of the eighty-one have resulted in operational facilities; two have permits to operate; two are under judicial review; and four failed to become operational owing to market factors (Condron and Sipher 1987). Yet the need for facilities persists.

Amplified hazards are hidden, then, in a more specialized sense than are the three preceding classes of hidden hazards. They remain hidden to the professional assessors until the consequences are upon us, at which point they become highly visible to society generally. Improved methods of assessment are needed to pinpoint hazards whose nature and magnitude are defined substantially by their interaction with social structures and processes. But even improved assessment will have difficulty securing acceptance for its findings, because making such hazards visible threatens the established regulatory and siting processes and alters the ground rules of whose values will prevail, which impacts are legitimate, and what evidence will count.

VALUE-THREATENING HAZARDS

Every technology, as Edward Wenk (1986) vividly recounted, plays Jekyll and Hyde. Among the most pervasive and resistant to formal assessment of the Hydeian effects are those that alter human institutions, life-styles, and basic values. Because the pace of technology outstrips the social institutions that husband it, disharmony occurs in purpose, economy, and social stability. Technology has promoted the growth and power of industrial enterprises, thereby necessitating a growth in governmental policies to regulate their potential hazards. While nurturing technical complexity, technological growth hinders public understanding and engenders feelings of anxiety. In trying to cope, some people seek spiritual explanation; others look to cults for the comfort of simplistic explanations (Wenk 1986, p. 8).

Hazards that threaten basic human values are as amorphous as they are profoundly troubling. Such values continually shape the directions of science, the unfolding of technological applications, and the institutions that manage both. Equally certain is that technology is a major driving force in shaping social values. Consider the impacts of the automobile on families, contraceptives on sexual mores, life-extending technologies on religious beliefs, and computers and informational banks on privacy. Like amplified hazards, value-threatening hazards are often not hidden, as the noisy conflicts over abortion, smoking, and motorcycle helmets make clear. Yet some technologies pose uncertain and dimly perceived threats to values, intermingled with substantial benefits. Others involve subtle effects, that in themselves are not dramatic but that add to the long-term erosions already in progress. Somehow, the wariness that often greets a new technology seldom includes an acknowledgment that threats to values have been instrumental to its innovation.

Why is it that nuclear power and genetic engineering appear to evoke such public apprehension and so confound formal analysis? Both were born amidst great promise for their benefits to humanity and the global environment. Nuclear power promised enough energy to bring electricity to the world's population without damaging the environment. Genetic engineering promises a new revolution in agriculture for the world's hungry and the eradication of diseases that have been the scourge of health and well-being for centuries. Yet both technologies generate great unease amidst the optimism. Is it that the power of nuclear energy may ultimately fuel destruction rather than progress? And for genetic engineering, have humans usurped the gods by intervening in the fundamental matter of our existence? The means to bring to the surface and debate these issues seem lacking in our institutional processes, in which the social conflicts over risk may be, at heart, projected worries over technologies that touch the wellsprings of our existence.

Then there are the more subtle cumulative effects on values that may already be deteriorating. Privacy, under constant assault from technologies ranging from computerized information banks to electronic eavesdropping, may be disappearing as a basic social value in America. The proliferation of video display terminals, with their far-reaching potential for social isolation and the routinization of labor, may in the future be categorized with piecework, factory assembly lines, and "sweatshops" as socially harmful innovations. Meanwhile, biological and genetic screening, with its potential for selecting among peoples in a host of institutional settings, threatens the basic notions of equity and equality of opportunity. The unrelenting portrayal of violence on television and videotapes erodes our sensitivity to the use of force in human relations and the settlement of arguments. Frequently, ideological trappings color our perceptions of such hazards.

How do such hazards make their way onto the public agenda? Hazards that threaten human values and that are incremental, uncertain, and cumulative appear esoteric and unsubstantiated in a world of scientific evidence and discourse. Hazards that pose issues that transcend the capability of existing institutions to confront and debate them remain unarticulated, are expressed as emotional outcries in a cost-benefit and risk-reduction optimizing world, or find their arena of debate outside established institutions. The means to incorporate these murky hazards into deliberative public policy and decision making continue to elude us.

MULTIPLE HIDDEN HAZARDS

This exploration of hidden hazards has concentrated on the archetypal hazards of each class. But it is surely the case that many hazards have multiple attributes or dimensions that, especially in their interaction, obscure hazards to experts or publics. Thus, occupational hazards involving reproductive effects may simultaneously be ideological (involving notions of the workplace in a market economy), marginal (particularly if construed as gender specific), and value threatening (if affecting beliefs about contraception and abortion). Nuclear power and genetic engineering elicit such strong concerns because they are both amplified and value-threatening hazards. Global poverty surely involves ideological interpretations of the causes and effects of poverty as well as its marginality to the more affluent advanced industrial societies. In short, probably few hazards are hidden purely because of a single attribute; rather, their multiattribute character conceals them from expert assessment or from more general societal recognition.

CONCLUSIONS

What can one conclude about this perplexing genus of hidden hazards? What is the prospect that such hazards will be better integrated over time into our overall knowledge of the epidemiology of natural and technological hazards?

There are few reasons to be optimistic, for such hidden hazards are deeply embedded in our belief systems, scientific practices, and social relations. Uncovering them not only will require new departures in the sciences and social sciences but also will constitute a political act and challenge. Because these hazards vary in their genesis and impacts, any observations must be specific to particular classes:

- *Global elusive* hazards will certainly command determined scientific efforts in the coming decade to delineate their causes, distribution, and global implications. In this sense, these hazards should increasingly assume their place on society's strand of worry beads. At the same time, it is unclear whether our understanding of their human causes will progress equally or whether such hazards will overcome the global political conflicts that their amelioration will certainly provoke. Disjunctures in assessment and global management could well flourish during the 1990s.
- *Ideological* and *marginal* hazards lie deeply embedded in the belief systems and social structures of different societies. They are not likely to become more visible with advances in science and assessment capabilities. Because they have the same roots as do society's institutions and values, they are not "manageable." Rather, they are likely to linger as sources of continuing social and technological controversies.
- *Amplified* and *value-threatening* hazards will continue their dichotomous opaqueness to anticipatory assessments of technology and overt stimulus to social clashes over technology deployment and governmental decisions. Through these encounters, both the assessment and the operation of the political economy will continue to discover, and define, their limits.
- *Multiple* hazards will continue to be a changing domain, as hazards ebb and flow in their social meaning in differing societies and cultural groups. But they will continue to furnish society with hazard surprises as well as deep challenges to assessment technologies.

REFERENCES

Bingham, E. (1985). "Hypersensitivity to Occupational Hazards." In *Hazards: Technology and Fairness,* ed. National Academy of Engineering, pp. 79–88. Washington, D.C.: National Academy Press.

Blaikie, P., and Brookfield, H. (1987), *Land Degradation and Society*. London: Methuen.

Bowonder, B. (1981). "Issues in Environmental Risk Assessment." *Technological Forecasting and Social Change* 19:99–227.

Chisolm, J. J., and O'Hara, D. J. (1982). *Lead Absorption in Children: Management, Clinical and Environmental Aspects*. Baltimore: Urban and Schwarenberg.

Condron, M. M., and Sipher, D. L. (1987). *Hazardous Waste Facility Siting: A National Survey*. Albany: New York State Legislative Commission on Toxic Substances and Hazardous Wastes.

Derr, P., Goble, R., Kasperson, R. E., and Kates, R. W. (1981). "Worker/Public Protection: The Double Standard." *Environment* 23 (September): 6–15, 31–36.

Derr, P., Goble, R., Kasperson, R. E., and. Kates, R. W. (1983). "Responding to the Double Standard of Worker/Public Protection." *Environment* 25 (July–August): 6–11, 35–36.

Downing, J., Berry, L., Downing, L., Downing, T. E., and Ford, R. (1987). *Drought and Famine in Africa, 1981–1986: The U.S. Response*. Worcester, Mass.: International Development Program, Clark University.

Freudenberg, N. (1984). *Not in Our Backyards: Community Action for Health and the Environment*. New York: Monthly Review Press.

Graham, J., and Shakow, D. (1981). "Risk and Reward: Hazard Pay for Workers." *Environment* 23 (October): 14–20, 44–45.

Greenberg, M. R., and Anderson, R. F. (1984). *Hazardous Waste Sites: The Credibility Gap*. New Brunswick, N.J.: Rutgers University Press.

Hagman, G., Beer, H., Benz, M., and Wijkman, A. (1984). *Prevention Better Than Cure: Report on Human and Environmental Disasters in the Third World*. 2nd ed. Stockholm: Swedish Red Cross.

Harrison, P., and Palmer, R. (1986). *News Out of Africa: Biafra to Band Aid*. London: Hilary Shipman.

Hilts, P. J. (1988). "Radon Tests Urged for All Houses: EPA, Surgeon General Note 'Urgency', Issue Joint Health Advisory." *Washington Post,* September 13, pp. A1, A16.

International Council of Scientific Unions (1986). *The International Geosphere–Biosphere Programme: A Study of Global Change*. Paris: The Council.

International Labor Office (ILO) (1977). *Occupational Exposure Limits to Airborne Toxic Substances*. Geneva: ILO.

International Trade Unions of Chemical, Oil and Allied Workers (ICPS) (n.d.). *Threshold Limit Values for Toxic Products in Workplaces*. Budapest: Alain Covet.

Jancar, B. (1987). *Environmental Management in the Soviet Union and Yugoslavia: Structure and Regulation in Federal Communist States*. Durham, N.C.: Duke University Press.

Kane, D. N. (1985). *Environmental Hazards to Young Children*. Phoenix. Oryx Press.

Kasperson, R. E., Renn, O., Slovic, P., Brown, H. S., Emel, J., Goble, R., Kasperson, J. X., and Ratick, S. (1988). "The Social Amplification of Risk: A Conceptual Framework." *Risk Analysis* 8(2): 177–87.

Kates, R. W. (1978), *Risk Assessment of Environmental Hazard*. SCOPE Report 8. Chichester: Wiley.

Kates, R. W. (1985). "Managing Technological Hazards: Success, Strain, and Surprise." In *Hazards: Technology and Fairness,* ed. National Academy of Engineering, pp. 206–20. Washington, D.C.: National Academy Press.

Kates, R. W., Chen, R. S., Downing, T. E., Kasperson, J. X., Messer, E., and Millman, S. (1988). *The Hunger Report: 1988*. Providence: Alan Shawn Feinstein World Hunger Program, Brown University.

Kates, R. W., Chen, R. S., Downing, T. E., Kasperson, J. X., Messer, E., and Millman, S. (1988). *The Hunger Report: Update 1989*. Providence: Alan Shawn Feinstein World Hunger Program, Brown University.

Li, L. M. (1987). "Famine and Famine Relief: Viewing Africa in the 1980s from China in the 1920s." In *Drought and Hunger in Africa: Denying Famine a Future*, ed. M. H. Glantz, pp. 415–34. Cambridge: Cambridge University Press.

Lindell, M. K., and Earle, T. C. (1983). "How Close Is Close Enough?: Public Perceptions of Risks of Industrial Facilities." *Risk Analysis* 3(4): 245–53.

Malone, T. (1986). "Mission to Planet Earth: Integrating Studies of Global Change." *Environment* 28 (October): 6–11, 39–42.

Mathews, J. T. (1989). "Redefining Security." *Foreign Affairs* 68 (Spring): 162–77.

Melville, M. (1981). "Risks on the Job: The Worker's Right to Know." *Environment* 23 (November): 12–19, 42–45.

Montreal Protocol (1987). *1987 Montreal Protocol on Substances That Deplete the Ozone Layer: Final Act*. Montreal: United Nations Environment Programme.

National Environmental Policy Act (NEPA) (1970). 42 USC 4341. Amended by P.L. 94–52, 3 July 1975; P.L. 94–83, 9 August 1985.

Nelkin, D., and Brown, M. S. (1984). *Workers at Risk: Voices from the Workplace*. Chicago: University of Chicago Press.

Nriagu, J. O. (1983). *Lead and Lead Poisoning in Antiquity*. New York: Wiley.

Occupational Safety and Health Administration (OSHA) (1983). "Hazard Communication: Final Rule." *Federal Register* 48, November 25, 1983, pp. 53279–348.

Office of Technology Assessment (OTA) (1985). *Reproductive Health Hazards in the Workplace*. Washington, D. C.: OTA.

Office of Technology Assessment (OTA) (1987). *Background Paper: Public Perceptions of Biotechnology*. Vol. 2 of *New Developments in Biotechnology*. Washington, D.C.: OTA.

O'Keefe, P., and Wisner. B. (1975). *The World Food Crisis: Issues*. Occasional Paper no. 1. Bradford: University of Bradford.

Popper, F. J. (1983). "LP/HC and LULUs: The Political Uses of Risk Analysis in Land-Use Planning." *Risk Analysis* 3(4): 255–63.

Rayner, S., and Cantor, R. (1987). "How Fair Is Safe Enough?: The Cultural Approach to Societal Technology Choice." *Risk Analysis* 7:3–13.

Roschin, A. V., and Timofeevskaya, L. A. (1975). "Chemical Substances in the Work Environment: Some Comparative Aspects of USSR and US Hygienic Standards." *Ambio* 4(1): 30–33.

Ross, L. (1987). "Environmental Policy in Post-Mao China." *Environment* 29 (May): 12–17, 34–39.

Sewell, W. R. D., and Foster, H. D. (1976). "Environmental Risk: Management Strategies in the Developing World." *Environmental Management* 1(1): 49–59.

Simon, P. J. (1983). *Reagan in the Workplace: Unraveling the Health and Safety Net*. Washington, D.C.: Center for Study of Responsive Law.

Slovic, P. (1987). "Perception of Risk." *Science* 236:280–89.

Susman, P., O'Keefe, P., and Wisner, B. (1983). "Global Disasters, a Radical Interpretation." In *Interpretations of Calamity from the Viewpoint of Human Ecology*, ed. K. Hewitt, pp. 263–77. Boston: Allen & Unwin.

Texler, J. (1986). *Environmental Hazards in Third World Development*. Studies on Developing Countries, no. 120. Budapest: Institute for World Economics, Hungarian Academy of Sciences.

United Nations Environment Programme (1986). *The State of the Environment, 1986: Environment and Health*. Nairobi: United Nations Environment Programme.

Viscusi, W. K. (1983). *Risk by Choice: Regulating Health and Safety in the Workplace.* Cambridge, Mass.: Harvard University Press.

von Winterfeldt, D., and Edwards, W. (1984). *Understanding Public Disputes About Risky Technologies.* New York: Social Science Research Council.

Wenk, E., Jr. (1986). *Tradeoffs.* Baltimore: Johns Hopkins University Press.

Winell, M. (1975). "An International Comparison of Hygienic Standards for Chemicals in the Work Environmnet." *Ambio* 4(1): 34–42.

Wisner, B. (1976). *Man-Made Famine in Eastern Kenya: The Interrelationship of Environment and Development.* Discussion Paper, no. 96. Brighton, Eng.: Institute of Development Studies, University of Sussex,

World Commission on Environment and Development (1987). *Our Common Future.* New York: Oxford University Press.

2

Acceptable Evidence
in a Pluralistic Society

SHEILA JASANOFF

In recent years policy analysts have come to recognize cross-national comparison as a technique for illuminating noteworthy or desirable elements of decisions in particular national contexts (Heidenheimer, Heclo, and Adams 1975). This approach has proved especially fruitful in studies of science-based regulation. Comparisons between countries have helped identify institutional, political, and cultural factors that condition decision makers' use of scientific knowledge. This research has demonstrated, for example, that the analysis of evidence, especially in fields characterized by high uncertainty, can be influenced by the participation of differing classes of professionals in the administrative process, the composition and powers of scientific advisory committees, and the legal and political processes by which regulators are held accountable to the public (Brickman, Jasanoff, and Ilgen 1985; Gillespie, Eva, and Johnston 1979; McCrea and Markle 1984; Jasanoff 1986, 1987). As a result, it is by no means uncommon to find decision makers interpreting the same scientific information in different ways in different countries.

Cultural variation appears to influence not only the way decision makers select among competing interpretations of data but also their methods of regulatory analysis and their techniques for coping with scientific uncertainty. For instance, in assessing both environmental and economic impacts, U.S. regulators place a higher value on formal analytical methods, whose validity can be publicly tested and verified, than do regulators in most European countries (Jasanoff 1983; Vogel 1986). One result of such methodological divergence is that evidence considered sufficient to trigger action in one country may fail to do so in another.

These findings should not be surprising to anyone familiar with the problems of policy-relevant science. When knowledge is uncertain or ambiguous, as is often the case in science bearing on policy, facts alone are inadequate to compel a choice. Any selection inevitably blends scientific with policy considerations, and policymakers accordingly are forced to look beyond science to legitimate their preferred reading of the evidence. In this quest, cross-national divergences can be expected to appear as policymakers fall back on established, possibly nation-specific, repertoires of institutional and procedural approaches to securing political legitimacy. The manner in which a decision maker views scientific uncertainty ultimately conforms to deeper patterns of political culture, incorporating a nation's norms and standards of governmental accountability. Comparative studies of science-based policy thus provide a window on interdependencies between politics and the evaluation of evidence in advanced industrial societies.

This chapter argues that the tendency of U.S. regulators to address uncertainty through quantitative analysis serves political needs that are not as keenly felt in other Western democracies, most notably Britain. I shall begin with an account of the treatment of uncertainty in U.S. policy, looking at three areas of health risk assessment in which quantitative analysis has gained momentum over the past decade: carcinogenic risk assessment, *de minimis* risk analysis, and the quantification of expert judgments (or judgmental probability encoding). I then shall contrast these analytical strategies with the sharply divergent methods used in Britain to evaluate the health risks of lead in the environment. Risk assessment in the two countries appears to reflect different understandings of what constitutes accountability in decisions grounded in insufficient evidence. To explain this, I shall examine factors lying outside science. The relatively low demand for quantitative risk analysis in Britain is attributed, at least in part, to the insulated position of British experts in the policy process. Subjected to far less intensive public scrutiny than are their American counterparts, Britain's expert committees preserve a degree of informality and flexibility in their technical analyses that would be unthinkable in the United States.

RISK IN NUMBERS

The use of quantitative analysis in the regulation of health hazards, especially those presented by toxic substances, is fairly new in the United States. According to one count, between 1978 and 1980 only eight chemicals were regulated on the basis of quantitative risk analysis, whereas between 1981 and 1985 the number of such decisions rose to fifty-three (Travis et al. 1987, p. 483). The purposes for which agencies use quantitative analysis also multiplied during this latter period. Numerical risk estimates are used not only to balance risk benefits and to set standards but also, increasingly, to compare risks and distinguish between those that merit regulation and those that do not. The nascent technique of judgmental probability encoding represents a still newer application of quantitative methodologies to the evaluation of scientific uncertainty, in this case to measure the extent of subjectivity in expert judgments. Despite their growing popularity, however, all of these methodologies continue to generate lively technical and political controversy.

Carcinogenic Risk Assessment

Quantitative risk analysis is now recognized as almost indispensable to estimating the risk of cancer to human populations. The method permits decision makers to aggregate data from disparate sources, including animal and epidemiological studies and exposure assessments. In 1980, however, federal agencies such as the Occupational Safety and Health Administration (OSHA) were sufficiently skeptical about this technique to recommend only a limited use of quantitative risk assessment in policymaking. In keeping with this philosophy, OSHA refrained from carrying out a risk assessment for benzene when proposing a new standard in 1978, a decision that eventually led to an embarrassing reversal by the Supreme Court. The current practice of the federal regulatory agencies stands in startling contrast with these earlier attitudes, and the growing number of regulatory decisions based on quantitative analysis attest to the central role of this methodology in legitimating policy.

Despite its centrality, however, risk assessment remains one of the most controversial aspects of public policymaking regarding carcinogens. Three charges are most frequently heard: first, that carcinogenic risk assessment is misleading; second, that it encourages the use of "bad science"; and, third, that it obscures distinctions between facts and values, thereby making it more difficult to hold risk assessors accountable to the public. These charges must be investigated further if we wish to understand why quantification nevertheless remains an attractive strategy for expressing the uncertainties inherent in risk decisions.

False Precision

The earliest applications of quantitative risk assessment to chemical carcinogens made it clear that the technique could be used to produce widely varying risk estimates. For example, in a study carried out for the Food and Drug Administration (FDA) in 1978, the National Research Council (NRC) found that the carcinogenic risk of saccharin varied by as much as six orders of magnitude, depending on the methods chosen to make high-to-low dose extrapolations and interspecies dose adjustments. Because of these uncertainties, the study concluded that "the quantitation of risks to humans cannot be made with confidence" (NAS–NRC 1978, pp. 3–72). Some years later, the differences in exposure estimates led to comparable variations in the risk estimates for ethylene dibromide produced by the Environmental Protection Agency (EPA) and by a consultant for the food industry (Hanson 1984, p. 13). Divergences like these fed fears among the public and some agency officials that risk assessment could be willfully manipulated to put a veneer of false precision on inherently arbitrary policy choices (Peterson 1985).

Agency officials were quick to realize that ill-considered uses of risk estimates could threaten the legitimacy of their policy decisions. Yet their response, ironically, was not to retreat from quantitative risk assessment but to insist on still greater precision. Agency risk analysts, for instance, embraced the idea that the uncertainty surrounding each risk estimate should be represented in numerical form. Accordingly, William Ruckelshaus, in his second term as administrator of the EPA, announced that risk calculations should be "expressed as distributions of estimates and not as magic numbers that can be manipulated without regard to what they really mean" (Ruckelshaus 1984, p. 161). And in an effort to explain to the public what such risk estimates "really mean," the EPA began placing great emphasis on the newly recognized field of "risk communication."

The "Bad Science" Dilemma

Besides worrying about the imprecision of quantitative risk analysis, members of the scientific community voiced concern that the pressure to perform such assessments might lead the regulatory agencies to overlook major qualitative deficiencies in the empirical studies used as a basis for risk assessment. During the mid–1980s, for example, the EPA's Scientific Advisory Panel (SAP), an expert committee associated with the agency's pesticide program, found fault with the agency on several occasions for glossing over flaws in the design and conduct of toxicological studies relevant to regulation. One impatient panelist described the quantitative risk assessments performed on the basis of such studies as "gimcrack mathematics" (Jasanoff 1990, p. 136).

Advances in scientific knowledge provided additional grounds for concern about the soundness of many quantitative risk assessments. For nearly twenty years, federal agencies had attempted to conduct carcinogenic risk assessments in accordance with generic principles or guidelines. But in spite of many attempts to ensure the technical validity of these principles, especially through repeated scientific peer review, controversies continued to arise over their applicability to specific cases. A major failing of the generic principles, in the view of most critics, was their inflexibility in the face of new knowledge, whether about a particular chemical or the general processes of carcinogenesis. The EPA's risk assessment of formaldehyde appeared to confirm these misgivings.

Formaldehyde emerged as a major policy dilemma for federal regulation when the Chemical Industry Institute of Toxicology (CIIT), an industry-financed research organization, reported in late 1979 that the substance caused a marked increase in nasal cancer among exposed rats. The CIIT's rat study was universally regarded as both well designed and well conducted. Yet the agency's efforts to incorporate its results into risk-assessment and regulatory policy ran into un-expected obstacles. At the core of the dispute was the shape of the dose response curve in the CIIT bioassay: Sharply nonlinear, it suggested that the severe carcinogenic response in rats at high doses might have a mechanistic explanation that was not applicable to other species, including humans. Not surprisingly, this position was forcefully defended by representatives of the formaldehyde industry, who pointed to the statistically insignificant increase of cancer incidence in mice, as well as to largely negative results in human epidemiological studies, as further confirmation that the rat data alone should not be taken as diagnostic of the risk to humans.

From the standpoint of quantitative risk assessment, the nonlinearity of the dose response greatly increased the uncertainties of deciding what extrapolation model should be used for estimating risk at lower doses. Even a multidisciplinary international workshop devoted to formaldehyde failed to develop a consensus on this issue (Report on the Consensus Workshop on Formaldehyde 1984, pp. 328–81). In the meantime, CIIT scientists carried out new research on re-actions between formaldehyde and DNA in the nasal cells of rats at varying exposure levels. The aim of this research was to elucidate the specific mechanism of formaldehyde-induced cancer in rats and to provide, if possible, a better scientific basis for selecting a low-dose extrapolation model for quantitative risk assessment.

According to the EPA's published risk-assessment guidelines, a linearized multistage model should be selected for low-dose extrapolation unless the evidence suggests that another approach would be more appropriate. The model is generally regarded as "conservative" in that it tends to overstate the degree of risk at low exposure levels. EPA believes this conservatism is in accord with the precautionary character of the legislation it administers. One of the key assumptions underlying the linearized multistage model is that the "administered dose" (i.e., the level inhaled by the test animals) is directly proportional to the "delivered dose" (i.e., the amount actually reaching the target cells). The CIIT experts, however, concluded that the evidence on formaldehyde did not bear out this assumption. Their research suggested that, at least in this particular case, a more accurate measure of the delivered dose could be derived from a count of the DNA adducts formed by formaldehyde in the rat nasal lining. This delivered dose, the CIIT scientists further argued, should be used in place of the administered dose to quantify the cancer risk of formaldehyde.

The EPA's draft risk assessment for formaldehyde, completed in 1985, failed to take note of these new developments and continued to use the linearized multistage model applied to the administered dose. In due course, this document was peer reviewed by a subcommittee of the agency's Science Advisory Board (SAB). The committee concluded that the EPA's risk assessment would not be scientifically adequate without a fuller investigation of the issues raised by the CIIT's research on the pharmacokinetics of formaldehyde. In response to the SAB's review, the EPA convened a panel of specialists on metabolism, statistics, and DNA adducts to consider, generally, whether pharmacokinetic information could reasonably be incorporated into risk assessments and, specifically, whether the CIIT's data on DNA adducts should have been factored into the formaldehyde risk assessment. These initiatives bolstered the EPA's scientific credibility but ultimately had little influence on the agency's decision making: The pharmacokinetics panel supported the EPA's judgment that the CIIT's research was not yet sufficiently validated for use in policy-relevant risk assessments (EPA 1987).

Although the review by the expert panel permitted the EPA to continue using its established principles of risk assessment, not all observers were persuaded of the correctness of the agency's final decision. The CIIT scientists, in particular, continued to believe that the agency should have paid greater heed to the emerging pharmacokinetic information on formaldehyde and refused to accept as dispositive the views of a possibly arbitrary grouping of seven outside experts. Furthermore, the agency's failure to change its position on this issue was construed by industry experts as evidence that the conservative and pro-regulatory values reflected in the multistage extrapolation model would not easily yield to rapid advances in scientific knowledge.

Facts and Values

A third criticism directed against quantitative policy analysis, whether scientific or economic, is that it blurs the distinctions that decision makers should ideally maintain between facts and values. Students of carcinogenic risk assessment, in particular, have convincingly argued that facts and values are intertwined at virtually every stage in the application of this controversial methodology (Rush-efsky 1986; Whittemore 1983). The conviction that risk assessment can never be a value-free exercise led an NRC committee to recommend in 1983 that the functions of risk assessment and risk management should not be institutionally separated in the regulatory process, even though agencies should seek as far as possible to prevent risk-management considerations from influencing their risk assessments (NAS–NRC 1983).

A number of writers have suggested that the use of statistical techniques to evaluate the risks of environmental chemicals increases the probability that the regulator's value judgments will not be fully exposed to public scrutiny and criticism. The problem stems in part from the way that people perceive numbers. Thus, psychological research on risk perception indicates that "both scientists and lay people may underestimate the error and unreliability in small samples of data, particularly when the results are consistent with preconceived, emotion-based beliefs" (Whittemore 1983, p. 28).

Another argument against quantitative risk assessment is that the professionals responsible for carrying out such analyses may not adequately inform policymakers about the values embedded in competing approaches to measuring and representing risk. There are, for example, different uncertainties associated with using population—as opposed to individual—risks in making decisions about toxic substances (Silbergeld 1986). Although professional risk assessors may be well aware of these uncertainties, they may not necessarily make them explicit in the course of assessing risk. The tendency not to state uncertainties about alternative methodological assumptions may even be exacerbated by the effort to separate risk assessment from risk management. The statisticians and modelers who develop the risk-assessment programs and the policymakers who deal with the "softer" social and political impacts of regulation may array themselves, like C. P. Snow's "two cultures," across an ever-widening communications gap.

De Minimis Risks

Displaying curiously little concern about possible abuses of quantitative methods, regulatory agencies have, if anything, intensified their efforts to displace qualitative judgments by means of mathematical formulations. One area of decision making in which this trend can be seen quite clearly is in the developing "discipline" of de minimis risk analysis. The term de minimis derives from the ancient legal maxim de minimis non curat lex ("the law does not concern itself with trifles"), which common law courts relied on to prevent litigants from seeking judicial remedies for trivial injuries. More recently, U.S. courts have invoked the de minimis principle to bar regulatory action against risks of no serious consequence to public health, safety, or welfare (U.S. v. Lexington Mill and Elevator Co., 232 U.S. 399 [1914]; Alabama Power Co. v. Costle, 636 F.2d 323 [D.C. Cir. 1979]; Monsanto v. Kennedy, 613 F.2d 947 [D.C. Cir. 1979]; Natural Resources Defense Council v. EPA, 824 F.2d 1146 [D.C. Cir. 1987]).

By the late 1970s, a combination of economic, scientific, and policy considerations led U.S. regulators to consider adopting explicit policies for designating certain risk levels as *de minimis,* that is, too low to merit regulation (Whipple 1984). Advances in analytical chemistry revealed trace quantities of hazardous substances in environments that had previously been regarded as safe. At the same time, active programs of chemical testing in government and industry identified an ever-more numerous array of chemicals as potentially dangerous to public health and the environment. In an increasingly antiregulatory political climate, powerful voices argued that proceeding against each of these newly revealed hazards would lead to the imposition of potentially crippling regulatory costs on society, and with few commensurate benefits. Under these conditions, regulatory agencies welcomed the opportunity to dismiss as *de minimis* those risks that fell below a certain probabilistic threshold.

As a quantitative decision rule, the *de minimis* strategy provided an apparently apolitical reason for deciding whether regulatory action was warranted and therein lay its appeal. As one analyst noted, "Without the *de minimis* cut-off, regulators may find it difficult to justify decisions to summarily dismiss small risks that come to official attention, even if there is reason to believe that better candidates for regulatory consideration await identification" (Mumpower 1986, p. 441). Both the EPA and the FDA indicated that from a policy standpoint, they would regard risks of less than one in a million as negligible.

Yet even as regulators displayed growing support for the *de minimis* principle, evidence accumulated that defining a numerical cutoff point for regulation does not absolve regulators of the need to make policy choices and that careless use of this strategy could even have negative consequences. An important limitation of the *de minimis* strategy is that it does not tell decision makers how to take account of distributive factors or, put differently, how to strike the balance between equity and efficiency. For example, an annual risk of one per million for the entire U.S. population would result in roughly the same number of fatalities (240) as a one-per-hundred risk exclusively affecting a small town of 24,000 people. The former case would conventionally be regarded as *de minimis.* The latter case, however, might warrant a positive regulatory response on the grounds that "subgroups should not be exposed to disproportionately high levels of risk, and that risks should be shared among as many people as possible" (Mumpower 1986, p. 439).

The superficial logic of the *de minimis* risk policy, moreover, may encourage uncritical case-by-case applications that result in the sanctioning of unacceptably high risks. Most policymakers would agree that a hazard with an annual risk of injuring one individual in a million can reasonably be dismissed as *de minimis.* Yet if numerous additional risks of this magnitude are introduced incrementally, the cumulative impact of all these new *de minimis* risks over an individual's lifetime could well rise above the level of acceptability (Mumpower 1986, p. 442). In the real world, of course, risks are not only introduced but also eliminated. To make an informed use of a *de minimis* policy, regulators would therefore have to make numerous assumptions about the dynamics of technological innovation, maturation, and decay. These assumptions, in turn, could generate more political controversy than could the *de minimis* principle itself.

Disciplined Judgment

An even more striking—and conceptually more problematic—use of quantitative analytic techniques was the EPA's effort in the mid–1980s to represent in statistical terms the variations in expert opinion concerning the health and environmental effects of air pollutants. The impetus for this process of "judgmental probability encoding" came from a recognition that in the field of health risk assessment—in which scientific evidence is unusually fragmentary and indirect—inferences invariably contain large measures of subjective judgment. Experts, moreover, differ in their assessments of the uncertainties associated with particular inferences. To get a balanced picture of expert opinion on any issue, therefore, the EPA concluded it would be necessary to get an accurate sense of the range of divergence among experts, as well as of the levels of confidence that each attached to his or her own judgments about the evidence.

The effort to find a rigorous way of representing variations in expert opinion advanced furthest in work done by the EPA's Office of Air Quality Planning and Standards (OAQPS) in establishing national ambient standards for air pollutants. In particular, consultants for OAQPS were asked to encode expert judgments with respect to two health effects associated with elevated blood-lead levels: reductions in hemoglobin and children's IQ. The exercise represented a radical intrusion into the thought processes of scientific experts. The encoder's task was not only to ascertain, in quantitative terms, the range of uncertainty within a single expert's mind but also to capture differences in the confidence that each expert had in his or her own predictions. In a system committed to separating fact and value, judgmental probability encoding represented the ultimate attempt to sort out the role of both constituents in the formation of expert opinion.

Proponents of this analytical approach maintain that "the only alternatives to encoding judgmental probabilities [when direct data are unavailable] are to treat the added uncertainty qualitatively, which currently cannot be done in a rigorous manner, or to ignore it altogether, which is indefensible" (Wallsten and Whitfield 1986, p. 5). But the encoding exercises carried out to date reveal a number of potential pitfalls in the application of this technique to policymaking. Most experienced users of this methodology concede that its validity depends on obtaining the judgments of experts who span the whole range of respectable scientific opinion. Failure to represent the entire range could give the final decision maker a significantly distorted picture of the extent of scientific disagreement on a given issue. Yet even the advocates of encoding admit that "the problem of establishing the range of opinion and selecting appropriate experts appears to be difficult and judgmental in nature" (Wallsten and Whitfield 1986, p. 5). Given the opportunities (and incentives) for manipulating evidence in the policy process, it may be naively optimistic to hope that a task requiring so much skill and judgment will always be carried out in a responsible, unbiased, and replicable manner.

Encoding the judgments of experts may provide a clearer picture of the extent and nature of their disagreements, but it cannot produce consensus where none exists. The limits of the technique were clearly illustrated in a study of professional judgments concerning the health effects of sulfur emitted from coal-fired power plants. The experts consulted in the study diverged so widely that it proved impossible to construct a single model of sulfur-related health damage based on their opinions (Morgan et al. 1984, p. 209). Indeed, the authors of the study noted that they had "resisted the temptation to construct a single result" because "we believe that with the diversity of opinion that exists such a result would have little meaning, might serve to mask the existing divergence of views, and might be picked up and used in risk assessment activities without appropriate qualifications" (Morgan et al. 1984, p. 214). While admiring the authors' self-restraint, one may wonder whether agency risk analysts, driven by political pressures and legal timetables, are likely to display an equally scrupulous sense of professional responsibility.

Yet another possible misuse of the encoding technique is to apply it to areas in which there is adequate direct evidence of health and environmental effects to permit the use of classic statistical techniques. In a recent case of this kind, the OAQPS staff tried to use judgmental probability encoding to derive a dose response function for human health effects caused by ozone. However, the Clean Air Scientific Advisory Committee (CASAC), the body charged by law with reviewing the scientific basis for the EPA's air pollution standards, pointed out that this approach was inappropriate because the available dose response information was strong enough and consistent enough for ordinary statistical analysis (Jasanoff 1990, p. 118).

These examples indicate that encoding expert judgments raises many of the same questions that have troubled critics of carcinogenic risk assessment. Will the method lead to consistent results? Will regulators be tempted to manipulate the outcome through, for example, a biased selection of experts? Will the quantification of essentially subjective judgments convey a false sense of precision regarding expert opinion? Will the use of a complex statistical technique enhance the policymaker's understanding of the relevant scientific issues, or will it induce undue deference and transfer too much control over risk management to technical experts?

A SYSTEM OF TRUST

Britain's philosophy for dealing with uncertainty in policy-relevant science offers in many respects a perfect counterfoil to that of the United States, and many of its distinctive attributes are reflected in the work of that country's prestigious Royal Commissions. Formally appointed by and answerable to the sovereign, these bodies are frequently called upon to advise the government on pressing issues of national policy. Their membership is usually heterogeneous, drawing on diverse professions and disciplines and including both noted academics and representatives of major political-interest groups. A Royal Commission thus is inherently multipartite, but it is at the same time a microcosm of Britain's establishment, and its credibility is very much a reflection of its members' collective credibility.

The Lead Report

Britain established a permanent Royal Commission on Environmental Pollution in 1970 in response to the growing worldwide demand for more governmental attention to environmental problems. Its mandate was broadly defined: "to advise on matters, both national and international, concerning the pollution of the environment; on the adequacy of research in this field; and the future possibilities of danger to the environment" (Royal Commission on Environmental Pollution 1971, p. iii). Since its creation, the commission has reported to Parliament at two- or three-year intervals on issues of particular concern. The commission's ninth report on lead in the environment (Royal Commission on Environmental Pollution 1983), prompted in part by the controversy over leaded gasoline and human health, provides revealing insights into the British approach to resolving scientific uncertainty.

The Royal Commission's objectives in the lead report were more wide-ranging than are those of the typical U.S. regulatory agency or advisory committee. The report consisted of four major components: a survey of the extent and nature of lead pollution, an analysis of the effects of lead on humans and animals, a discussion of specific problems connected with lead in gasoline, and a set of recommendations for further research and regulatory policy. Thus, within the framework of a single report, the commission members were concerned with issues of risk assessment as well as risk management, science as well as public policy.

One area of uncertainty that concerned both the Royal Commission and the EPA's air office, OAQPS, was the effect of low-level exposure to airborne lead on the intelligence and behavior of children. In the United States, as we noted, the OAQPS saw this issue as central to policy. EPA consultants accordingly tried to address the uncertainties in the data by eliciting and encoding expert judgments concerning mean IQ reductions at varying levels of exposure. The Royal Commission, however, adopted an entirely different strategy that significantly downplayed the significance of this body of evidence.

Lead and IQ

By the time the Royal Commission took up the issue, numerous British and international expert bodies had already investigated the question of lead and children's IQ. The relevant scientific literature, for example, had been reviewed by a working group appointed by the Department of Health and Social Security (DHSS) and by independent scholars. By its own admission, the commission did not feel it had much to add to these surveys; hence it undertook no independent reanalysis of the evidence connecting lead to children's behavior and academic performance. Rather, the commission's primary concern was to establish whether this impressive volume of data delivered any clear message about risk, and on this point its conclusions were extremely equivocal:

> In our view the accumulated evidence may indicate a causal association between the body burden of lead and psychometric indices, or the effects of confounding factors,

or both. On present evidence we do not consider it possible to distinguish between these possibilities. We consider it unlikely that population surveys alone will settle the issue in the near future, particularly as the effect of lead on behavior and intelligence, as suggested by the studies mentioned above, is at the most small at the concentrations found in the general population. (Royal Commission on Environmental Pollution 1983, pp. 61–62)

But although the commission found no persuasive evidence of a risk to children's health, it departed from prior British policy in recommending, first, that lead additives be phased out altogether from gasoline (Royal Commission on Environmental Pollution 1983, p. 143) and, second, that the British government should make strenuous efforts to reduce all manufactured sources of lead in the environment. How did it justify this highly precautionary posture?

Sir Richard Southwood, the commission's chair during the preparation of the lead report, provides illuminating glimpses into the role of scientific evidence and uncertainty in the group's deliberations (Southwood 1985, pp. 346–50). The commission, in Southwood's account, determined from the outset "that scientific uncertainty over the effects of low levels of lead were [sic] likely to remain; the recently published MRC report has underlined that and as Sir James Gowans said in relation to the same problem 'The things we would like to know may be unknowable' " (Southwood 1985, p. 349). Accordingly, the commission made a value decision that it would address the problem not with the bias of "innocent until proved guilty" (a bias that Southwood considers fairly typical of professional scientists) but in accordance with what Southwood, himself an ecologist, saw as the basic evolutionary principle for assessing risk: "dangerous until proved safe."

Faced with the limitations of the scientific evidence, the commission adopted what it termed the "approach of the Manager." This entailed a four-step analysis: First, the commission looked at the basic physiology of lead and, second, at its persistence in the environment. Both these steps indicated grounds for concern, as lead affects RNA with no threshold and remains undegraded in the environment for long periods of time. Third, the commission considered the average blood-lead concentrations in the general population in relation to levels at which undisputed health effects may occur. This comparison revealed that the safety margin between the two levels was uncomfortably small, particularly for urban populations. Finally, the commission took note of the probable economic impacts of eliminating lead from gasoline and determined that a phaseout would not be unreasonably costly if manufacturers were given a period of time to adjust to the use of unleaded fuels.

ACCOMMODATING UNCERTAINTY

The Royal Commission's treatment of evidence relating to the environmental and health risks of lead differed in notable respects from American agencies' risk-assessment practices for toxic pollutants, including lead. The most striking difference, perhaps, was in the attitude of the experts in the two countries toward the apparent limitations of the evidence. In analyzing the risks of lead, the Royal Commission seemed much more willing than the EPA was to accept some uncertainties as irreducible and hence not worth further study and investigation. Thus, with respect to the lead–IQ link, the commission expressed skepticism that additional research would lead to greater scientific certainty, given the confounding factors inherent in any study focusing on this type of causation. The commission's framework for decision making was geared instead toward making policy recommendations on the basis of what was known or readily knowable: in this case, the genotoxicity, pervasiveness, and environmental persistence of lead.

The British and U.S. experts also differed in the manner in which they chose to express their risk determinations. The Royal Commission's approach to assessing the effects of lead was entirely qualitative; it described the risk to children, for instance, in words ("at the most small"), not numbers. Translated into American terms, this was equivalent to finding that the risk of damage to behavior or intelligence was *de minimis*. But notably, the British experts made no attempt to quantify the "small" effect even in the aggregate, let alone to make more finely tuned estimates of IQ decrements at different exposure levels. In contrast, all of the major developments in U.S. risk-assessment methodology— carcinogenic risk assessment, *de minimis* risk analysis, and judgmental probability encoding—were oriented toward representing the probability of harm in quantitative terms.

A third significant difference between the British and American analytical approaches is the latter's intense preoccupation with isolating and, if possible, measuring the degree of subjectivity in the judgments of technical experts. In the area of carcinogen risk assessment, this concern was manifested in the increasing demand for expressing the uncertainties surrounding any numerical estimate of risk. Risk estimates, as U.S. regulators now insist, must be represented as bands rather than single numbers so as to communicate an accurate sense of the reliability of these projections to the final decision maker. Judgmental probability encoding represents an even more drastic attempt to separate expert assessments of uncertainty into subjective and objective components.

The Royal Commission's lead report, by contrast, made no effort to sort out where and to what extent subjective elements entered into its authors' reading of the evidence. Writing in the first person, the commission made it clear that the report as a whole was indeed a subjective document. Given the commission's view that much of the evidence was inherently limited and incapable of definite resolution, the report could hardly claim to be anything else. The commission explicitly recognized that it had to leap beyond the state of available knowledge. To the extent that the results could claim to be authoritative, it was not because they were "objective" but because they represented the collective judgment of a group whose expertise and disinterest were beyond question.

Finally, the regulatory treatment of scientific uncertainty seemed to invite very different levels of public debate and controversy. In Britain, the Royal Commission's method of resolving the uncertainties with respect to lead called forth hardly any public challenge. Members of the commission were not questioned about key elements in their reasoning or regulatory philosophy. For example, no one objected either to the commission's judgment that IQ decrements associated with lead were "small" or to the determination that there was nonetheless an adequate scientific basis for regulating environmental lead. A paper by lead additive manufacturers criticizing some aspects of the commission's economic analysis received virtually no publicity and had no effect on policy.

The picture on the U.S. side was, of course, vastly different. All of the formal techniques used by U.S. regulators to assess risk or express uncertainty were widely debated both within and outside the administrative process. The validity of the EPA's approach to carcinogen risk assessment, for example, was tested both in a succession of formal administrative proceedings, including peer review by the agency's Science Advisory Board, and in scientific journals, congressional hearings, and even lawsuits. Through these repeated cycles of explanation, revision, and refinement, the EPA was finally able to persuade its varied public and private constituencies—industry, environmentalists, labor, Congress, the Office of Management and Budget (OMB), the courts, and independent scientists—that its analytical principles were consistent with both its regulatory mission and current scientific knowledge. The process of persuasion, however, was infinitely more costly than any risk-assessment exercise undertaken by a British agency.

EVIDENCE AND ACCOUNTABILITY

The differences between the British and American approaches to dealing with uncertain scientific evidence appear, at one level, paradoxical. Both countries, after all, are liberal democratic societies, committed to the principle that political power must be wielded rationally and without caprice by officials answerable to the public. Both subscribe to certain common procedural strategies for ensuring political accountability. In each country, moreover, the principle of accountability encompasses the use of technical evidence by regulators. Both nations recognize that a decision-making process incapable of exposing arbitrariness, error, or bias in an agency's handling of science does not lend itself to effective democratic control. Why, then, have risk assessment practices in the two countries diverged so markedly?

A closer comparison between the administrative and political cultures of Britain and the United States begins to reveal the reasons. The relatively closed, consensual, and nonlitigious British regulatory approach presents a stark, and well-documented, contrast with the open, adversarial, and adjudicatory style of policy development in the United States (Badaracco 1985; Brickman, Jasanoff, and Ilgen 1985; Irwin 1985; Vogel 1986; Wilson 1985). In Britain, as in most European countries, the authority to make technically grounded decisions is given to relatively apolitical and autonomous bodies of civil servants, who are reasonably well insulated against political pressure. In the United States, by contrast, the relatively low status of the civil service is coupled with a system of political appointments that diminishes the independence of the regulatory agencies, though it increases their responsiveness to presidential policies. British advisory committees, too, operate in a less-exposed environment than do their American counterparts, and they are less publicly accountable for the advice they give (Brickman, Jasanoff, and Ilgen 1985, pp. 162–64; Collingwood and Reeve 1986).

American regulators, moreover, are subject to an elaborate system of checks and balances that places them under the supervision of Congress and the courts, as well as the White House. Because of these multiple channels of accountability, U.S. agencies are answerable to a more diverse array of political pressures than are their British counterparts. Any interest group with a sufficient stake in regulation can usually count on support from one or another branch of government in pressing challenges against agency policy. During the Reagan administration, for example, business and industry forged a strong alliance with the Office of Management and Budget, whereas environmental, labor, and consumer groups continued to rely on Congress and the courts. By contrast, public participation in Britain is structured in the neocorporatist manner, so that access to government is limited to a few powerful "peak organizations" representing organized interests such as business and labor.

Given the highly critical and competitive character of U.S. politics, it is not surprising that all three governmental branches have insisted at one time or another that the rationality and objectivity of regulatory decisions should be demonstrated, if possible, through quantitative analysis. Starting in the Nixon administration, successive presidents responded to industry concerns about regulation by requiring agencies to quantify the economic impacts of their major regulatory proposals. Subsequently, the White House's efforts to hold down the costs of social regulation centered on revising and coordinating the techniques of quantitative risk assessment developed by individual administrative agencies. These activities were spearheaded in the Carter years by the Interagency Regulatory Liaison Group (IRLG) and in the Reagan years by the Office of Science and Technology Policy (OSTP).

The U.S. Congress, too, has displayed an active interest in asking agencies to quantify the basis for their regulatory decisions. Quantification of economic impacts became standard practice under provisions written into many health and safety statutes that agency decisions should be supported by explicit analyses of costs and benefits. Such mandates, in turn, provided a strong impetus for quantifying the assessment of risk. Using numerical risk estimates, agencies were able to calculate the expected costs of different levels of risk reduction and to select regulatory options that satisfied statutory requirements of cost effectiveness and economic feasibility.

More generally, however, American regulatory statutes place agencies in a political environment in which quantitative analysis becomes a lifeline to legitimacy. The unique U.S. propensity to define an ambitious legislative agenda without regard to the scientific, economic, and political realities (Brickman, Jasanoff, and Ilgen 1985, pp. 35–39) in many cases pushed agencies toward predicting consequences that were empirically unverifiable. The requirement that the EPA establish health-based air quality standards for pollutants like lead offers one such example. In order to meet the legislative demand for a uniform national standard, the EPA was virtually compelled to operate beyond the limits of available knowledge and to draw inferences about effects that were not directly measurable. Quantitative risk assessment filled the vacuum between what was known and what had to be known in order to implement legislation. There are no comparable standard-setting obligations under British law, and policymakers in Britain are free, for the most part, to base their decisions on pragmatic evaluations of what was already known and how much it would cost to know more.

Activist judicial review is another distinctive feature of the American regulatory process, and the courts too have played their part in pressing agencies to quantify their technical determinations. In the early years of modern environmental policymaking, federal courts declared that they were prepared to take a "hard look" at the basis for agency actions no matter how technically complex (Leventhal 1974, pp. 511–12). The "hard look" doctrine put the agencies on notice that their decisions would satisfy a reviewing court only if they were clearly reasoned, thoroughly documented, and responsive to the concerns of all affected parties. To meet these demanding requirements, agencies took unusual pains to build massive technical dossiers before promulgating major regulatory decisions. The costs of generating such review-proof records were generally deemed worthwhile in order to forestall potentially embarrassing reversals in court.

Paradoxically, however, the very comprehensiveness of the administrative record emerged as a threat to the credibility of federal decision makers. Obliged to canvass all of the ambiguities, complexities, and question marks presented by science, agencies were seldom in a position to argue that the evidence displayed in the record clearly supported their chosen course of action. Indeed, by revealing all the weaknesses in the underlying evidence, regulators were virtually doomed to justifying their decisions by means of arguments that appeared, at best, subjective and, at worst, tainted by economic or political bias.

The seeming arbitrariness of the scientific rationale underlying expensive regulatory measures made the courts in the 1980s increasingly skittish about accepting only the agencies' unsupported judgments on how to interpret complex risk evidence. An era of relatively sustained deference to agency expertise gave way to a period of judicial skepticism. In a number of cases involving toxic substances, the courts either refused to accept administrative assessments of risk or asked that more compelling technical support be provided for agency decisions (*Gulf South Insulation* v. *Consumer Product Safety Commission,* 701 F.2d 1137 [5th Cir. 1983]).

One of the most influential markers of this shift in judicial attitudes was a 1980 decision of the U.S. Supreme Court, in which a majority of the justices concluded that OSHA's proposed workplace exposure standard for benzene could not be upheld (*Industrial Union Department. AFL-CIO* v. *American Petroleum Institute,* 448 U.S. 607 [1980]). Especially relevant to our discussion was the court's endorsement of quantitative risk assessment. A plurality of the justices held that OSHA could tighten the benzene standard only if it could demonstrate, by means of convincing numerical projections, that there was a "significant" risk to health at existing exposure levels and that this risk would be reduced by adopting the proposed new standard.

In the United States, then, the use of quantitative risk analysis can be seen as a response to the exceptionally exposed position in which regulatory agencies are placed. Reducing scientific uncertainty to mathematical terms offers decision makers a means of rationalizing actions that might otherwise seem insupportably arbitrary and subjective. Evidence translated into quantitative terms appears to speak for itself rather than through a distorting filter of political interpretation. Numbers carry particular appeal when decision makers are answerable to a multiplicity of lay critics, both inside and outside the government. Even the most carefully crafted risk estimates, presented with due attention to uncertainty, appear to transcend politics, effacing the underlying discretionary judgments of experts. Numerical assessments possess a kind of symbolic neutrality that is rarely attained by qualitative formulations about the "weight" or "sufficiency" of the evidence.

But although the strategy of quantification seems on the surface both more transparent and more amenable to public control than are other forms of regulatory analysis, the debate over risk assessment in the United States suggests that these apparent advantages are offset by serious practical and philosophical costs. There are, to start with, the investments of time, money, and personnel that must be made in order to develop risk-assessment methodologies and to create an evidentiary base (e.g., bioassays or clinical studies) that lends itself to such analysis. There is, moreover, little evidence that these expenditures shorten the regulatory process, facilitate dispute resolution, or otherwise lead to "better" decisions. As the formaldehyde case suggests, the use of quantitative methods may close some areas of scientific controversy only by shifting debate into other areas. In this case, the initial dispute about the appropriate model for low-dose extrapolation was eventually displaced by an equally intransigent debate over the validity of the CIIT's research on delivered dose.

The phenomenon of protracted technical controversy, accompanied by continual shifts from one to another locus of uncertainty, supports the view that quantitative risk assessment is far less an independent decision technique than a surrogate for deeper political divergences that choose (or sometimes are forced by law) to express themselves as disputes about evidence. Yet if risk analysis merely converts political arguments into technical ones, the cause of accountability is not likely to be well served. Because of their esoteric nature, such methodologies threaten to elude the traditional processes of democratic control. Scientific peer review provides a partial remedy against crass abuses of quantitative techniques, but this is a remedy that tends, if anything, to consolidate power within the expert community. It is in its own way a threat to the democratic legitimation of public policies, not least because peer review in the regulatory environment can so easily be subverted by politics (Jasanoff 1985).

It is a particularly ironic twist that in striving to counter possible misuses of quantitative analysis, U.S. agencies have been driven to develop analytical methods that still more precisely capture the uncertainties embedded in risk assessment. Techniques such as judgmental probability encoding and some of the refinements in carcinogenic risk assessment may indeed provide truer pictures of what experts really believe and how far their evaluations rest on subjective considerations. In the process of characterizing these beliefs, however, some of the benefits initially sought through quantitative analysis could well be lost. If the agencies succeed, the numbers they generate will merely reflect more and more perfectly the uncertainties of the underlying evidence. The ultimate decision maker will still be confronted with the problem of cutting the knot of uncertainty, and it is by no means clear that better quantitative characterizations of the range of political choice will enhance the legitimacy of the final decision.

CONCLUSION

I have suggested that risk assessment in the United States is the product of a political process that combines large policy expectations with little trust in those called upon to formulate specific policy outcomes. Such a system threatens the credibility both of regulators and science as an institution (Weinberg 1985, pp. 209–22). Pressing the evidence to produce levels of precision that it cannot support augments controversy and may feed the disenchantment with science and the scientific community already endemic in the U.S. political environment. Simultaneously, the effort to bring an unattainable level of technical rationality to decisions that are fundamentally subjective and political may weaken trust in government. The suspicion that risk assessment serves merely a political function in federal policymaking is likely to fuel the public demand for radical alternatives, such as the discharge prohibitions and right-to-know requirements of Proposition 65 in California. Approved by California voters in 1986, this law requires businesses to stop dumping listed carcinogens into drinking water and to provide clear warnings when they dispose of these substances elsewhere. Experts decried the provision as populist excess, fed by limited public understanding of science. But one could equally read Proposition 65 as an impatient response to the obfuscation and political stalemate induced by proliferating, yet inconclusive, uses of quantitative risk assessment.

Comparisons between the United States and Britain suggest, however, that the British methods of dealing with technical uncertainty, despite their many attractions, cannot readily be assimilated into the U.S. policy process. The tensions and paradoxes that mark the analysis of risk in America are to a large extent unavoidable by-products of a legal and political environment that prizes openness and diversity along with scientific rationality. In this pluralistic society, political arguments, including esoteric arguments about scientific evidence, must necessarily be resolved through negotiation and debate among many competing interests. The politics of regulation in the United States leaves little room for appealing to a single higher authority, which, like Britain's Royal Commission on Environmental Pollution, can speak as the voice of reason and expertise for a whole society.

REFERENCES

Badaracco, J. (1985). *Loading the Dice.* Boston: Mass. Harvard Business School Press.
Brickman, R., Jasanoff, S., and Ilgen, T. (1985). *Controlling Chemicals: The Politics of Regulation in Europe and the United States.* Ithaca, N.Y.: Cornell University Press.

Collingwood, D., and Reeve, C. (1986). *Science Speaks to Power*. London: Frances Pintner.

Environmental Protection Agency (EPA), Office of Pesticides and Toxic Substances (1987). *Assessment of Health Risks to Garment Workers and Certain Home Residents from Exposure to Formaldehyde*. Appendix I: *Expert Panel Report on HCHO Pharmacokinetic Data and CIIT Response*. Washington, D.C.: EPA.

Gillespie, B., Eva, D., and Johnston, R. (1979). "Carcinogenic Risk Assessment in the United States and Great Britain: The Case of Aldrin/Dieldrin." *Social Studies of Science* 9:265–301.

Hanson, D. J. (1984). "Agricultural Uses of Ethylene Dibromide Halted." *Chemical and Engineering News*, March 5, p. 13.

Heidenheimer, A. J., Heclo, H., and Adams, C. T. (1975). *Comparative Public Policy: The Politics of Social Choice in Europe and America*. New York: St. Martin's Press.

Irwin, A. (1985). *Risk and the Control of Technology*. Oxford: Oxford University Press.

Jasanoff, S. (1983). "Negotiation or Cost–Benefit Analysis: A Middle Road for U.S. Policy?" *Environmental Forum* 2:37–43.

Jasanoff, S. (1985). "Peer Review in the Regulatory Process." *Science, Technology, and Human Values* 10:20–32.

Jasanoff, S. (1986). *Risk Management and Political Culture*. New York: Russell Sage Foundation.

Jasanoff, S. (1987). "Cultural Aspects of Risk Assessment in Britain and the United States." In *The Social and Cultural Construction of Risk*, ed. B. B. Johnson and V. T. Covello, pp. 359–97. New York: Reidel.

Jasanoff, S. (1990). *The Fifth Branch: Science Advisors as Policymakers*. Cambridge, Mass.: Harvard University Press.

Leventhal, H. (1974). "Environmental Decisionmaking and the Role of the Courts." *University of Pennsylvania Law Review* 122:511–12.

McCrea, F., and Markle, G. (1984). "The Estrogen Replacement Controversy in the USA and UK: Different Answers to the Same Question?" *Social Studies of Science* 14:1–26.

Morgan, M. G., Morris, S. C., Henrion, M., Amaral, D. A. L., and Rish W. R. (1984). "Technical Uncertainty in Quantitative Policy Analysis—A Sulfur Air Pollution Example." *Risk Analysis* 4:209.

Mumpower, J. (1986). "An Analysis of the *de Minimis* Strategy for Risk Management." *Risk Analysis* 6:441.

National Academy of Sciences–National Research Council (NAS–NRC) (1978). *Saccharin: Technical Assessment of Risks and Benefits*. Pt. I. National Technical Information Series, no. PB292 695. Washington, D.C., November 6.

National Academy of Sciences–National Research Council (NAS–NRC) (1983). *Risk Assessment in the Federal Government: Managing the Process*. Washington, D.C.: National Academy Press.

Peterson, C. (1985). "How Much Risk Is Too Much?" *Washington Post National Weekly Edition*, February 4, p. 8.

"Report on the Consensus Workshop on Formaldehyde" (1984). *Environmental Health Perspectives* 58:328–81.

Royal Commission on Environmental Pollution (1971). *First Report*. London: Her Majesty's Stationery Office.

Royal Commission on Environmental Pollution (1983). *Ninth Report: Lead in the Environment*. London: Her Majesty's Stationery Office.

Ruckelshaus, W. (1984). "Risk in a Free Society." *Risk Analysis* 4:161.

Rushefsky, M. E. (1986). *Making Cancer Policy*. Albany: State University of New York Press.

Silbergeld, E. (1986). "The Uses and Abuses of Scientific Uncertainty in Risk Assessment." *Natural Resources and Environment* 2:17–20, 57–59.

Southwood, T. R. E. (1985). "The Roles of Proof and Concern in the Work of the Royal Commission on Environmental Pollution." *Marine Pollution Bulletin* 16:346–50.

Travis, C. C., Richter, S. A., Crouch, E. A. C., Wilson, R., and Klema, E. D. (1987). "Risk and Regulation." *Chemtech August 1987:* 478–83.

Vogel, D. (1986). *National Styles of Regulation*. Ithaca, N.Y.: Cornell University Press.

Wallsten, T. S., and Whitfield, R. G., (1986). "Assessing the Risks to Young Children of Three Effects Associated with Elevated Blood-Lead Levels." Report prepared for EPA Office of Air Quality Planning and Standards, Argonne National Laboratory, Argonne, Ill., December.

Weinberg, A. (1985). "Science and Its Limits: The Regulator's Dilemma." *Issues in Science and Technology* 2:209–22.

Whipple, C. (1984). "Application of the *de Minimis* Concept in Risk Management." Paper presented to the Joint Session of the American Nuclear Society and Health Physics Society, New Orleans, June 6.

Whittemore, A. S. (1983). "Facts and Values in Risk Analysis for Environmental Toxicants." *Risk Analysis* 3:23–33.

Wilson, G. (1985). *The Politics of Safety and Health*. Oxford: Oxford University Press.

3

Beyond Numbers: A Broader Perspective on Risk Perception and Risk Communication

PAUL SLOVIC

> To effectively manage . . . risk, we must seek new ways to involve the public in the decision-making process. . . . They [the public] need to become involved early, and they need to be informed if their participation is to be meaningful.
>
> Ruckelshaus (1983, p. 1028)

In a bold and insightful speech before the National Academy of Sciences at the beginning of his second term as the administrator of the Environmental Protection Agency (EPA), William Ruckelshaus called for a governmentwide process for managing risks that involved the public. Arguing that the government must accommodate the will of the people, he quoted Thomas Jefferson's famous dictum to the effect that "if we think [the people] not enlightened enough to exercise their control with a wholesome discretion, the remedy is not to take it from them, but to inform their discretion" (Ruckelshaus 1983, p. 1028). Midway into his tenure as the EPA administrator, Ruckelshaus's experiences in attempting to implement Jefferson's philosophy led him to a more sober evaluation: "Easy for *him* to say. As we have seen, informing discretion about risk has itself a high risk of failure" (Ruckelshaus, 1984, p. 160).

This chapter attempts to illustrate why the goal of informing the public about risk issues—which in principle seems easy to attain—is surprisingly difficult to accomplish. To be effective, risk communicators must recognize and overcome a number of obstacles that have their roots in the limitations of scientific risk assessment and the idiosyncrasies of the human mind. Doing an adequate job of communicating means finding comprehensible ways of presenting complex technical material that is clouded by uncertainty and inherently difficult to understand. Greater awareness of the nature of risk perceptions, the fundamental values and concerns that underlie those perceptions, and the difficulties of communicating about risk should enhance the chances of designing an environment in which all parties can cooperate in solving the common problems of risk management.

LIMITATIONS OF RISK ASSESSMENT

Risk assessment is a complex discipline, not fully understood by its practitioners, much less by the lay public. At the technical level, there is still much debate over terminology and techniques, and technical limitations and disagreements among experts inevitably affect communication in the adversarial climate that surrounds many risk issues. Those conveying these issues to the public must be aware of the strengths and limits of the methods they use to generate this information. In particular, such risk communicators need to understand that risk assessments are constructed from theoretical models that are based on assumptions and subjective judgments. If these assumptions and judgments are deficient, the resulting assessments may be quite inaccurate.

Nowhere are these problems more evident than in the assessment of chronic health effects caused by low-level exposures to toxic chemicals and radiation. The typical assessment uses studies of animals exposed (relatively briefly) to extremely high doses of the substance, in order to draw inferences about the risks to humans exposed to very low doses (sometimes over long periods of time). The models designed to extrapolate the results from animals to humans and from high doses to low doses are controversial. For example, some critics have argued that mice may be from 3×10^4 to 10^9 times more prone to cancer than are humans (Gori 1980). Different models for extrapolating from high-dose exposures to low-dose exposures produce estimated cancer rates that can differ by factors of 1000 or more at the expected levels of human exposures (which themselves are often subject to a great deal of uncertainty). Difficulties in estimating synergistic effects (interactions between two or more substances, such as between cigarette smoking and exposure to asbestos) and effects on particularly sensitive people (e.g., children, pregnant women, the elderly) further compound the problems of risk assessment. In light of these various uncertainties, one expert concluded that "discouraging as it may seem, it is not plausible that animal carcinogenesis experiments can be improved to the point where quantitative generalizations about human risk can be drawn from them" (Gori 1980, p. 259).

In addition, in the adversarial climate of risk discussions, these limitations of assessment are brought forth to discredit quantitative risk estimates. To be credible and trustworthy, a communicator must know enough to acknowledge valid criticisms and to discern whether the available risk estimates are valid enough to help the public gain perspective on the dangers they face and the decisions that must be made.

On the positive side, there are some hazards (e.g., radiation, asbestos) whose risks are relatively well understood. Moreover, for many other hazards, risk estimates are based on a chain of conservative decisions at each choice point in the analysis (e.g., studying the most sensitive species, using the extrapolation model that produces the highest risk estimate, giving benign tumors the same weight as malignant ones, etc.). Despite the uncertainties, one may have great confidence that the "true risk" is unlikely to exceed the estimate resulting from such a conservative process. In other words, uncertainty and subjectivity do not imply chaos. Communicators thus must know when this point is relevant and how to make it when it applies.

Parallel problems exist in engineering risk assessments designed to estimate the probability and severity of rare, high-consequence accidents in complex systems such as nuclear reactors or liquefied natural gas (LNG) plants. The risk estimates are calculated from theoretical models (in this case, fault trees or event trees) that attempt to depict all possible accident sequences and their (judged) probabilities. Limitations in the quality or comprehensiveness of the analysis and the quality of the judged risks for individual sequences, or improper rules for combining estimates, can seriously compromise the validity of the assessment.

LIMITATIONS OF PUBLIC UNDERSTANDING

Just as they must understand the strengths and limitations of risk assessment, communicators must appreciate the wisdom and folly in public attitudes and perceptions. Among the important research findings and conclusions are the following.

People's Perceptions of Risk Are Sometimes Inaccurate

Risk judgments are influenced by the memorability of past events and the imaginability of future events. As a result, any factor that makes a hazard unusually memorable or imaginable, such as a recent disaster, heavy media coverage, or a vivid film, can seriously distort the public's perceptions of risk. In particular, studies by Lichtenstein, Slovic, Fischhoff, Layman, and Combs (1978), Morgan and colleagues (1983), and others have found that risks from dramatic or sensational causes of death, such as accidents, homicides, cancer, and natural disasters, tend to be greatly overestimated. Risks from undramatic causes such as asthma, emphysema, and diabetes, which take one life at a time and are common in nonfatal form, tend to be underestimated. News media coverage of hazards has been found to be biased in much the same direction, thus contributing to the difficulties of obtaining a proper perspective on risks (Combs and Slovic 1978).

Risk Information May Frighten and Frustrate the Public

The fact that perceptions of risk are often inaccurate points to the need for warnings and educational programs. However, to the extent that misperceptions are due to reliance on imaginability as a cue for riskiness, such programs may run into trouble. Merely mentioning possible adverse consequences (no matter how rare) of some product or activity can enhance their perceived likelihood and make them appear more frightening. The anecdotal observation of attempts to inform people about recombinant DNA hazards supports this hypothesis (Rosenburg 1978), as does a controlled study by Morgan and colleagues (1985).

In the latter study, people's judgments of the risks from high-voltage transmission lines were assessed before and after they read a brief and rather neutral description of the findings from studies of possible health effects caused by such lines. The results clearly indicated a shift toward greater concern in three separate groups of subjects who read the description. Whereas the mere mention and refutation of potential risks raises concerns, the use of conservative assumptions and "worst-case scenarios" in risk assessment causes people to react extremely negatively because of their difficulty appreciating the improbability of such extreme but imaginable consequences. The fact that imaginability may blur the distinction between what is (remotely) possible and what is probable obviously poses a serious obstacle to risk-information programs.

Other psychological research shows that people may have great difficulty making decisions about gambles when they are forced to resolve conflicts generated by the possibility of experiencing both gains and losses, and uncertain ones at that (Slovic 1982; Slovic and Lichtenstein 1983). As a result, whenever possible, people try to reduce the anxiety generated by uncertainty by denying that uncertainty, thus making the risk seem either so small that it can safely be ignored or so large that it clearly should be avoided. They object to being given statements of probability rather than fact; they want to know exactly what will happen.

Given a choice, people would rather not have to confront the gambles inherent in life's dangerous activities. They would prefer being told that risks are managed by competent professionals and are thus so small that they need not worry about them. However, if such assurances cannot be given, they will want to be informed of the risks, even though finding out about them might make them feel anxious and confused (Alfidi 1971; Fischhoff 1983; Weinstein 1979).

Strong Beliefs Are Hard to Modify

It would be comforting to believe that polarized positions respond to informational and educational programs. Unfortunately, psychological research demonstrates that people's beliefs change slowly and are extraordinarily persistent in the face of contrary evidence (Nisbett and Ross 1980). Once formed, initial impressions tend to structure the way that subsequent evidence is interpreted. That is, new evidence will appear reliable and informative if it is consistent with one's initial beliefs, and conversely, contrary evidence will be dismissed as unreliable, erroneous, or unrepresentative.

Naive Views Are Easily Manipulated by Presentation Format

When people lack strong opinions, the opposite situation exists—they are at the mercy of the way that the information is presented to them. Subtle changes in the way that risks are expressed can have a major impact on their perceptions and decisions. One dramatic example of this comes from a study by McNeil, Pauker, Sox, and Tversky (1982), who asked people to imagine that they had lung cancer and had to choose between two therapies, surgery or radiation. The two therapies were described in some detail. Then, some of the subjects were presented with the cumulative probabilities of surviving for varying lengths of time after the treatment. Other subjects received the same cumulative probabilities framed in terms of dying rather than surviving (e.g., instead of being told that 68 percent of those having surgery will have survived after one year, they were told that 32 percent will have died). Framing the statistics in terms of dying increased the percentage of subjects choosing radiation therapy over surgery, from 18 percent to 44 percent. Moreover, the effect was as strong for physicians as for laypersons.

Numerous other examples of "framing effects" have been demonstrated by Tversky and Kahneman (1981) and Slovic, Fischhoff, and Lichtenstein (1982). The fact that subtle differences in how risks are presented can have such marked effects suggests that those responsible for information programs have considerable latitude to manipulate perceptions and behavior. This possibility raises ethical problems that must be addressed by any responsible risk-information program.

PLACING RISKS IN PERSPECTIVE

Choosing Risk Measures

When we know enough to be able to describe risks quantitatively, we face a wide choice of options regarding the specific measures and statistics used to describe the magnitude of risk. Fischhoff, Lichtenstein, Slovic, Derby, and Keeney (1981) point out that choosing a risk measure involves several steps: (1) defining the hazard category, (2) deciding what consequences to measure (or report), and (3) determining the unit of observation. The way in which the hazard category is defined can have a major effect on risk statistics.

Crouch and Wilson (1982) provide some specific examples of how different measures of the same risk can sometimes give quite different impressions. For example, they show that the number of accidental deaths per million tons of coal mined in the United States has decreased steadily over time. In this respect, the industry is getting safer. However, they also report that the rate of accidental deaths per one thousand coal-mine employees has increased. Neither measure is the "right" measure of mining risk. Each tells part of the same story.

The problem of selecting measures is complicated even more by the framing effects described earlier. Thus not only do different measures of the same hazard give different impressions, but also the same measure, differing only in (presumably) inconsequential ways, can lead to vastly different perceptions.

A case study of communicating information about the risks of the pesticide ethylene dibromide (EDB) points to an important distinction between macro- and micromeasures of risk (Sharlin 1986). The Environmental Protection Agency, which was responsible for regulating EDB, broadcast information about the aggregate risk of this pesticide to the exposed population. Although the media accurately transmitted this macroanalysis, newspaper editorials and public reaction clearly were unable to translate it into a microperspective regarding the

Table 3.1. Annual Fatality Rates per 100,000 Persons at Risk

Risk	Rate
Motorcycling	2000
Aerial acrobatics (planes)	500
Smoking (all causes)	300
Sport parachuting	200
Smoking (cancer)	120
Firefighting	80
Hang gliding	80
Coal mining	63
Farming	36
Motor vehicles	24
Police work (nonclerical)	22
Boating	5
Rodeo performer	3
Hunting	3
Fires	2.8
1 diet drink/day (saccharin)	1.0
4 T. peanut butter/day (aflatoxin)	0.8
Floods	0.06
Lightning	0.05
Meteorite	0.000006

Source: Adapted from Crouch and Wilson (1982).

risk to the exposed individual. In other words, the newspaper reader or TV viewer had trouble inferring an answer to the question, Can I eat the bread? from the aggregate risk analysis.

Basic Statistical Presentations

In this section, I shall describe a few of the statistical displays most often used to educate people about general and specific risks. I do not mean to endorse these presentations as effective; in fact, they have some serious limitations, which shall be discussed later.

Among the few "principles" in this field that seem to be useful is the assertion that comparisons are more meaningful than are absolute numbers or probabilities, especially when those absolute values are quite small. Sowby (1965) argued that to decide whether or not we are responding adequately to radiation risks we need to compare them with "some of the other risks of life," and Rothschild (1978) observed that "there is no point in getting into a panic about the risks of life until you have compared the risks which worry you with those that don't, but perhaps should."

Familiarity with annual mortality risks for the population as a whole or as a function of age may be one standard for evaluating specific risks. Sowby (1965) took advantage of such data to observe that one hour riding a motorcycle was as risky as one hour of being seventy-five years old. Table 3.1 provides annual mortality rates from a wide variety of causes.[1]

Mortality rates fail to capture the fact that some hazards (e.g., pregnancy, motorcycle accidents) cause death at a much earlier age than do others (e.g.,

Table 3.2. Estimated Loss of Life Expectancy Due to
Various Causes

Cause	Days
Cigarette smoking (male)	2,250
Heart disease	2,100
Being 30% overweight	1,300
Being a coal miner	1,100
Cancer	980
Stroke	520
Army in Vietnam	400
Dangerous jobs, accidents	300
Motor vehicle accidents	207
Pneumonia, influenza	141
Accidents in home	95
Suicide	95
Diabetes	95
Being murdered (homicide)	90
Drowning	41
Job with radiation exposure	40
Falls	39
Natural radiation (beir)	8
Medical x-rays	6
Coffee	6
All catastrophes combined	3.5
Reactor accidents (UCS)	2*
Radiation from nuclear industry	0.02*

*These items assume that all U.S. power is nuclear. UCS is the Union of Concerned Scientists,
the most prominent group of critics of nuclear energy.
Source: Cohen and Lee (1979).

lung cancer due to smoking). One way to provide a perspective on this consideration is to calculate the average loss of life expectancy due to the exposure to the hazard, based on the distribution of deaths as a function of age. Some estimates of loss of life expectancy from various causes are shown in Table 3.2.

Yet another innovative way to gain perspective was devised by Wilson (1979), who displayed a set of activities (Table 3.3), each of which was estimated to increase one's chance of death (during any year) by one in a million.

Comparisons within lists of risks such as those in Tables 3.1, 3.2, and 3.3 have been advocated not just to gain some perspective on risks but also to guide decision making. Thus Cohen and Lee (1979) argued that "to some approximation, the ordering [in Table 3.2] should be society's order of priorities" (1979, p. 720). However, Slovic, Fischhoff, and Lichtenstein (1980b) contended that such claims could not be logically defended. Although carefully prepared lists of risk statistics can provide some degree of insight, they are only a small part of the information needed for decisions. As a minimum, such inputs should include a detailed account of the costs and benefits of the available options, as well as an indication of the uncertainty in these assessments. As we have seen, uncertainties in risk estimates are often quite large. Failure to indicate uncertainty not only deprives the recipient of needed information but it also can lead to distrust and rejection of the analysis.

Table 3.3. Risks Estimated to Increase Chance of Death in Any Year by 0.000001 (1 part in 1 million)

Activity	Cause of Death
Smoking 1.4 cigarettes	Cancer, heart disease
Spending 1 hour in a coal mine	Black lung disease
Living 2 days in New York or Boston	Air pollution
Traveling 10 miles by bicycle	Accident
Flying 1000 miles by jet	Accident
Living 2 months in Denver on vacation from New York	Cancer caused by cosmic radiation
One chest x-ray taken in a good hospital	Cancer caused by radiation
Eating 40 T of peanut butter	Liver cancer caused by aflatoxin B
Drinking 30, 12-oz. cans of diet soda	Cancer caused by saccharin
Drinking 1000, 24-oz. soft drinks from recently banned plastic bottles	Cancer from acrylonitrile monomer
Living 150 years within 20 miles of a nuclear power plant	Cancer caused by radiation
Risk of accident by living within 5 miles of a nuclear reactor for 50 years	Cancer caused by radiation

Source: Wilson (1979).

TOWARD A BROADER PERSPECTIVE ON RISK PERCEPTION AND COMMUNICATION

A stranger in a foreign land would hardly expect to communicate effectively with the natives without knowing something about their language and culture. Yet risk assessors and risk managers have often tried to communicate with the public under the assumption that they and the public share a common conceptual and cultural heritage in the domain of risk. That assumption, however, is false and has led to failures of communication and rancorous conflicts.

The Psychometric Paradigm

Evidence against the "commonality assumption" comes from sociological, psychological, and anthropological studies directed at understanding the determinants of people's risk perceptions and behaviors. In psychology, research within what has been called the "psychometric paradigm" has explored the ability of psychophysical scaling methods and multivariate analysis to produce meaningful representations of risk attitudes and perceptions (see, e.g., Brown and Green 1980; Gardner et al. 1982; Green 1980; Green and Brown 1980; Johnson and Tversky, 1984; Lindell and Earle 1983; MacGill 1983; Renn 1981; Slovic, Fischhoff, and Lichtenstein, 1980a, 1984; Vlek and Stallen 1981; von Winterfeldt, John, and Borcherding 1981).

Researchers employing the psychometric paradigm typically ask people to judge the current riskiness (or safety) of diverse sets of hazardous activities, substances, and technologies and to indicate their desires for the risk reduction and regulation of these hazards. These global judgments have then been related to judgments about the hazard's status on various qualitative characteristics of risk, some of which are shown in Table 3.4.

Table 3.4. Characteristics Examined in Psychometric Studies of Perceived Risk

Voluntary–Involuntary
Chronic–Catastrophic
Common–Dread
Injurious–Fatal
Known to those exposed–Not known to those exposed
Known to Science–Not known to science
Controllable–Not controllable
Old–New

In the following section, I shall briefly review some of the results obtained from psychometric studies of risk perception and outline some of the implications of these results for risk communication and risk management.

Revealed and Expressed Preferences

The original impetus for the psychometric paradigm came from the pioneering effort of Starr (1969) to develop a method for weighing technological risks against benefits in order to answer the fundamental question: How safe is safe enough? His "revealed preference" approach assumed that by trial and error society has arrived at an "essentially optimum" balance between the risks and benefits associated with any activity. One may therefore use historical or current risk and benefit data to reveal patterns of "acceptable" risk–benefit trade-offs. Examining such data for several industries and activities, Starr concluded that (1) the acceptability of risk from an activity is roughly proportional to the third power of the benefits for the activity and (2) the public will accept risks from voluntary activities (such as skiing) that are roughly one thousand times as great as it would tolerate from involuntary hazards (such as food preservatives) that provide the same level of benefits.

The merits and deficiencies of Starr's approach have been debated at length (see, e.g., Fischhoff et al. 1981). I shall not elaborate on them here, except to note that the concern about the validity of the many assumptions inherent in the revealed preferences approach stimulated Fischhoff and associates (1978) to conduct an analogous psychometric analysis of questionnaire data, resulting in "expressed preferences." More recently, numerous other studies of expressed preferences have been carried out within the psychometric paradigm.

These studies have shown that perceived risk is both quantifiable and predictable. Psychometric techniques seem well suited to identifying similarities and differences among people with regard to risk perceptions and attitudes. They have also demonstrated that the concept of risk means different things to different people. When experts judge risk, their responses correlate highly with technical estimates of annual fatalities. Lay people can assess annual fatalities if they are asked to (and produce estimates somewhat like the technical estimates). However, their judgments of risk are related more to other hazard characteristics (e.g., catastrophic potential, threat to future generations) and, as a result, tend to differ from their own (and experts') estimates of annual fatalities.

Another consistent finding from psychometric studies of expressed preferences is that people tend to view current risk levels as unacceptably high for most activities. The gap between perceived and desired risk levels suggests that people are not satisfied with the way that market and other regulatory mechanisms have balanced risks and benefits. Across the domain of hazards, there seems to be little systematic relationship between perceptions of current risks and benefits. However, studies of expressed preferences do seem to support Starr's claim that people are willing to tolerate higher risks from activities seen as highly beneficial. But whereas Starr concluded that voluntariness of exposure was the key mediator of risk acceptance, expressed preference studies have shown that other (perceived) characteristics such as familiarity, control, catastrophic potential, equity, and level of knowledge also seem to influence the relationship among perceived risk, perceived benefit, and risk acceptance (Fischhoff et al. 1978; Slovic, Fischhoff, and Lichtenstein 1980a).

Various models have been advanced to represent the relationships among perceptions, behavior, and these qualitative characteristics of hazards. As we shall see, the picture that emerges from this work is both orderly and complex.

Factor-analytic Representations

Many of the qualitative risk characteristics are highly correlated with one another, across a wide range of hazards. For example, hazards judged to be "voluntary" tend also to be judged as "controllable"; hazards whose adverse effects are delayed tend to be seen as posing risks that are not well known; and so on. Investigation of these relationships by means of factor analysis has shown that the broader domain of characteristics can be condensed into a small set of higher-order characteristics or factors.

The factor space presented in Figure 3.1 has been replicated across groups of lay people and experts judging large and diverse sets of hazards. Factor 1, *dread risk,* is defined at its high (right-hand) end by perceived lack of control, dread, catastrophic potential, fatal consequences, and the inequitable distribution of risks and benefits. Nuclear weapons and nuclear power score highest on the characteristics that make up this factor. Factor 2, *unknown risk,* is defined at its high end by hazards judged to be unobservable, unknown, new, and delayed in their manifestation of harm. Chemical technologies score particularly high on this factor. A third factor, reflecting the number of people exposed to the risk, has been obtained in several studies. Making the set of hazards more or less specific (e.g., partitioning nuclear power into radioactive waste, uranium mining, and nuclear reactor accidents) has had little effect on the factor structure or its relationship to risk perceptions (Slovic, Fischhoff, and Lichtenstein 1985).

Research has shown that lay people's risk perceptions and attitudes are closely related to the position of a hazard within this type of factor space. Most important is the horizontal factor dread risk. The higher a hazard's score is on this factor (the farther to the right it appears in the space), the higher its perceived risk will be, the more people will want to see its current risks reduced, and the more they will want strict regulation used to achieve the desired reduction in risk (Figure 3.2). In contrast, experts' perceptions of risk are not closely related to any of the various risk characteristics or factors derived from these characteristics (Slovic, Fischhoff, and Lichtenstein 1985). Instead, as we noted earlier,

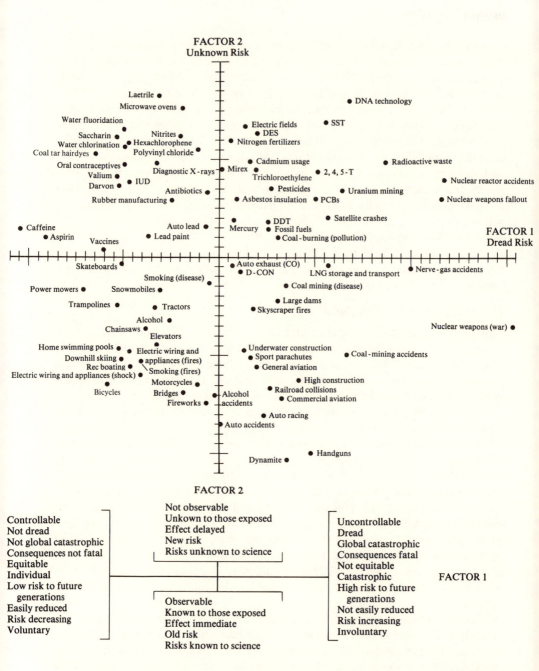

Figure 3.1 Location of eighty-one hazards on Factors 1 and 2 derived from the relationships among eighteen risk characteristics. Each factor is made up of a combination of characteristics, as indicated by the lower diagram. (Slovic, Fischhoff, and Lichtenstein 1985)

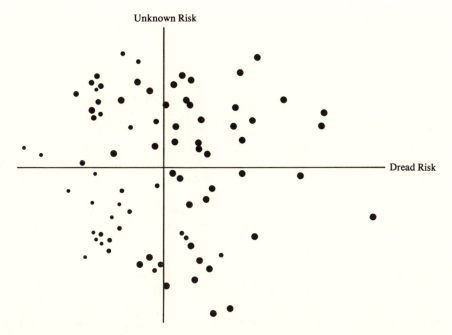

Figure 3.2 Attitudes toward regulation of the hazards in Figure 3.1. The larger the point, the greater the desire for strict regulation to reduce risk. (Slovic, Fischhoff, and Lichtenstein 1985)

experts appear to see riskiness as synonymous with expected annual mortality (Slovic, Fischhoff, and Lichtenstein 1979). As a result, conflicts over "risk" may result from experts and lay people having different definitions of the concept.

The representation shown in Figure 3.1, though robust and informative, is by no means a universal cognitive mapping of the domain of hazards. Other psychometric methods (such as multidimensional-scaling analysis of hazard-similarity judgments), applied to quite different sets of hazards, produce different spatial models (Johnson and Tversky 1984). The utility of these models for understanding and predicting behavior remains to be determined.

Accidents as Signals

Risk analyses typically model the impacts of an unfortunate event (such as an accident, a discovery of pollution, sabotage, and product tampering) in terms of direct harm to victims: deaths, injuries, and damages. The impacts of such events, however, sometimes extend far beyond these direct harms and may include significant indirect costs (both monetary and nonmonetary) to the responsible government agency or private company that far exceed the direct costs. In some cases, all companies in an industry are affected, regardless of which company was responsible for the mishap. In extreme cases, the indirect costs of a mishap may extend past industry boundaries, affecting companies, industries, and agencies whose business is minimally related to the initial event. Thus, an unfortunate event can be thought of as analogous to a stone dropped in a pond. The ripples spread outward, encompassing first the directly affected victims, then the responsible company or agency, and, finally, reaching other companies, agencies, and industries.

Some events make only small ripples; others make larger ones. The challenge is to discover the characteristics associated with an event and the way that it is managed that can predict the breadth and seriousness of those impacts (see Figure 1.3). Early theories equated the magnitude of impact to the number of people killed or injured, or to the amount of property damaged. However, the accident at the Three Mile Island (TMI) nuclear reactor in 1979 provides a dramatic demonstration that factors besides injury, death, and property damage impose serious costs. Despite the fact that not a single person died at TMI and few if any latent cancer fatalities are expected, no other accident in our history has produced such costly societal impacts. The accident at TMI devastated the utility that owned and operated the plant. It also imposed enormous costs on the nuclear industry and on society, through stricter regulation (resulting in increased construction and operation costs), reduced operation of reactors world-wide, greater public opposition to nuclear power, and reliance on more expensive energy sources. It may even have led to a more hostile view of other complex technologies, such as chemical manufacturing and genetic engineering. The point is that traditional economic and risk analyses tend to neglect these higher-order impacts, and so they greatly underestimate the costs associated with certain kinds of events.

Although the TMI accident is extreme, it is by no means unique. Other recent events resulting in enormous higher-order impacts include the chemical manufacturing accident at Bhopal, India; the pollution of Love Canal, New York, and Times Beach, Missouri; the disastrous launch of the space shuttle *Challenger;* and the meltdown of the nuclear reactor at Chernobyl. Following these extreme events are myriad mishaps varying in the breadth and size of their impacts.

An important concept that has emerged from psychometric research is that the seriousness and higher-order impacts of an unfortunate event are determined in part by what that event signals or portends (Slovic, Lichtenstein, and Fischhoff 1984). The informativeness or "signal potential" of an event, and thus its potential social impact, appears to be systematically related to the characteristics of the hazard and the location of the event within the factor space described earlier (Figure 3.3). An accident that takes many lives may produce relatively little social disturbance (beyond that experienced by the victims' families and friends) if it occurs as part of a familiar and well-understood system (such as a train wreck). However, a small accident in an unfamiliar system (or one perceived as poorly understood), such as a nuclear reactor or a recombinant DNA laboratory, may have immense social consequences if it is perceived as a harbinger of further and possibly catastrophic mishaps.

The concept of accidents as signals was eloquently expressed in an editorial addressing the tragic accident at Bhopal:

> What truly grips us in these accounts [of disaster] is not so much the numbers as the spectacle of suddenly vanishing competence, of men utterly routed by technology, of fail-safe systems failing with a logic as inexorable as it was once—indeed, right up until that very moment—unforseeable. And the spectacle haunts us because it

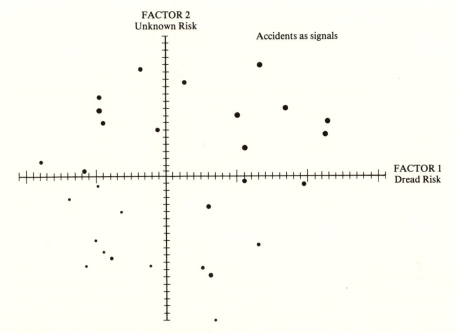

Figure 3.3 Relation between signal potential and risk characterization for thirty hazards in Figure 3.1. The larger the point, the greater the degree to which an accident involving that hazard was judged to "serve as a warning signal for society, providing new information about the probability that similar or even more destructive mishaps might occur within this type of activity." Media attention and the higher-order costs of a mishap are likely to be correlated with signal potential. (Slovic, Lichtenstein, and Fischhoff 1984)

seems to carry allegorical import, like the whispery omen of a hovering future. (*New Yorker* 1985, p. 29)

One implication of the signal concept is that effort and expense beyond that indicated by a cost–benefit analysis might be warranted to reduce the possibility of "high-signal accidents." Unfortunate events involving hazards in the upper-right quadrant of Figure 3.1 appear particularly likely to be able to produce large ripples. As a result, risk analyses involving these hazards need to be sensitive to these possible higher-order impacts, in order to bring greater protection to potential victims as well as to companies and industries.

Other Paradigms

Other important contributions to our current understanding of risk perception have come from geographers, sociologists, and anthropologists. The geographical research focused originally on understanding human behavior in the face of natural hazards, but it has since broadened to include technological hazards as well (Burton, Kates, and White 1978). The sociological work (Moatti, Stemmeling, and Fagnani 1984; Short, 1984) and the anthropological studies (Douglas and Wildavsky 1982) have shown that the perceptions of risk that have been identified within the psychometric paradigm may have their roots in social and cultural factors. Short argues that in some instances the response to hazards is caused by social influences transmitted by friends, family, fellow workers, and respected public officials. In these cases, risk perception may form afterwards, as part of one's *post hoc* rationale for his or her behavior. In a similar vein, Douglas and Wildavsky assert that people acting within social organizations downplay certain risks and emphasize others as a means of maintaining the organization's viability.

Implications for Risk Communication

Risk-perception research has a number of direct implications for communication efforts. In addition to the problems noted earlier regarding the comparative evaluation of risk statistics (such as those in Tables 3.1, 3.2, and 3.3), psychometric studies provide further insight into the reasons that comparisons are rejected as guides to personal or public decision policies. Risk perceptions and risk-taking behaviors appear to be determined not only by accident probabilities, annual mortality rates, or mean loses of life expectancy but also by other characteristics of hazards such as uncertainty, controllability, catastrophic potential, equity, and threat to future generations. Within the perceptual space defined by these and other characteristics, each hazard is unique. To many persons, statements such as "the annual risk from living near a nuclear power plant is equivalent to the risk of riding an extra three miles in an automobile" give inadequate consideration to the important differences in the nature of the risks from these two technologies.

Recognition of the limitations of such cross-hazard risk comparisons led Covello, Sandman, and Slovic (Chapter 4, this volume) to search for types of comparisons that may be better able to put results in perspective. They describe a dozen types of comparisons that may be more useful than a comparison of unrelated risks.

Another concept from risk-perception research that affects communication is the notion that accidents are signals. This concept implies that when informed about a particular hazard, people's concerns will generalize beyond the immediate problem to other related hazards. For example, with regard to the EDB scare, two newspaper editors wrote: "The cumulative effect—the 'body burden count' as scientists call it—is especially worrisome considering the number of other pesticides and carcinogens humans are exposed to" (*Honolulu Sunday Star-Bulletin and Advertiser,* February 5, 1984). And "Let's hope there are no cousins of EDB waiting to ambush us in the months ahead" (*San Francisco Examiner,* February 10, 1984).

As a result of this broad (and legitimate) perspective, communications from risk managers pertaining to the risk and control of a single hazard, no matter how carefully presented, may fail to alleviate people's fears, frustrations, and anger. If people trust the ability of the risk manager to handle the broader risk problems, these general concerns will probably not surface.

Whereas the psychometric research implies that risk debates are not merely about risk statistics, the sociological and anthropological work implies that some of these debates may not even be about risk. That is, risk may be a rationale for actions taken on other grounds, or it may be a surrogate for social or ideological concerns. When this is the case, communication about risk is simply irrelevant to the discussion, and so if possible, hidden agendas need to be brought to the surface for open discussion (Edwards and von Winterfeldt 1987).

Perhaps the most important message from the research done to date is that there is wisdom as well as error in public attitudes and perceptions. Lay people sometimes lack certain basic information about hazards. Their basic conceptualization of risk, however, is much richer than that of the experts and reflects legitimate concerns that are typically omitted from expert risk assessments. As a result, risk-communication efforts are destined to fail unless they are structured as a two-way process (Renn, in press). Each side, expert and public, has something valid to contribute, and each side must respect the insights and intelligence of the other.

ACKNOWLEDGMENT

Support for writing this chapter was provided by the National Science Foundation under grants SES–8517411 and SES–8722109 to Decision Research. The text of this chapter draws heavily on the author's joint work with Baruch Fischhoff and Sarah Lichtenstein and also on previous material:

Slovic, P. (1986). "Informing and Educating the Public About Risk." *Risk Analysis* 4:403–15.
Slovic, P. (1987). "Perception of Risk." *Science* 236:280–85.

NOTE

1. The accuracy of the statistics presented in this section is not guaranteed, as they have been taken directly from various published studies. The problems in using and interpreting risk data were pointed out earlier.

REFERENCES

Alfidi, R. J. (1971). "Informed Consent: A Study of Patient Reaction." *Journal of the American Medical Association* 216:1325.
Brown, R. A., and Green, C. H. (1980). "Precepts of Safety Assessments." *Journal of the Operational Research Society* 11:563–71.
Burton, I., Kates, R. W., and White, G. F. (1978). *The Environment as Hazard*. Oxford: Oxford University Press.
Cohen, B., and Lee, I. (1979). "A Catalog of Risks." *Health Physics* 36:707–22.
Combs, B., and Slovic, P. (1978). "Newspaper Coverage of Causes of Death." *Journalism Quarterly* 56:837–43.
Crouch, E. A. C., and Wilson, R. (1982). *Risk/Benefit Analysis*. Cambridge, Mass.: Ballinger.
Douglas, M., and Wildavsky, A. (1982). *Risk and Culture*. Berkeley: University of California Press.

Edwards, W., and von Winterfeldt, D. (1987). "Public Values in Risk Debates." *Risk Analysis* 7:141–58.

Fischhoff, B. (1983). "Informed Consent for Transient Nuclear Workers." In *Equity Issues in Radioactive Waste Management,* ed. R. Kasperson, pp. 301–28. Cambridge, Mass.: Oelgeschlager, Gunn & Hain.

Fischhoff, B., Lichtenstein, S., Slovic, P., Derby, S. L., and Keeney, R. L. (1981). *Acceptable Risk.* Cambridge: Cambridge University Press.

Fischhoff, B., Slovic, P., Lichtenstein, S., Read, S., and Combs, B. (1978). "How Safe Is Safe Enough? A Psychometric Study of Attitudes Towards Technological Risks and Benefits." *Policy Sciences* 8:127–52.

Gardner, G. T., Tiemann, A. R., Gould, L. C., DeLuca, D. R., Doob, L. W., and Stolwijk, J. A. J. (1982). "Risk and Benefit Perceptions, Acceptability Judgments, and Self-reported Actions Toward Nuclear Power." *Journal of Social Psychology* 116:179–97.

Gori, G. B. (1980). "The Regulation of Carcinogenic Hazards." *Science* 208:256–61.

Green, C. H. (1980). "Risk: Attitudes and Beliefs." In *Behavior in Fires,* ed. D. V. Canter, Chichester: Wiley.

Green, C. H., and Brown, R. A. (1980). "Through a Glass Darkly: Perceiving Perceived Risks to Health and Safety." Research paper, School of Architecture, Duncan of Jordanstone College of Art, University of Dundee, Scotland.

Johnson, E. J., and Tversky, A. (1984). "Representations of Perceptions of Risks." *Journal of Experimental Psychology: General* 113:55–70.

Lichtenstein, S., Slovic, P., Fischhoff, B., Layman, M., and Combs, B. (1978). "Judged Frequency of Lethal Events." *Journal of Experimental Psychology: Human Learning and Memory* 4:551–78.

Lindell, M. K., and Earle, T. C. (1983). "How Close Is Close Enough? Public Perceptions of the Risks of Industrial Facilities." *Risk Analysis* 3:245–54.

MacGill, S. M. (1983). "Exploring the Similarities of Different Risks." *Environment and Planning B: Planning and Design* 10:303–29.

McNeil, B. J., Pauker, S. G., Sox, H. C., Jr., and Tversky, A. (1982). "On the Elicitation of Preferences for Alternative Therapies." *New England Journal of Medicine* 306:1259–62.

Moatti, J. P., Stemmelen, E., and Fagnani, F. (1984). "Risk Perception Social Conflicts and Acceptability of Technologies (An Overview of French Studies)." Paper presented at the Annual Meeting of the Society for Risk Analysis, Knoxville, Tenn.

Morgan, M. G., et al. (1983). "On Judging the Frequency of Lethal Events: A Replication." *Risk Analysis* 3:11–16.

Morgan, M. G., Slovic, P., Nair, I., Geisler, D., MacGregor, D., Fischhoff, B., Lincoln, D., and Florig, K. (1985). "Powerline Frequency and Magnetic Fields: A Pilot Study of Risk Perception." *Risk Analysis* 5:139–49.

New Yorker. (1985). "The Talk of the Town." 50(53):29.

Nisbett, R., and Ross, L. (1980). *Human Inference: Strategies and Shortcomings of Social Judgment.* Englewood Cliffs, N.J.: Prentice-Hall.

Renn, O. (1981). "Man, Technology and Risk: A Study on Intuitive Risk Assessment and Attitudes Towards Nuclear Power." Report Jul-Spez, 115, Nuclear Research Center, Jülich, Federal Republic of Germany, June.

Renn., O. (in press). "Premises of Risk Communication: Results of the West German Planning Cell Experiments." In *Risk Communication,* ed. R. Kasperson and P. J. Stallen, New York: Reidel.

Rosenburg, J. (1978). "A Question of Ethics: The DNA Controversy." *American Educator* 2(27):

Rothschild, N. (1978, November). "Coming to Grips with Risk." Address on BBC television, reprinted in *Wall Street Journal,* March 13, 1979.

Ruckelshaus, W. D. (1983). "Science, Risk, and Public Policy." *Science* 221:1026–28.

Ruckelshaus, W. D. (1984). "The Regulation of Carcinogenic Hazards." *Risk Analysis* 4:157–62.

Sharlin, H. I. (1986). "EDB: A Case Study in the Communication of Health Risk." *Risk Analysis* 6:61–68.

Short, J. F., Jr. (1984). "The Social Fabric at Risk: Toward the Social Transformations of Risk Analysis." *American Sociological Review* 49:711–25.

Slovic, P. (1982). "Toward Understanding and Improving Decisions." In *Human Performance and Productivity.* Vol. 2: *Information Processing and Decision Making,* ed. W. C. Howell and E. A. Fleishman, pp. 157–83. Hillsdale, N.J.: Erlbaum.

Slovic, P., Fischhoff, B., and Lichtenstein, S. (1979). "Rating the Risks." *Environment* 21(3):14–20, 36–39.

Slovic, P., Fischhoff, B., and Lichtenstein, S. (1980a). "Facts and Fears: Understanding Perceived Risk." In *Societal Risk Assessment: How Safe Is Safe Enough?* ed. R. Schwing and W. A. Albers, Jr., pp. 181–214. New York: Plenum.

Slovic, P., Fischhoff, B., and Lichtenstein, S. (1980b). "Informing People About Risk." In *Product Labeling and Health Risks: Banbury Report* 6, ed. L. Morris, M. Mazis, and I. Barofsky, pp. 165–81. Cold Spring Harbor, N.Y.: Banbury Center.

Slovic, P., Fischhoff, B., and Lichtenstein, S. (1982). "Response Mode, Framing, and Information-processing Effects in Risk Assessment." In *New Directions for Methodology of Social and Behavioral Science: Question Framing and Response Consistency,* ed. R. Hogarth, pp. 21–36. San Francisco: Jossey-Bass.

Slovic, P., Fischhoff, B., and Lichtenstein, S. (1984). "Behavioral Decision Theory Perspectives on Risk and Safety." *Acta Psychologica* 56:183–203.

Slovic, P., Fischhoff, B., and Lichtenstein, S. (1985). "Characterizing Perceived Risk." In *Perilous Progress: Managing the Hazards of Technology,* ed. R. W. Kates, C. Hohenemser, and J. X. Kasperson, pp. 91–125. Boulder, Colo.: Westview, Press.

Slovic, P., and Lichtenstein, S. (1983). "Preference Reversals: A Broader Perspective." *American Economic Review* 73:596–605.

Slovic, P., Lichtenstein, S., and Fischhoff, B. (1984). "Modeling the Societal Impact of Fatal Accidents." *Management Science* 30:464–74.

Sowby, F. D. (1965). "Radiation and Other Risks." *Health Physics* 11:879–87.

Starr, C. (1969). "Social Benefit Versus Technological Risk." *Science* 165:1232–38.

Tversky, A., and Kahneman, D. (1981). "The Framing of Decisions and the Psychology of Choice." *Science* 211:452–58.

Vlek, C. A. J., and Stallen, P. J. (1981). "Judging Risk and Benefit in the Small and in the Large." *Organizational Behavior and Human Performance* 28:235–71.

von Winterfeldt, D., John, R. S., and Borcherding, K. (1981). "Cognitive Components of Risk Ratings." *Risk Analysis* 1:277–88.

Weinstein, N. D. (1979). "Seeking Reassuring or Threatening Information About Environmental Cancer." *Journal of Behavioral Medicine* 2:125–39.

Wilson, R. (1979). "Analyzing the Daily Risks of Life." *Technology Review* 81(4): 40–46.

4

Guidelines for Communicating Information About Chemical Risks Effectively and Responsibly

VINCENT T. COVELLO, PETER M. SANDMAN, AND PAUL SLOVIC

The Emergency Planning and Community Right-to-Know Act of 1986 (Title III of the Superfund amendments) and many state and local laws are imposing much more openness on the chemical industry. During the next few years, industry officials, government officials, and representatives from public-interest groups will increasingly be called upon to provide and explain information about chemical risks to the general public and to people living near chemical plants.

This chapter presents and discusses guidelines for communicating information about chemical risks effectively, responsibly and ethically. A basic assumption of the chapter is that discussing risk, when done properly, is always better than withholding information. In the long run, more effective, responsible, and ethical risk communication will be better for communities, industry, government, and society as a whole.

The chapter consists of four parts: (1) guidelines for communicating risk information, (2) guidelines for presenting and explaining risk-related numbers and statistics, (3) guidelines for presenting and explaining risk comparisons, and (4) problems frequently encountered in communicating risk information. Most of the material in this chapter deals with health risks, not the risks of accidents. In some cases, accidents raise similar communication issues, especially when most of the expected adverse health effects are long term rather than immediate. However, when considering the risks of accidents, it is generally best to focus on preventive measures, emergency response procedures, containment and re-mediation procedures, and the extent of the possible damage.

GUIDELINES FOR COMMUNICATING
RISK INFORMATION

There are no easy prescriptions for communicating risk information effectively, responsibly, and ethically (Table 4.1). But those who have studied and partic-ipated in debates about risk do generally agree on seven principles that underlie

Table 4.1. Factors Important to Risk Perception and Evaluation

Factor	Conditions Associated with Greater Public Concern	Conditions Associated with Less Public Concern
Catastrophic potential	Fatalities and injuries grouped in time and space	Fatalities and injuries, scattered and random
Familiarity	Unfamiliar	Familiar
Understanding	Mechanisms or process not understood	Mechanisms or process understood
Uncertainty	Risks scientifically unknown or uncertain	Risks known to science
Controllability (personal)	Uncontrollable	Controllable
Voluntariness of exposure	Involuntary	Voluntary
Effects on children	Children specifically at risk Delayed effects	Children not specifically at risk Immediate effects
Manifestation of effects on future generations	Risk to future generations	No risk to future generations
Identification of victim	Identifiable victims	Statistical victims
Dread	Effects dreaded	Effects not dreaded
Trust in institutions	Lack of trust in responsible institutions	Trust in responsible institutions
Media attention	Much media attention	Little media attention
Accident history	Major and sometimes minor accidents	No major or minor accidents
Equity	Inequitable distribution of risks and benefits	Equitable distribution of risks and benefits
Benefits	Unclear benefits	Clear benefits
Reversibility	Irreversible effects	Reversible effects
Personal stake	Individual personally at risk	Individual not personally at risk
Origin	Caused by human actions or failures	Caused by acts of nature or God

effective risk communication (Covello and Allen 1988). These principles apply equally well to both the public and the private sectors; (Covello and Allen 1988; Covello, McCallum, and Pavlova 1989; Covello, Sandman, Slovic 1987; Covello, von Winterfeldt, and Slovic 1989; Hance, Chess, and Sandman 1987; Krimsky and Plough 1988). Although many of these principles may seem obvious, they are continually and consistently violated. Thus a useful way to read them is try to understand why they are frequently not followed.

Accept and Involve the Public as a Legitimate Partner

Two basic tenets of risk communication in a democracy are, first, that people and communities have a right to participate in decisions that affect their lives, their property, and the things they value; and second, that the goal of risk communication should not be to diffuse public concerns or avoid action. Risk communication should produce an informed public that is involved, interested, reasonable, thoughtful, solution oriented, and cooperative.

The guidelines for this principle are to demonstrate sincerity and respect for the public by involving the community early, before important decisions are made. Make clear the appropriateness of basing decisions about risks on factors other than the magnitude of the risk. Involve all parties that have an interest or a stake in the risk in question.

Plan Carefully and Evaluate Performance

Different goals, audiences, and media require different risk communication strategies. Risk communication will be successful only if it is carefully planned.

The guidelines for this principle are to begin with clear, explicit objectives, such as providing information to the public, motivating individuals to act, stimulating an emergency response, or contributing to conflict resolution. Evaluate the information about risks and know its strengths and weaknesses. Classify the different subgroups in the relevant audience, and aim your message at them. Recruit spokespersons who are good at presentation and interaction. Train the staff—including the technical staff—in communication skills, and reward outstanding performance. Whenever possible, pretest your message; carefully evaluate it; and learn from its shortcomings.

Listen to the Audience

People in the community are often more concerned about issues such as trust, credibility, control, competence, voluntariness, fairness, caring, and compassion than they are about mortality statistics and the details of quantitative risk assessment. If they are not listened to, they will not listen. Communication is a two-way activity.

The guidelines for this principle are not to make assumptions about what people know, think, or want done about risks. Take time to find out what people are thinking, using techniques such as interviews, focus groups, and surveys. Let all parties that have an interest or a stake in the issue be heard. Recognize people's emotions. Let people know that you understand what they have said, and address their concerns. Look for "hidden agendas," symbolic meanings, and broader economic or political considerations that often underlie and complicate the task of risk communication.

Be Honest, Frank, and Open

Trust and credibility are your most precious assets when communicating risk information. Difficult to obtain, they are almost impossible to regain once they have been lost.

The guidelines for this principle are to state your credentials but not to ask or expect to be trusted. If you do not know the answer to a question or are uncertain about it, say so. Get back to people with correct answers. Admit your mistakes. Disclose risk information as soon as possible (while emphasizing any appropriate reservations about reliability). Do not minimize or exaggerate the level of risk. Speculate only with great caution. If in doubt, lean toward sharing more information, not less, or people may think you are hiding something. Discuss any uncertainties, strengths, and weaknesses regarding data, including those identified by other credible sources. Identify worst-case estimates as such, and cite ranges of risk estimates when appropriate.

Coordinate and Collaborate with Other Credible Sources

Allies can help communicate risk information; conflicts or public disagreements with other credible sources make risk communication more difficult.

The guidelines for this principle are to take time to coordinate all interorganizational and intraorganizational communications. Devote the bulk of your effort and resources to the slow, hard work of building bridges with other organizations. Use credible and authoritative intermediaries. Consult with others to determine who is best able to answer questions about risk. Try to issue communications jointly with trustworthy sources such as credible university scientists, physicians, trusted local officials, and opinion leaders.

Meet the Needs of the Media

The media are prime transmitters of information on risks. They play a critical role in setting agendas and determining outcomes. The media are generally more interested in politics than in risk, more interested in simplicity than in complexity, and more interested in danger than in safety.

The guidelines for this principle are to be open with and accessible to reporters. Respect their deadlines. Provide information tailored to the needs of each medium, such as graphics and other visual aids for television. Prepare in advance and provide background material on complex risk issues. Follow up on stories with praise or criticism, as warranted. Try to establish long-term relationships of trust with specific editors and reporters.

Speak Clearly and with Compassion

Technical language and jargon are useful as professional shorthand, but they are barriers to successful communication with the public.

The guidelines for this principle are to use simple, nontechnical language. Be sensitive to local norms, such as speech and dress. Use vivid, concrete images that communicate on a personal level. Use examples and anecdotes that make technical risk data come alive. Avoid distant, abstract, unfeeling language about deaths, injuries, and illnesses. Acknowledge and respond (both in words and with actions) to people's emotions—anxiety, fear, anger, outrage, helplessness. Acknowledge and respond to the distinctions that the public views as important to evaluating risks. Compare risks to help put them in perspective, but avoid comparisons that ignore distinctions that people consider important. Always try to include a discussion of actions that are under way or can be taken. Tell people what cannot be done; promise only what can be done, and do what you have promised. Never let your efforts to inform people about risks prevent the acknowledgment—and statement—that any illness, injury, or death is a tragedy.

Although these principles can be helpful, they are no substitute for good judgment. Much of the work needed to make research on risk communication relevant to practitioners has not been done (for details, see Covello, von Winterfeldt, and Slovic 1987). Therefore, practitioners need to rely heavily on a combination of intuition and experience. Only they can know whether a particular piece of advice is valid and applies to their situations. Risk communication can never be made risk free.

Actions Versus Words

Communications about risks should include a discussion of what control measures and precautionary actions are being taken. In most risk situations, actions speak louder than words. People want to know what is being done to prevent an accident and how a company or government agency is preparing for the possibility, not how likely it thinks one is. People want to know what is being done to reduce toxic emissions and discharges, not just how much is being emitted and how many deaths or illnesses may result. People often care about managerial competence and conscientiousness more than about the risk itself. Many people perceive mismanagement, incompetence, and lack of conscientiousness as the central issues in risk assessment and communication. Explaining what has been done, what is being done, and what is planned to reduce and manage the risk is at least as important as explaining how small the risk is. People may be much more willing to listen to what is said about how small a risk is after they have been given information about what is being done to make it still smaller. At the same time, they will want relevant information about plant safety records, emergency preparedness drills, and other efforts to protect people and the environment, for example, programs to prevent accidents and to reduce or eliminate chemical odors.

Besides outlining measures that can be taken, it is important to tell people what cannot be done. Unkept promises can destroy credibility. If certain measures cannot be taken—for example, because they are too expensive, because they are against the law, because the technology or data are not available, or because the home office has not given approval—it is better to say so.

The Role of Risk Comparisons

Because risk comparisons help put risks into perspective, they are a powerful tool in risk communication. Still, they cannot take the place of long-term, sensitive interaction with the local community. Risk comparisons are useful only in a continuing, sound community relations program. Almost any accurate and appropriate risk comparison will work if the people in the community trust the spokesperson. But if the official has low credibility, inspires no trust, or has no relationship with the community, even the best risk comparison will fail (for more details about establishing trust, see Hance, Chess, and Sandman 1987).

Risk comparisons, although only a part of the answer, should be a fundamental component of any risk-communication program. In the next few years, officials will constantly be asked to explain chemical risks and to put them into perspective. And as Title III of the Superfund Act goes into effect (not to mention various state laws and referenda, such as California's Proposition 65), the demand for information about risk will increase sharply. Many officials and representatives from community and public-interest groups are looking for better ways to explain risk information accurately, responsibly, ethically, and in a way that will make sense to people with no technical training. Finding better ways to compare risks is thus an important part of improving risk communication.

Public Perceptions of the Chemical Industry

The chemical industry worldwide faces the problem of explaining risk to a public that is sometimes fearful, often hostile, and almost always skeptical. Industry officials are understandably concerned about the public's distrust of their industry.

Public distrust of the chemical industry is grounded in the belief that chemical companies have been remiss in many ways; that is, they have been insensitive to plant neighbors, unwilling to acknowledge problems, opposed to regulation, closed to dialogue, unwilling to disclose risk-related information, and negligent in fulfilling their responsibilities for health, safety, and the environment. Having determined (with some justification) that chemical companies are not always to be trusted, many people have been slow to notice that the industry is changing, that chemical companies are working harder to gain the public's trust and to attain credibility. But one of the costs of this heritage of mistrust is the public's willingness—and sometimes eagerness—to believe that chemical plants represent one of the greatest risks posed by modern technology. When people appear to be ignoring evidence that chemical risks are small, they may well be responding to evidence that the chemical industry has failed to act responsibly in the past.

Fortunately, the prospects for overcoming this distrust are better locally than they are globally. People stereotype less and scapegoat less when they are dealing with someone they know. Therefore, part of the solution to the global problem is local. When chemical companies have built up a track record of dealing openly, fairly, and safely with their employees, customers, and neighboring communities, this general distrust may diminish. In the meantime, many people—especially those who have had no direct experience with a chemical plant or have had a negative experience—are likely to view chemical plants as a microcosm of an industry that they believe has been arrogant and careless, that has killed children, and that has destroyed the environment for profit. Building bridges locally, and explaining risk credibly, may be difficult. But it is necessary because of the initial mistrust that faces most officials from the chemical industry.

Public Perceptions of Government Agencies

People often distrust those government agencies responsible for regulating the chemical industry. Trust and credibility can be undermined by numerous factors, including the public perception that government agencies are unduly influenced by the chemical industry; that government agencies are inappropriately biased in favor of particular chemical risk management policies or approaches; that the managers and staff in government agencies are neither technically nor managerially competent to deal with chemical risk issues; that government officials often lack adequate skills to communicate risk information and interact with concerned citizens; that government officials have mismanaged regulatory programs and have made highly questionable regulatory decisions; that experts and officials in government agencies have lied, presented half-truths, or made serious errors; that equally prominent government experts have taken diametrically opposed positions on chemical risk issues; that financial, personnel, and other resources for chemical risk regulation have been inadequate; and that approaches to chemical risk assessment and management by different authorities have often been inconsistent.

Factors Affecting Risk Acceptability

Understanding the distinction between risk and risk acceptability is critical to overcoming mistrust and communicating effectively. Even though the level of risk is related to risk acceptability, it is not a perfect correlation. Two factors affect the way people assess risk and evaluate acceptability, and they modify the correlation.

First, the level of risk is only one of several variables that determines acceptability. Among the other variables that matter, and should matter, are fairness, benefits, alternatives, control, and voluntariness (see Slovic 1987). In general, a fairly distributed risk is more acceptable than is an unfairly distributed one. A risk entailing significant benefits to the parties at risk is more acceptable than is a risk with no such benefits. A risk for which there are no alternatives is more acceptable than a risk that could be eliminated by using an alternative technology. A risk over which the parties concerned have some control is more acceptable than is a risk that is beyond their control. A risk that the parties concerned can assess and decide to accept is more acceptable than is a risk that is imposed on them. These statements are true in the same sense that it is true that a small risk is more acceptable than a large risk is. Risk is multidimensional, and so size is only one of the relevant dimensions.

If one grants the validity of these points, then a whole range of risk-management approaches becomes possible. If factors such as fairness, familiarity, and voluntariness are as relevant as size is in judging the acceptability of a risk, then any efforts by industry or government officials to make a risk more fair, more familiar, and more voluntary are as appropriate as are efforts to make the risk smaller. Similarly, if control is indeed important to determining the acceptability of a risk, then any efforts by industry and government to share their power, such as establishing and assisting community advisory boards or supporting third-party monitoring activities, will help make a risk more acceptable.

Second, deciding what level of risk ought to be acceptable is not a technical question but a value question. People vary in how they assess risk acceptability. That is, they weigh the various factors according to their own values, sense of risk, and stake in the outcome. Because acceptability is a matter of values and opinions and because values and opinions differ, debates about risk are often really debates about accountability and control. The real issue is whose values and opinions will decide the outcome. Although the standpoint of industry and government are still crucial, the views of the general public count for a great deal, which is one reason that industry and government officials are being asked to explain risks in the first place: Public values and opinions about acceptability do matter.

When industry and government officials explain and compare the risks associated with a particular chemical emission, they are providing information that they think people want to have in order to decide whether the risk is acceptable. But data on risk levels are only one of many kinds of information on which people base decisions about risk acceptability. Risk comparisons cannot preempt those decisions. No risk comparison will be successful if it appears to be trying to settle the question of whether a risk is acceptable. For example, it is often tempting for industry or government officials to use the following argument when they meet with community groups or members of the public, even though, it should always be avoided. The argument is as follows:

> The risk of *a* (emissions from the plant) is lower than the risk of *b* (driving to the meeting or smoking during breaks). Because you (the audience) find *b* acceptable, you should also find *a* acceptable.

This argument has a basic flaw in its logic, and so its use can severely damage trust and credibility. This is because some listeners will analyze the argument in this way:

> I do not have to accept the (small) added risk of living near a chemical plant just because I accept the (perhaps larger, but voluntary and personally beneficial) risk of sunbathing, bicycling, smoking, or driving my car. In deciding about the acceptability of risks, I consider many factors, only one of them being the size of the risk—and I prefer to do my own evaluation. Your job is not to tell me about what I should accept but to tell me about the size of the risk and what you are doing about it.

Closely tied to the issue of risk acceptability is the question of what the goal of risk communication should be. It should not be to avoid responsible action or simply to pacify local citizens. Instead, risk communication should produce an involved, informed, interested and fair-minded public, so that public opinions and concerns will be (or remain) reasonable, thoughtful, calm, solution oriented, and cooperative.

Disclosing and Providing Risk-related Information

Given the public's increasing demands to participate and the increasing number of community right-to-know laws, industry and government officials cannot afford to hold back relevant data about risks. But making such information available is not the same as packing it all into one incomprehensible speech, pamphlet, or news release. It is better to provide and explain two or three numbers, carefully selected, than to inundate an audience with meaningless facts.

When explaining risk-related numbers, keep in mind that such numbers are not the whole story. Generally, the community is more interested in trustworthiness and credibility than in data and details of quantitative risk assessment.

People want industry or government officials to acknowledge, respect, and share their concerns. By focusing too much on "the data," officials can easily fall into the trap of reinforcing the stereotype of industry and government as uncaring. To many industry and government officials, risk means risk statistics. But to the audience, risk means illness, suffering, and (possibly) death—not just for them, but also for their families and children. Language (including numbers) that distances the official from that reality defeats its purpose.

GUIDELINES FOR PROVIDING AND EXPLAINING
RISK-RELATED NUMBERS AND STATISTICS

Levels of risk are only one among several numbers and statistics that need to be conveyed when communicating risk to the public. Other numbers and statistics include (1) concentrations (such as parts per million or billion), (2) probabilities (the likelihood of the event), and (3) quantities (such as how many tons of air toxic x were emitted into the air near the plant, how much effluent was released into the water, and how much soil was contaminated).

This list is neither complete nor precise. The main point is that industry and government officials often think they are providing information about risk when they actually are providing risk-related numbers and statistics. There is nothing wrong with the latter. Indeed, because the most difficult number to conceptualize, present, and explain is the risk itself, sometimes it does makes sense to talk about other numbers or statistics. And, discussing risk-related numbers and statistics often contributes to a clearer understanding of how a risk is calculated. For example:

> The plant releases z pounds of air toxic x per month (quantity) into the air, causing y parts per billion of air toxic x in the air at the plant gate (concentration) on a bad day. A population exposed to that dose of air toxic x for seventy years would experience a w percent increase in the rate of lung cancer (risk).

Using Comparisons to Explain Risk-related
Numbers and Statistics

Risk-related numbers and statistics can be presented as part of a comparison, but such comparisons often create their own problems. For example, "concentration" comparisons ("a drop of vermouth in a million-gallon martini"), though common, are seldom helpful. Most concentration comparisons have the same drawbacks as do the worst-risk comparisons: They seem to minimize and trivialize the problem and also to prejudge its acceptability. Therefore concentration comparisons—usually designed to convince the audience to stop worrying—are likely to fail. Furthermore, concentration comparisons can also be misleading, as chemicals vary widely in their potency (one drop of chemical x in a swimming pool might kill everyone, whereas one drop of chemical y would have no effect whatsoever). In view of these reservations and others, it is best to use concentration comparisons sparingly and sensitively.

Like "concentration" comparisons, "probability" comparisons can be misleading. For example, some analysts suggest that instead of comparing the probability of one unlikely (chemically hazardous) event with the probability of another unlikely but negative event, such as getting hit by lightning, officials would do better to compare it with an unlikely positive event, such as winning the Irish Sweepstakes (Bean 1987).

"Quantity" comparisons, on the other hand, tend to be more useful. For example, it can be quite helpful to translate tons of ash into Astrodomes–full or swimming pools–full. Quantity comparisons usually help the audience visualize how much is there.

Alternative Ways to Express Risk-related Numbers and Statistics

Changes in the law now require industry and government officials to offer the public a variety of numbers and statistics related to plant inventories and emissions. There are many ways to express such data. To communicate them clearly, spokespeople should not simply pass on unprocessed data to the public. Instead, they should select the best way(s) to present and explain them.

Industry and government officials often are attacked for "lying with statistics." In arguments about risks, no number can be "the right one." Each way of expressing a risk "frames" the risk a little differently and thus has a different impact. Doctors, for example, prescribe medications that save the lives of 60 percent of seriously ill patients, not medications that lose 40 percent of the patients, even though the results are exactly the same (Slovic 1987).

Risk is usually presented as the expected number of deaths per unit of something per years exposed. The various ways of expressing this number, however, are not equivalent. For example, consider the issue of how many people will die annually as a result of emissions of air toxic x. This risk information can be expressed as

- Deaths per million people in the population.
- Deaths per million people within n miles of the facility.
- Deaths per unit of concentration (LD–50, for example, or any toxicity measure).
- Deaths per facility.
- Deaths per ton of air toxic x released.
- Deaths per ton of air toxic x absorbed by people.
- Deaths per ton of chemical produced.
- Deaths per million dollars of product produced.

Depending on the circumstances, these different expressions strike an audience as being more or less appropriate, more or less frightening, more or less comprehensible, and more or less credible. Consider and compare the first two. People living or working near a facility are obviously at much greater risk from emissions of air toxic x than are people farther away. Therefore, dividing the expected number of deaths from air toxic x—virtually all of them near the plant—by the U.S. population, yields a reassuringly low risk figure, but a grossly misleading one. It is like including nonsmokers in a calculation of the risk of smoking.

Transformations of scale are also important to presenting risk-related numbers and statistics. Should you describe the amount of effluent in pounds or tons (or barrels or truckloads)? Six parts per billion sounds like a good deal more than 0.006 parts per million.

Risk data can be expressed in various ways for different levels of exposure. One possibility, for example, is the period of exposure that is equal to some unit risk: *n* years at the plant gate, for example, equals a one-in-a-million chance of cancer, or *n* years within ten miles of the plant (a less alarming number), or *n* years at the highest exposure so far detected (a more alarming number).

One unit for expressing risk data is lost life expectancy (based on a conservative presumed normal life expectancy of seventy years). This measure gives more weight to early deaths and less to deaths in old age. Risk can also be expressed in terms of lost workdays. This measure cannot be used, however, for estimating risks to children, nonworking parents, and retired people.

If the pertinent data are available—which is seldom the case—risk can and should be calculated for health consequences other than deaths. Deaths are often used by analysts as a surrogate for other health consequences, including cancers, birth defects, genetic damage, immune-system damage, neurological damage, fertility problems, behavioral disorders, liver disorders, kidney disorders, blood problems, and cardiovascular problems. All are adverse chronic health consequences. Adverse acute health consequences include trauma, burns, rashes, and eye problems. One way to express such acute consequences is to count hospital days or medical costs. Risks can also be calculated for adverse consequences for wildlife, endangered species, resources, plants, agriculture, air quality, water quality, habitat, aesthetics, climate, and so on.

Having chosen a unit and calculated a number, one can choose among a variety of ways of expressing it. Suppose emissions of the air toxic is expected to kill 1.4 people in 1000 over a lifetime. Which of the following should be told to people?

1. The lifetime risk is 0.0014.
2. The lifetime risk is 0.14 percent.
3. The lifetime risk is 1 in 710.
4. In a community of 1000 people, we could expect
    1.4 to die as a result of exposure.

Although these alternatives are equivalent, their meaning to and effect on audiences are not. Some terms may make the risk seem small, and others may make the risk seem larger. The terms of expression, therefore, need to be selected responsibly and ethically.

When presenting risk-related numbers and statistics to the public, it is important to be aware that (1) most risk numbers are estimates, and (2) these estimates are often based on data from animals exposed to very high doses and on modeling of one sort or another. We know how many people die each year from auto accidents and from being struck by lightning. We can measure these rates. But we can only calculate and estimate—not measure—how many people die each year from chronic exposure to trace quantities of chemicals. Given this context, industry and government officials have several options. They can offer their best estimate of the risk. They can (as some federal agencies do) offer an upper-bound or "worst-case" estimate. In such an estimate, each step in the calculation moves toward greater risk, thus attempting to assess the maximum risk that could reasonably be expected. Or they can offer the most likely estimate, along with the highest and lowest estimates. This latter option is recommended whenever possible, as most people want complete information and may resent the choice being made for them.

With this background, what can we say about how best to provide and explain risk-related numbers and statistics?

First, do not just present the risk-related numbers and statistics as you find them. The first step in selecting numbers and statistics wisely is realizing that selection is necessary.

Second, pick a risk number for which the data are good. Some of the data manipulations that have been discussed are just simple transformations of the same number. However, some transformations (for example, lost-life-expectancy data) require additional information. Similarly, if there are data concerning deaths and illnesses but not environmental effects, talk about deaths and illnesses. If, however, people are concerned about environmental effects such as odors or haze, be prepared to talk about the partial data that are available and about efforts to gather more complete information.

Third, use whole numbers and simple fractions whenever possible, such as 6 parts per billion instead of 0.006 parts per million. However, in a table with several values, it is best to express all of them with the same denominator, even if it means violating this rule. For example:

Substance	Risk per Million Persons Exposed
Air toxic x	3.0
Air toxic y	0.6
Air toxic z	4000.0

Fourth, pick a number that will be easy for the audience to understand; thus avoid unfamiliar units or overly complex concepts. Explain the number in words that help clarify its meaning. For example, "a risk of 0.047" is comprehensible to only a few experts, but everyone can understand that roughly 5 people in an auditorium of 100 would be affected. Similarly, "a cancer risk of 4.7×10^{-6}" is hard going even for an expert. Another way of expressing this number is as follows:

Imagine ten cities of 100,000 people each, all exposed to n amount of air toxic x. In five of these ten cities, probably no one would be affected. In each of the other five cities, there will be one additional cancer, on the average.

Fifth, use graphs, charts, and other visual aids.

Sixth, choose a number considered fair and relevant by knowledgeable others. "Fair" means that the number does not give a misleading impression of the risk; rather, it makes serious risks look serious and modest risks look modest. "Relevant" means that the number speaks to the issue at hand. For example, to an audience of nearby residents, it is more relevant to discuss risk at the plant gate than average risk to the larger community.

Seventh, pick a number that the audience will also consider fair and relevant. Find out which health or environmental consequences the audience cares about. Use various methods to obtain this information, including interviews, surveys, and focus groups. Explaining cancer risks to an audience concerned about miscarriages or property values or diminished elk hunting is pointless. Similarly, if an audience is angry or distrustful (a not uncommon state of affairs), bend over backwards to frame the risk in a way that the audience will not resent, even if doing so gives an impression of greater risk than the data appear to justify. It is pointless to offer a risk estimate that the audience will reject as not being credible.

Eighth, strive for comprehensibility and clarity, but do not oversimplify. Most people, if sufficiently motivated, are capable of understanding quite complex quantitative information. For example, people who have not been trained in probability and statistics can still understand mortgage rates and the complex odds for poker games and at racetracks. If people do not or cannot seem to understand information about risk, it may be that they are being overloaded with technical jargon and details, or they may simply not trust the person providing the data enough to want to understand what is being said.

Ninth, pay close attention to the numbers that others are using. Although not everyone is obliged to use the same risk numbers (especially if they think some numbers are not fair or relevant), confusion will result if everyone is using a different measure of the same risk. If terms must be changed, you should explain why. This tactic can be risky, however. Other participants in the debate may have more credibility with the public, providing all the more reason to use their measures or explain why you are not.

Finally, tenth, if there is sufficient opportunity, consider offering several different estimates of the same risk. For example:

> Our best estimate is a. Our cautious, worst-case estimate is b. The highest estimate we have heard, from the ABC organization, is c.

Credibility is improved when higher calculated estimates of the risk are acknowledged.

Personalizing Risk-related Numbers and Statistics

Technical experts see risk as a statistical number, whereas laypersons view risk in much more emotional terms, bound up with prospects of individual suffering and death. For many people, the term *low expected mortality* is fraught with emotional meaning; it could mean the death of the baby of the woman in the third row.

These different views are not just a matter of rational versus emotional. They also involve a matter of "macro" versus "micro." Risk calculations are on the macro level; that is, what will happen to the community as a whole. But citizens' concerns are micro; what might happen to me and those I love.

One solution to this dilemma is to "personalize" the risk information, especially the quantitative risk information, which most needs personalizing. Several approaches are available:

- Using examples and anecdotes—hypothetical if necessary, real if possible—to make the risk data come alive.
- Talking about oneself—for example, about risks that are personally unacceptable, about perceptions of the risk under discussion, or about feelings toward the audience at that moment.
- Using concrete images to give substance to abstract risk data.
- Avoiding distant, abstract, and unfeeling language about death, injury, and illness.
- Listening to people express their concerns.

GUIDELINES FOR PROVIDING AND EXPLAINING RISK COMPARISONS

Assume that as part of the overall risk communication effort, an industry or government official has decided to explain a particular risk number. As part of that explanation, the official wants to compare the number with some other number. The first step is to be clear about why the risk comparison is being presented.

Risk comparisons help put risks into perspective. Both research and experience show that people are typically less familiar with quantitative risk data than they are, for example, with quantitative length data (Covello 1989). Most people have a solid grasp of how long three feet is. But they do not have a solid grasp of how risky a lifetime risk of 0.047 is. The job of risk comparisons is to make the original risk number more meaningful by comparing it with other risks.

In many cases, risk comparisons are intended to be reassuring. A company or agency may, for example, believe that a particular risk is small, often so small that the company or agency believes that no action to reduce it further is needed. The purpose of the risk comparison, therefore, is to help people understand why the company or agency believes that the risk is small, so that people can make informed decisions about whether they want to accept it.

Less typical, but equally important, are cases involving risks that a chemical company or government agency believes are large. The purpose of the risk comparison is to help people understand why the company or agency believes that the risk is large. When faced with such a risk, the chemical company or agency should acknowledge that the risk is serious, offer comparisons that show it is serious, and focus on what is being done to reduce it.

Problems arise in presenting risk comparisons because people in the community often have good reasons to distrust efforts to minimize risks. Most people recognize the interest of a chemical company in minimizing public perceptions of chemical risks in general and of specific chemical risks in particular. Comparisons aimed at minimizing a risk frequently irritate people, because they see, and resent, the comparison as an effort to persuade people that the risk is smaller than they thought.

With this as a background, how can a company or agency present a minimizing risk comparison effectively, responsibly, and ethically?

First, acknowledge and state the company's or agency's stake in the issue and why the comparisons are being presented. If the purpose is to convince people in the community that the risk is tolerably small, explain this as clearly as possible.

Second, do not ask, or expect, to be trusted. Instead, point out why people need not put all their trust in the company or agency. Cite various institutions and procedural safeguards or mechanisms—including inspections, union safety committees, community oversight committees (including the new ones established under Title III), media investigations, and risk assessments by independent experts (even the assessments of opposing groups)—to make the point. Urge the community to seek out various sources of information with differing biases. The less the community feels it has to trust, the more the community will feel it can trust.

Third, whether people will see a "minimizing" risk comparison as useful or as misleading depends on whether it seems to be prejudging the acceptability of the risk. People will be more willing to learn from a risk comparison—even a minimizing risk comparison—if they are confident that the community's right to participate fully in decisions that affect their lives, their property, and the things they value is accepted and that decisions about the acceptability of a risk will be based on factors other than the size of the risk.

A Categorization and Ranking System for Risk Comparison

Some kinds of risk comparisons are more likely than others to be perceived as an effort to preempt judgments about the acceptability of a risk. The highest-ranking comparisons are those that put the least strain on the trust between an industry or government official and the public. These comparisons tend to strike even skeptical listeners as relevant, appropriate, and helpful information. The lowest-ranking comparisons, on the other hand, are those that have no obvious claim to relevance, appropriateness, or helpfulness. Such comparisons are more likely to be seen as manipulative or misleading, that is, as efforts to preempt judgments about the acceptability of a risk. Thus, the lowest-ranking comparisons are much more difficult to present and require much more effort to communicate effectively. That difficulty, though, does not mean that the lowest-ranking risk comparisons should never, under any circumstances, be used. Indeed, sometimes the only way to make a point is to use one. The general rule of thumb is to select from the highest-ranking risk comparisons whenever possible and to use a low-ranking risk comparison cautiously.

First-Rank Risk Comparisons (most acceptable)

The following are examples of comparisons of the same risk at two different times:

The risk from air toxic x is 40 percent less than it was before the plant installed scrubbers last October.

With the new procedures in place, the risk will be cut in half by this time next year.

Examples of comparisons with a standard:

Exposure of plant workers to air toxic x is well below the level that the Occupational Safety and Health Administration considers safe.

Plant emissions of air toxic x are 10 percent of what is permitted under the old EPA standard, and slightly under the level established by the new, stricter EPA standard.

Examples of comparisons with different estimates of the same risk:

Our best estimate of the risk is x, although you should be aware that the ABC organization has calculated an upper-bound or worst-case risk estimate of y.

Our best estimate of risk is x on the basis of methodology alpha and y on the basis of methodology beta.

Our best estimate of the risk is x, whereas that of the ABC organization is y and of the XYZ organization is z.

Second-Rank Risk Comparisons (less desirable)

The following are examples of comparisons of the risk of doing something versus not doing it:

If the newest and most advanced emission control equipment is installed, the risk will be x, whereas if this equipment is not installed, the risk will be y.

Examples of comparisons of alternative solutions to the same problem:

The risk associated with incinerating waste is x, and the risk associated with using a landfill is y.

When using the preceding type of comparison, be careful not to leave out alternatives with lower risks than the one being advocated.

Examples of a comparison with the same risk as that experienced in other places:

The most serious air toxic x problems have been encountered in the Denver area; the air toxic x problem in this community is only one-fifth as serious as Denver's.

Third-Rank Risk Comparisons (even less desirable)

The following are examples of comparisons of average risk with peak risk at a particular time or location:

The risk posed by emissions of air toxic x on an average day is one-thousandth as great as the risk last Wednesday, when a valve malfunctioned.

The risk posed by emissions of air toxic x to the nearest home is 90 percent less than the risk at the plant gate, and the risk two miles from the plant gate is 90 percent less than the risk at the nearest home.

The risk posed by emissions of air toxic x to the average community resident is 90 percent less than the risk to plant workers.

Examples of comparisons of the risk from one source of a particular adverse effect with the risk from all sources of that same adverse effect:

The risk of lung cancer posed by emissions of air toxic x is roughly three-hundredths of 1 percent of the total lung cancer risk in this community.

The risk of lung cancer posed by emissions of air toxic x would increase the total number of cases of lung cancer expected in a community of this size in a typical year from 500 to 500.15.

Fourth-Rank Risk Comparisons (marginally acceptable)

The following are examples of comparisons of risk with cost, or of one cost–risk ratio with another cost–risk ratio.

Reducing the risk posed by air toxic x by half would cost y dollars.

Saving one life by controlling emissions of air toxic x would cost y dollars, whereas saving a life by improving particulate control would cost only z dollars.

Example of a comparison of risk with benefit:

The chemical product whose waste by-product is air toxic x is used by hospitals to sterilize surgical instruments and thus helps save many lives and reduce many risks.

Risk–benefit comparisons tend to be more acceptable when the benefits accrue to the same people. Even this type of comparison, however, may strike people as bribery (the community will receive x amount of tax revenue from the facility) or blackmail (the community will lose x number of jobs if the plant is shut down). The general rule of thumb is to explain the benefits of the chemical product or facility separately from its risks, as people generally prefer to make their own risk–benefit comparisons.

Example of a comparison of occupational risks with environmental risks:

The community is exposed to far less air toxic x than plant workers are, and medical tests of plant workers show no evidence of adverse health effects.

Example of a comparison with other risks from the same source:

Emissions of air toxic x from chemical plant a are no more serious a problem than are emissions of air toxic y from plant b (which the community has long found acceptable).

Example of a comparison with other specific causes of the same disease, illness, or injury:

Air toxic x produces far less lung cancer than does exposure to natural background levels of the same chemical.

This last example violates at least one of the distinctions that the public considers important to evaluating risks: manufactured versus natural. For that reason and others, such comparisons fall into the fourth rank.

Fifth-Rank Risk Comparisons (use with extreme caution!)

All of the preceding types of comparisons have some claim to relevance and legitimacy: a strong claim in the top ranks but a much weaker claim in the bottom ranks. In the fifth rank are all those risk comparisons that have little or no claim to relevance or legitimacy. Central among these are comparisons of two or more completely unrelated risks. Even in the fifth rank, however, we can make distinctions. For example, the more that a risk comparison disregards factors that people consider important to evaluating risks, the more likely it is to be ineffective (see Table 4.1). The most important concerns are (1) catastrophic potential, (2) dread, (3) voluntariness, (4) newness, (5) familiarity, and (6) controllability.

The classic example of a comparison that violates these distinctions is telling people at a public meeting that their risk from air toxic x is lower than the risk they took when they drove their cars to the meeting or when they enjoyed a cigarette during a break. Unless there is already a high level of trust between the speaker and the audience, this sort of comparison is almost guaranteed to provoke outrage because it seems to make the following claim:

> Because the risk of emissions of air toxic x is less than that of driving or smoking, two conclusions follow: (1) The risk of emissions of air toxic x must logically be more acceptable, and (2) people who drive or smoke have surrendered their right to object to the plant's emission of air toxic x.

This is a false argument based on a flawed premise, and so to seem to be advancing such a claim is to invite resentment from the audience.

Comparing the risk from air toxic x with such things as the risk of food additives is also far from ideal, as there is no special reason to believe that the risks of food additives are relevant to risks of air toxic x. But at least the comparison does not appear to violate any of the most important risk distinctions. By contrast, comparing the risk of air toxic x with the risk from driving without a seat belt violates most of the major risk distinctions. The latter risk is voluntary, familiar, and controlled by the individual, and so comparing this risk with that of emissions of air toxic x—which is likely to be perceived as involuntary, unfamiliar, and beyond the citizen's control—is bound to infuriate the audience.

A comparison that is just as bad (if not worse) is to relate the risk of air toxic x to an unfamiliar risk from a familiar activity, such as the risk from ingesting the aflatoxin in peanut butter. At least people recognize that driving without a seat belt is genuinely risky, though they see little analogy between that and plant emissions. But people do not widely recognize eating peanut butter as risky, and they see even less analogy between it and plant emissions. Thus an industry or government official who tells an audience of mothers that air toxic x is less dangerous than the peanut butter they feed their children is likely to be seen as patronizing and not believable.

The conclusion from this discussion of risk comparisons is not that they are impossible or useless. Rather, it is that industry and government officials must take responsibility for the appropriateness of their risk comparison, just as they must take responsibility for the accuracy of officially published risk data.

PROBLEMS FREQUENTLY ENCOUNTERED WHEN
EXPLAINING RISK NUMBERS

Various objections can be anticipated in response to explanations of risk-related information, including risk comparisons. Some have already been discussed. Others are related to (1) data uncertainties, (2) information disclosure, and (3) demands for zero risk.

Data Uncertainties

Because risk data are often highly uncertain, some people are likely to object that they cannot be trusted. When addressing this issue, therefore, the following guidelines can be helpful (for more detail, see Hance, Chess, and Sandman 1987).

Acknowledge—do not hide—uncertainty. Sounding more certain than the data can justify is a sure way to lose trust and credibility. Uncertainties caused by generalizing from animal data to humans, gaps in the data, and differences of interpretation should be acknowledged up front, not "admitted" belatedly when an opposing interest group points them out. But acknowledging uncertainty is not the same as asserting total randomness or declaring ignorance.

Degree of certainty can be represented along a scale or continuum, from "total confidence" in the data to "guesswork." When there is total confidence in the data, do not hesitate to say so. Most data produced by risk assessments, however, are somewhere in the middle of the scale. Such data are characterized by uncomfortably large margins of error; often they are based on disputable assumptions, models, and extrapolations, such as those from animals to people and from high doses to low doses. Nonetheless, such data are a much better basis for decision and action than guesswork is. In these cases, the information is helpful but far from certain. Risk assessments deal with this uncomfortable level of uncertainty by making assumptions in the direction of overestimating the risk.

Somewhere near the guesswork end of the scale are risk assessments based on very inadequate data. When this is the case, say so. State as much as you know and explain (1) what will be done to get better risk data and (2) what will be done in the meantime to reduce the risk or to protect people against it. This attitude is crucial. Under Title III, communities may begin asking about many possible health and environmental consequences for thousands of different chemicals, and in many cases few or no data are available to answer these questions. When there is no evidence one way or the other, such as in the case of data on the relationship between air toxic x and miscarriages, never respond with "there is no evidence to show that air toxic x causes miscarriages." The statement is technically correct—but profoundly misleading. If similar compounds have been tested and found not to cause miscarriages, say so. But if there are theoretical reasons for doubting that air toxic x could interfere with pregnancies, say that. If there are records of female plant workers with no miscarriage problems, say that. And if the possible connection has not yet been tested and deserves to be studied, say that.

In some cases, risk estimates are very uncertain because the risk is very low and measurement is very difficult, such as in cases of the chronic effects of chemicals with extremely low toxicity. In these cases it is best to explain that the data are uncertain because it is so hard to measure such improbable effects.

When acknowledging and explaining data uncertainties, talk about how the risk estimate was obtained and by whom. Demystifying the risk assessment process is a public benefit in its own right and will help the audience understand the simultaneous claim that (1) the risk estimate is uncertain and (2) the estimate is based on the best available scientific data. Explaining this process will also enable two points of special relevance and importance in any risk communication to be made.

First, the presence of a toxic chemical in the environment, such as an emission from a chemical manufacturing plant, does not necessarily signify a significant health or environmental risk. This point is often extremely difficult to communicate, especially to those worried about long-term, low-dose effects. Even if hundreds or thousands of tons of air toxics are emitted into the air each year, they may pose no significant health or environmental risk to people. For a toxic chemical to be a risk, there must be an exposure. That is, there must be a way for the air toxic to get from where it is (the manufacturing plant) to where people or the things they value are. Routes of exposure are as important to risk assessment as toxicity measurements are.

Second, when a route of exposure exists, the next important question is the concentration of the chemical that may reach people. This concentration amount, typically far lower than the concentration at the source, will often become even lower with the passage of time as the substance breaks down. Risk assessors should consider not only whether a toxic substance is in the air or groundwater but also how much is in the air or groundwater. This distinction becomes clearer as the measurement equipment improves. By being able to measure parts per trillion and parts per quadrillion, scientists are now able to detect "low but measurable" chemical concentrations that were unknown before. Thus the proper question is not, How risky is a toxic substance? but, rather, How risky is the concentration of the toxic substance to which people are being exposed?

Regardless of the explanation of uncertainty in risk estimates, some people still will be dissatisfied. The demand for certainty is a natural human trait, especially in stressful situations when people feel helpless, angry, or at the mercy of others. Often community members will suggest ways of improving the quality of risk-assessment data. These suggestions are worth taking seriously, not only for their technical information, such as information about exposure routes that may have been overlooked, but also as a means for reassuring the community that the agency or company is listening, that it cares about their concerns, and that it is willing to be responsive. Although absolute certainty can never be achieved, community proposals for striving for that goal deserve respectful attention. But remember never to promise something that cannot or will not be done.

Information Disclosure

In some cases, risk data are quite preliminary, and officials may want to delay releasing the information until there has been further testing. When addressing this issue, the following two guidelines may be helpful (for more details, see Hance, Chess, and Sandman 1987):

Uncertainty is seldom, if ever, an acceptable excuse for waiting to communicate risk information. Especially if the information is alarming, it should be released promptly (with appropriate reservations about its reliability). This disclosure may be a legal obligation in some situations and an ethical obligation in nearly all situations.

Almost without exception, it is better to announce a possible problem promptly than to be accused later of having covered one up. Of course, common sense is the rule. Often the preliminary data are highly unreliable, and emergency action would not be warranted. In such cases, it may be preferable to wait a week or two for retesting and quality control, rather than to start a furor over nothing. But officials or managers with unsettling risk data are not likely to be tempted to announce the data prematurely. Rather, the temptation—a very dangerous one—is to wait too long. The worst of all possible worlds is to have investigative reporters or disaffected plant workers discover and announce the problem. Do not wait for that to happen (it will). Do not wait until the problem is solved or until all the scientific data are in. As soon as it is reasonable, announce the problem and discuss what can be done about it.

Demands for Zero Risk

One of the predictable things that will happen when industry or government officials try to explain a risk number at a public meeting is that someone in the audience will point out that it is not zero, following up with the assertion that it should be. It is important in such situations to explain that zero risk does not exist; that we all habitually—if ignorantly—tolerate quite substantial and avoidable risks, including smoking and driving without seat belts fastened; and that it is an impossible demand to expect zero risk from a chemical company.

How one responds to the demand for zero risk depends on its source. Consider five possibilities:

The demand for zero risk may be a demand for action. The facility may be able to (in effect) eliminate virtually all the risk in question through a change in procedure. If so, the costs and benefits of doing so should be assessed. If the situation warrants, action should be taken.

The demand for zero risk may be an exaggerated way of making the point that the risk is too high. In these cases, the demand is in effect a negotiating position. The response to this demand may be complicated by the fact that although further risk reduction is entirely feasible, the remaining risk is insignificant, and such action simply makes no economic sense. For several reasons, however, relatively little is gained by citing the economic benefits that compensate for risks that will not be eliminated. First, given the high value that people attach to good health, people will generally reject the argument that the reduced risk to the community does not justify the added costs to the company. Second, to the contention that the benefits compensate for the risks, people may accuse officials of bribery and blackmail. Finally, because economic benefits are typically distributed differently from health and environmental risks, such arguments often raise serious problems of fairness and equity.

The demand for zero risk may be sincere but ill informed. The people making it may simply not understand why something that is known to be risky can be allowed to continue. If so, respond gently with some fundamental risk education. Explain that zero risk does not exist. Point out that all of life's activities carry some risk, which we usually ignore if the risk is small enough and if the activity is beneficial enough. Be sure to agree that the risk should be made as low as possible, and discuss what is being done to achieve the lowest possible level.

The demand for zero risk may be politically motivated. The demand may be designed to win followers, influence politicians, or attract the interest of journalists. Such political activity is perfectly legitimate in a democracy. Environmental and public-interest groups opposed to the chemical industry—whether they are staffed by volunteers or paid organizers—are often highly dedicated and committed to their cause. Just as industry and government are entitled to build their case, environmental and public-interest groups are also entitled to build their political case. With occasional exceptions, the strategies used by environmental and public-interest groups are legitimate political strategies. For example, an environmental group may appeal to the community's health, safety, and environmental concerns, just as a company or agency may appeal to the community's economic concerns. An environmental group may recruit and mobilize experts to support its view, just as a company or agency may recruit and mobilize experts to support its view. Given that such activities are entirely legitimate in a democracy, it is unwise for officials to attack the sincerity and legitimacy of activities by environmental and public-interest groups. Industry and government officials have a right to disagree with the views of environmental and public-interest groups. Disagreeing, however, is not the same as questioning the integrity of environmental and public-interest groups or the right of such groups to use legitimate political strategies to promote their views. If the demand for zero risk is politically motivated, industry or government officials are unlikely to make much headway in large meetings held in public settings. One of the best options is to try to arrange a small meeting, in which frank discussion and negotiation might be possible.

Finally, the demand for zero risk may be a reflection of emotional distress stemming from outrage, anger, and distrust. For example, some people in a community may have decided that the company or agency is the enemy; that the company or agency is uncaring, arrogant, and dishonest; or that chemical manufacturing is an industry that does society little benefit at great cost. Risk, then, may not be the central issue. If a small number of community residents adopt a hostile view, an official can probably do little to change their minds. But if the outrage, anger, and distrust are widespread, these reactions should be considered as symptoms of serious problems in the overall communication effort or in the overall health and environmental protection effort. The underlying antagonism must be addressed, and the slow, hard work of building bridges, trust, and credibility must begin. In short, talking about risk to an audience that wants to believe that the risk is high is a pointless undertaking. The essential task is to figure out why the audience feels as it does and to pay more attention to those issues than to the battle over zero risk.

CONCLUSION

As we stated at the outset, there are no easy prescriptions for effective, responsible, and ethical risk communication. There are also limits on what people trying to communicate about risks can do, no matter how skilled, committed, and sincere they are. The guidelines, advice, and examples offered here may help. Over the next few years, risk communication will become increasingly important. Industry, government, and public-interest groups will continually be asked to provide information about chemicals and chemical risks to interested and affected individuals and communities. How they answer this challenge will affect the public's response to chemicals, chemical risks, and the chemical industry for years to come.

ACKNOWLEDGMENT

This chapter is based in part on a report written by us entitled "Risk Communication, Risk Statistics, and Risk Comparisons" (CMA, Washington, D.C.). We would like to thank the following people for their comments on previous versions and sections of this chapter: Frederick Allen, David Baird, Caron Chess, B. J. Hance, Richard N. Knowles, Brent McGinnis, Timothy O'Leary, John E. Slavick, Richard Symuleski, Detlof von Winterfeldt, and Ben Woodhouse.

REFERENCES

Bean, M. (1987). "Tools for Environmental Professionals Involved in Risk Communication at Hazardous Waste Facilities Undergoing Siting, Permitting, or Remediation." Report no. 87–30.8. Air Pollution Control Association, Reston, Va.

Covello, V. T. (1983). "The Perception of Technological Risks: A Literature Review." *Technological Forecasting and Social Change* 23:285–97.

Covello, V. T. (1988). "Informing the Public About Health and Environmental Risks: Problems and Opportunities for Effective Risk Communication." In *Risk Communication: A Symposium,* ed. N. Lind, pp. 5–15. Waterloo, Ontario: University of Waterloo Press.

Covello, V. T. (1990). "Risk Comparisons and Risk Communication." In *Risk Perception and Risk Communication,* ed. R. Kasperson and O. Renn, pp. 100–21. Boston: Reidel.

Covello V. T., and Allen F. (1988). *Seven Cardinal Rules of Risk Communication.* Washington, D.C.: Office of Policy Analysis, Environmental Protection Agency.

Covello, V., McCallum, D., and Pavlova, M. (eds.) (1989). *Effective Risk Communication: The Role and Responsibility of Government.* New York: Plenum Press.

Covello, V. T., Sandman, P., and Slovic P. (1988). *Risk Communication, Risk Statistics, and Risk Comparisons.* Washington, D.C.: Chemical Manufacturers' Association.

Covello, V. T., von Winterfeldt, D., and Slovic, P. (1987). "Communicating Risk Information to the Public." In *Risk Communication,* ed. J. C. Davies, V. Covello, and F. Allen, pp. 173–92. Washington, D.C.: Conservation Foundation.

Covello, V., von Winterfeldt, D., and Slovic, P. (1989). *Risk Communication: Research and Practice.* Cambridge: Cambridge University Press.

Davies, J. C., Covello, V. T., and Allen, F. W. (eds.) (1987). *Risk Communication.* Washington, D.C.: Conservation Foundation.

Fessenden-Raden, J., Fitchen, J., and Heath, J. (1987). "Risk Communication at the Local Level: A Complex Interactive Progress." *Science, Technology, and Human Values,* December, pp. 94–101.

Fischhoff, B. (1985a). "Managing Risk Perception." *Issues in Science and Technology* 2:83–96.

Fischhoff, B., (1985b). "Protocols for Environmental Reporting: What to Ask the Experts." *The Journalist,* Winter, pp. 11–15.

Fischhoff, B., Lichtenstein, S., Slovic, P., Derby, S. L., and Keeney, R. L. (1981). *Acceptable Risk.* Cambridge: Cambridge University Press.

Fischhoff, B., Slovic, P., and Lichtenstein, S. (1979). "Rat the Risks." *Environment* 21: 14–20, 36–39.

Hance, B., Chess, C. and Sandman, P. (1987). *Improving Dialogue with Communities: A Risk Communication Manual for Government.* Trenton, NJ: Office of Science and Research, New Jersey Department of Environmental Protection, December.

Johnson, B., and Covello, V. (eds.) (1987). *The Social and Cultural Construction of Risk: Essays on Risk Selection and Perception.* Boston: Reidel.

Kahneman, D., Slovic, P., and Tversky, A. (eds.) (1982). *Judgment Under Uncertainty: Heuristics and Biases.* Cambridge: Cambridge University Press.

Kasperson, R. (1986). "Six Propositions on Public Participation and Their Relevance to Risk Communication." *Risk Analysis* 6:275–82.

Kasperson, R., and Kasperson, J. (1983). "Determining the Acceptability of Risk: Ethical and Policy Issues." In *Risk: A Symposium,* ed. J. Rogers and D. Bates, Royal Society of Canada.

Klaidman, S. (1985). *Health Risk Reporting.* Washington, D.C.: Institute for Health Policy Analysis, Georgetown University.

Krimsky, S., and Plough, A. (1988). *Environmental Hazards: Communicating Risks as a Social Process.* Dover, Mass.: Auburn House.

Mazur, A. (1981). "Media Coverage and Public Opinion on Scientific Controversies." *Journal of Communication* 31:106–15.

Nelkin, D. (1984). *Science in the Streets.* New York: Twentieth Century Fund.

President's Commission on the Accident at Three Mile Island (1979). *Report of the Public's Right to Information Task Force.* Washington, D.C.: Government Printing Office.

Press, F. (1987). "Science and Risk Communication." In *Risk Communication,* ed. J. C. Davies, V. T. Covello, and F. W. Allen, pp. 11–17. Washington, D.C.: Conservation Foundation.

Ruckelshaus, W. D. (1983). "Science, Risk, and Public Policy." *Science* 221:1026–28.

Ruckelshaus, W. D. (1984). "Risk in a Free Society." *Risk Analysis,* September, pp. 157–63.

Ruckelshaus, W. D. (1987). "Communicating About Risk." In *Risk Communication,* ed. J. C. Davies, V. T. Covello, and F. W. Allen, pp. 3–9. Washington, D.C.: Conservation Foundation.

Sandman, P. M. (1986). *Explaining Environmental Risk.* Washington, D.C.: Office of Toxic Substances, Environmental Protection Agency.

Sandman, P. M. (1987). "Getting to Maybe: Some Communications Aspects of Hazardous Waste Facility Siting." In *Resolving Locational Conflict,* ed. R. Lake, pp. 324–44. New Brunswick, N.J.: Butgers University Press.

Sandman, P., Sachsman, D., and Greenberg, M. (1987). *Risk Communication for Environmental News Sources.* New Brunswick, N.J.: Industry/University Cooperative Center for Research in Hazardous and Toxic Substances.

Sandman, P. M., Sachsman, D., Greenberg, M., and Gotchfeld, M. (1987). *Environmental Risk and the Press.* New Brunswick, N.J.: Transaction Books.

Sharlin, H. (1986). "EDB: A Case Study in the Communication of Health Risk." *Risk Analysis* 6:61–68.

Slovic, P. (1986). "Informing and Educating the Public About Risk." *Risk Analysis* 4:403–15.

Slovic, P. (1987). "Perception of Risk." *Science* 236:280–85.

Slovic, P., and Fischhoff, B. (1982). "How Safe Is Safe Enough? Determinants of Perceived and Acceptable Risk." In *Too Hot to Handle,* ed. L. Gould and C. Walker, pp. 112–50. New Haven, Conn.: Yale University Press.

Slovic, P., Fischhoff, B., and Lichtenstein, S. (1982). "Facts Versus Fears: Understanding Perceived Risk." In *Judgment Under Uncertainty: Heuristics and Biases,* ed. D. Kahneman, P. Slovic, and A. Tversky, pp. 463–89. Cambridge: Cambridge University Press.

Thomas, L. M. (1987). "Why We Must Talk About Risk." In *Risk Communication,* ed. J. C. Davies, V. T. Covello, and F. W. Allen, pp. 19–25. Washington, D.C.: Conservation Foundation.

Wilson, R., and Crouch, E. (1987). "Risk Assessment and Comparisons: An Introduction." *Science* 236:267–70.

II
UNCERTAIN EVIDENCE IN RISK MANAGEMENT

The thesis of this volume is that we can become more sophisticated about what kind of evidence is important and how to count acceptable evidence in the assessment and management of risks. The thesis of Part II is that sophistication about acceptable evidence requires sophistication about uncertainty in decision-making contexts. At the same time, sophistication about uncertainty requires acknowledgment of the values that color decisions about uncertainty in risk assessment and its management.

Procedures for risk management in the United States in the past twenty-five years have not adequately recognized this role of values. In an attempt to ensure scientific authority, quantitative risk assessment was selected as the principal basis for setting standards and deciding on regulations in environmental policymaking, medicine, waste siting, and other areas of risk management. Ideally such a method would provide systematic standards to evaluate and manage such risky technologies. It would answer the question of when evidence is sufficient to show the existence or absence of specific risks and benefits. Answers to questions about the acceptability of risks, for example, when a risk is substantial enough for various regulatory measures, clearly involve social values. On the other hand, the information in quantitative risk assessments was supposed to incorporate scientific knowledge uncolored by such values. Given answers to questions of acceptable level of risk, then the question of acceptable evidence (in showing the existence or absence of such a risk) would be scientific and not political.

Part II examines a major source of difficulty in this view. Even with answers to questions about acceptable levels of risk, determining the acceptability of risk evidence is not free of values. An important reason for this difficulty is the inevitable uncertainties themselves. Uncertainties arise because risks typically have probabilistic effects. There is usually limited knowledge of the causal processes and also limited theoretical understanding. There are practical and ethical constraints on obtaining the data needed to improve the evidence on which to base decisions that can reduce risks. All of these uncertainties leave room to introduce value-laden decisions in collecting and interpreting evidence. Questions about what methods to use in estimating risks, how to extrapolate from animals to humans, and when clinical trials should be stopped do not allow unequivocal scientific answers, at least not at present. Therefore, answers to questions about acceptable evidence contain "extrascientific" values. The recognition that no stage of risk assessment is free of values has been a major theme in interdisciplinary work in science studies.

Although resolving questions of uncertainty is thus intertwined with values, a key theme of Part II is that it is a mistake to suppose that these questions can be answered by policymakers alone. The chapter in this section show that when policymakers, ethicists, and others pass judgment on the acceptability of evidence without understanding the difficult and often technical uncertainties involved, they defeat the very purpose they intend to accomplish, namely, to avoid bias and arrive at decisions that are more accountable to various individual and social values. Understanding the technical uncertainties thus is a prerequisite for developing systematic criteria for judging the acceptability of evidence. Thus, this understanding is a prerequisite if decisions are to be accountable.

Evidence may be incorrectly deemed acceptable because of linguistic and conceptual confusion, mistakes in collecting or interpreting the data, lack of awareness of the most recent evidence, or conscious and unconscious biases in identifying or evaluating evidence. Unearthing these errors requires identifying and understanding how they occur. The contributors to Part II use examples from relevant literature, policymaking episodes, and public events in order to identify these problems and to demonstrate why experts, policymakers, and citizens will have to cooperate to overcome these mistakes.

PROBLEMS IN SEPARATING RISK ASSESSMENT
FROM RISK MANAGEMENT

During the 1980s, regulatory agencies (e.g., the Environmental Protection Agency, EPA) adopted policies that separate the "science" of risk assessment from other activities related to regulation, categorized as *risk management*. Proponents of this separation assume that risk assessment is a field of objective, scientific analysis that can be divorced from the political values that are part of the subsequent management of risks. This assumption, however, is untenable, thus leading to the "uneasy divorce" in Ellen Silbergeld's "Risk Assessment and Risk Management: An Uneasy Divorce."

Silbergeld reviews the political history at the EPA of attempting to separate risk assessment from management. Far from improving the already contentious arena of environmental policy, Silbergeld argues, such separation encourages abuses of science and inappropriate interpretation of scientific data. Her discussion centers on issues of uncertainty, as these are problems that are exacerbated by divisions between assessment and management decisions. Those who endorse such a separation may believe that the resolution of uncertainty is a policy issue that can be handled by policymakers in a vacuum of science. But the resolution of uncertainty in risk assessment requires value-laden choices, and the uncertainty itself is often so difficult to explain that this division of expertise and responsibility results in inadequate decisions. If risk managers then claim "scientific" support for policy decisions based on these value-laden choices, the separation may undermine ethical discussion.

Silbergeld uses two issues pertaining to uncertainty to illustrate this problem: statistical representation of risk assessment data, and methods for allocating weight to scientific evidence. She then examines specific cases at the EPA during the 1980s of abusing and misinterpreting uncertainty.

UNCERTAINTIES IN MEDICINE

In Chapter 6, Valerie Miké demonstrates how uncertainties in the evidence underlying medical treatment arise and become exacerbated in the clinical setting. Regulatory policies and practices concerning treatments must take into account these difficulties and shortcomings. Physicians themselves find it hard to understand and accept technical and statistical evidence and to use it as a basis for treatment recommendations. Patients can be expected to have at least as much difficulty, and their view of doctors as authority figures may also impede their understanding and evaluating the evidence underlying recommendations. If they are unaware of what is and is not known about the risks, they will not be able to make a truly informed choice. The epistemological value of knowing about a risk is necessary for the ethical value of patients' rights not to be subjected to harm without consent. But knowing what is the risk is requires being able to judge critically the acceptability of the available evidence. This, in turn, requires understanding something of the weaknesses and uncertainties in the evidential basis. Miké exposes a number of such uncertainties in a discussion of specific medical practices (e.g., the case of diethylstilbesterol).

Moral philosophy also cannot deal adequately with ethical issues in medicine without understanding the uncertainties in the evidence. With the development of the field of bioethics, moral philosophers have increasingly discussed ethical issues in biomedical research, for example, the ethics of randomized clinical trials. However, as Miké explains, some of the most complex ethical issues in biomedical research turn on some of the most controversial aspects of the statistical reasoning underlying biomedical evidence. When ethical arguments are uninformed by such statistical complexities, distortions result. When charges of unethical conduct in biomedical research are based on such distortions, however inadvertent, further confusion results.

What Miké is arguing, then, is that responsible medical practices, as well as responsible arguments about the morality of such practices, require close collaboration among medical researchers, including laboratory scientists, clinical investigators, epidemiologists, and especially biostatisticians. They must assume the responsibility of communicating the results of their work in ways that medical practitioners, lawyers, ethicists, policymakers, the media, and the consumers of health care can understand. Only then will everyone gain insight into the nature and extent of the evidence underlying medical decisions and practices.

DISPUTES ABOUT EVIDENCE IN RADIOACTIVE WASTE MANAGEMENT

Chapter 7, by E. William Colglazier, "Evidential, Ethical and Policy Disputes: Admissible Evidence in Radioactive Waste Management," examines the epistemic and ethical issues that can and have arisen in the case of a major policy dispute over risk management in the United States. Through a minihistory of the social, political, and socioscientific disputes that have characterized this case, this chapter demonstrates that normative issues enter into policy disputes in a number of ways, that they are not limited to disputes over uncertainty.

Colglazier divides these issues into three kinds. *Procedural* issues refer to who should make what decisions for whom and by what process, and *distributional* issues refer to what distribution of costs, benefits, and risks is fair to the parties affected. But the disputants' positions on these issues also depend on their responses to *evidential* issues, how the costs, benefits, and risks are to be determined. What types and levels of scientific or other evidence are sufficient and admissible for making societal decisions? How can competing scientific claims be fairly assessed? Current disputes over radioactive waste management, for instance, deciding on waste repository sitings, revolve around the disagreements among experts and laypersons as to how to answer these questions. They provide fertile ground for unearthing how values in science and society influence different positions on criteria for acceptable evidence.

This chapter delineates the positions taken by key stakeholders at the national, state, and local levels. It describes these influences and identifies the values that seem to underlie the stakeholders' arguments. Like Chapters 1, 3, and 4, the identification of values in Chapter 7 shows why certain conditions help and others hinder in resolving actual policy disputes.

UNCERTAINTY AND RESPONSIBILITY

The final chapter in Part II, "Expert Claims and Social Decisions: Science, Politics, and Responsibility," by Rachelle D. Hollander, calls attention to the history of the influence of science and technology in Western society, using as an example Henrik Ibsen's play *Enemy of the People,* written in 1882. It develops the thesis that our new understanding of moral responsibility requires scientists, engineers, policymakers, interested scholars, and public- and private-sector representatives to determine not just acceptable risk but acceptable evidence as well.

Drawing on work by contributors to this volume and others, Hollander explains why morally responsible professional action requires acknowledging the limits of scientific evidence in reaching decisions. This is particularly true now, she believes, for what may be called the scientific state, in which decisions and actions based on scientific evidence are made in settings of inherent uncertainty. Choosing a scientific test that will select more false positives,—in order to minimize false negatives, or vice versa is not, or is not just, a scientific decision.

Rather when human beings will be affected, it is a moral decision. Furthermore, assumptions and interpretations of scientific studies are fallible; scientists assume that future work will show prior mistakes. Even when there is social agreement or societal concensus about ethical matters—that a pay premium is a good way to compensate for risk on the job, for instance—flaws in scientific studies of worker compensation may mean that evidence regarding whether the market functions to do this is faulty.

Neither agreement about values nor agreement about evidence assures us of final answers to our ethical questions; with each, disagreements can remain about the other. In a scientific state, the resolution of all of these kinds of questions requires that people who are not scientists or engineers become informed and involved in the discussions. Scientists, engineers, and other scholars must develop and apply methods for examining scientific knowledge and its limitations and point out both epistemic and ethical concerns that will be relevant to the decisions. While the choices that will be made will not be scientific ones, behaving responsibly requires professionals and others to work together for the common good.

5

Risk Assessment
and Risk Management:
An Uneasy Divorce

ELLEN K. SILBERGELD

Over the past decade, the concept of risk has become central to environmental policy. Environmental decision making has been recast as reducing risk by assessing and managing it. Risk assessment is increasingly employed in environmental policymaking to set standards and initiate regulatory consideration and, even in epidemiology, to predict the health effects of environmental exposures. As such, it standardizes the methods of evaluation used in dealing with environmental hazards. Nonetheless, risk assessment remains controversial among scientists, and the policy results of risk assessment are generally not accepted by the public.

It is not my purpose to examine the origin of these controversies, which I and others have considered elsewhere (see, e.g., EPA 1987), but rather to consider some of the consequences of the recent formulation of risk assessment as specific decisions and authorities distinguishable from other parts of environmental decision making. The focus of this chapter is the relatively new policy of separating certain aspects of risk assessment from risk management, a category that includes most decision-making actions. Proponents of this structural divorce contend that risk assessment is value neutral, a field of objective scientific analysis, while risk management is the arena where these objective data are processed into appropriate social policy. This raises relatively new problems to complicate the already contentious arena of environmental policy.

This separation has created problems that interfere with the recognition and resolution of both scientific and transscientific issues in environmental policymaking. Indeed, both science and policy could be better served by recognizing the scientific limits of risk-assessment methods and allowing scientific and policy judgment to interact to resolve unavoidable uncertainties in the decision-making process.

This chapter will discuss the forces that encouraged separating the performance of assessment and management at the EPA in the 1980s, which I characterize as an uneasy divorce. I shall examine some scientific and policy issues, especially regarding uncertainty, that have been aggravated by this policy of deliberate separation. Various interpretations of uncertainty have become central, and value-laden issues in decision making and appeals to uncertainty have often been an excuse for inaction.

In many environmental issues, values influence the resolution of uncertainty issues as well as other aspects of policy (Whittemore 1983). To the public, protection of the environment, particularly when considered as protecting human health and guaranteeing the natural world for future generations, is an ethical obligation of society (Hays 1987). Thus changing the structure of analysis and decision making has ethical aspects.

When he was administrator of the EPA for the second time, William Ruckelshaus raised concerns that allowing a technical elite to exercise the rights of intervention and protection may transgress the right of participatory democracy, and on these grounds, among others, he supported the separation between risk assessment and risk management in order to prevent a takeover by an elitist cult of experts (Ruckelshaus 1985; see also the responses by Silbergeld 1985, 1987a).

Uncertainty has major political and ethical impacts, as the need to decide whether or not to act, as well as what kind of action to take, requires deciding about uncertainty. Environmentalists and others who value prevention and protection tend to accept a correlative view of prevention and intervention in the face of uncertainty. And uncertainty is often invoked by industry, and more and more frequently by government, as grounds for inaction. For instance, in resisting petitions by the public to take action under Section 6 of the Toxic Substances Control Act, the EPA has invoked the requirement for reducing uncertainty before taking action, although in most cases, some action is needed to reduce uncertainty by eliciting information from potentially regulated industries.

Because uncertainty is always present in science-based decisions, what causes controversy and divisions among the interested parties is determining the amount or nature of the uncertainty that can be tolerated in order to resolve the issue; that is, establishing some minimal (usually unspecified) amount of certainty is necessary in order to enforce regulation (Finkel and Evans 1987). Successful legal challenges to proposed regulations have attacked the amount or the stability of the certainty that agencies have asserted (e.g., Judge Robert Bork's first opinion on vinyl chloride, which used a complicated concept of distinguishing between carcinogens on the basis of certainty). The Supreme Court rejected the Occupational Safety and Health Administration's (OSHA's) benzene rule on the grounds that it lacked rules for evaluating uncertainty. In contrast, the preventive view of uncertainty supports action to identify and reduce risks—to intervene—even in the face of considerable uncertainty.

Environmentalists often argue that experience supports lowering the barriers to action by limiting the dispositive power of uncertainty. Delaying action until some desired amount of uncertainty has been reduced often means that in the interval there is considerable damage, much of which may be irreparable, such as the release of a highly persistent chemical into the environment (see Rosner and Markowitz 1984 for an excellent discussion of the controversy over adding lead to gasoline in the 1920s). Although no strict analysis has ever been made of the social and economic costs of not regulating similar to those of the costs of allegedly inappropriate regulation (e.g., Nichols and Zeckhauser 1985), the values implicit in the differing views of what constitutes tolerable amounts of uncertainty transcend utilitarianism.

I shall limit my discussion of environmental decision making to those areas in which human health is the major concern, usually involving chemicals (or physical agents such as radiation) and, for the most part, to exposures to actual or putative chemical carcinogens.[1]

Scientific uncertainty is an unavoidable limit that is inherent in scientific knowledge and in the methods by which scientific facts are established. Because scientific knowledge is basically probabilistic rather than absolute and provisional rather than final, it can never be devoid of uncertainty or the possibility of inaccuracy or incompleteness. Systematic uncertainties are those limits on knowledge that are derived from the types of knowledge being sought: in this case, information about likely human health risks based on data from nonhuman species. Specific uncertainties are those arising from limitations of specific datasets, for instance, problems in experimental protocols, limits of statistical power, and the failure to control important covariates.

There are unavoidable systematic uncertainties in using animal data to infer human health risks. In the United States, it is not ethical to regard only human data as sufficient for regulatory action. Such a position is inconsistent with the EPA's statutory obligations under the Toxic Substances Control Act and the Resource Conservation and Recovery Act amendments of 1984. Further, placing a high value on prevention implies a willingness to tolerate considerable uncertainty. In contrast, placing higher value on repairing damage than on preventing it—as in traditional tort law that works after the fact of injury,—the amount of allowable uncertainty (in causation) is usually quite small. As stated by Judge Jack Weinstein in the Agent Orange case regarding the health effects associated with the exposure of soldiers in Vietnam to dioxin:

> The distinction between avoidance of risk through regulation and compensation for injuries after the fact is a fundamental one. In the former, risk assessments may lead to control of a toxic substance even though the probability of harm to any individual is small and the studies necessary to assume the risk are incomplete; society as a whole is willing to pay the price as a matter of policy. In the latter, a far higher probability (greater than 50%) is required since the law believes it unfair to require an individual to pay for another's tragedy unless it is shown that it is more likely than not that he caused it. (*In re Agent Orange*, pp. 111–12)

RISK ASSESSMENT AND RISK MANAGEMENT: STRUCTURING THE PROCESS

The view that risk assessment and risk management are and should be separate processes has been most clearly laid out by Russell and Gruber (1987). In this description of the optimal structure, by two practitioners at the EPA, the two processes are to be separated on three accounts: first, methodological (what goes on during each process); second, typological (which methods are applied in each process); and third, ownership (who does each process). In terms of methodology, Russell and Gruber assign to risk assessment the steps of hazard identification, dose response evaluation, and analysis of exposure. Taken together, the products of these methods lead to a quantitative risk assessment. Into risk management, they place all the processes of judgment, including the consideration of acceptability, feasibility, equity, and economics. Typologically, the assessment process is considered "hard" science, incorporating relevant biomedical sciences, statistics, and engineering. The management process is considered social science, with an emphasis on political science and economics and an increasing incorporation of decision theory. In terms of ownership, the assessment process is the preserve of scientists and technicians, who are to be insulated from policymakers and the public; the management process is the arena for policymakers and the public to negotiate an acceptable range of decision options within the constraints identified by the technicians who control the assessment process.

Expressed in these terms, the separation seems so extreme as to be unrealistic. Indeed, most former advocates of purity—whose views were first expressed in the National Academy of Sciences–National Research Council (NAS–NRC) report on risk assessment in the federal government, in 1983—have modified these divisions to allow more "joint custody" of the outcome. These modifications are, however, mostly an acknowledgment of the rights of crossover, that is, the right of the public to participate in both assessment and management, rather than a reconsideration of the fundamental assumption that the process can and should be so divided. As I noted at the beginning of this chapter, some people believe that quantitative risk assessment is inherently flawed and liable to produce defective policy, and that even if those flaws were to be corrected, additional problems would be introduced by the decision to divorce technique from judgment.

REASONS FOR THE DIVORCE:
A HISTORY OF BAD MANAGEMENT AND
UNSCIENTIFIC ASSESSMENT

There are understandable reasons that the EPA came to favor this separation. The collapse of the Reagan–Burford (Ann Burford, the former administrator of the EPA) EPA by the end of 1983 was caused by the failure of a deliberate campaign to mix politics and science in order to support specific "regulatory reform" agendas. The architects of this campaign sought to politicize the agency and purge it of technical and scientific staff and advisers who were deemed unsympathetic to deregulatory policies. The "black list," a rating of academic and other scientists—some of them with international reputations—was overtly used soon after Ronald Reagan's inauguration in 1981 by the new political appointees to prevent the EPA's Science Advisory Board[2] from criticizing the scientific basis for new ventures in policymaking (see Lash, 1984; as one of these blacklisted scientists, my bias should be made known).

Along with the attempt to oust mainstream scientists from advising the agency, the heads of programs in the first Reagan administration used dubious science to support some of their new directions. Dr. John Todhunter, the assistant administrator for pesticides and toxic substances, attempted to revise the methods for evaluating carcinogens in order to relieve the regulatory burden on pesticides, particularly those showing up in groundwater (see Perera 1984; OTA 1987). Rita Lavelle, the assistant administrator for solid waste, adopted nonstandard approaches to exposure assessment in order to avoid taking action at the Missouri dioxin sites. In 1982, on orders from George Bush's regulatory reform group in the White House, the Office of Air Quality and Radiation initiated a proposal to put lead back into gasoline and used unconventional arguments (including the notion that lead was an essential trace element in human nutrition) to counter a recently completed national survey of lead exposure, the second National Health and Nutrition Examination Survey. These events, which were widely publicized through congressional oversight hearings in 1982, supported the widespread perception that science had become completely "contaminated" by politics in the Reagan EPA.

Ruckelshaus and his housecleaners, in 1983, dealt directly with this contamination charge. First, scientists of unprecedented stature were recruited to take assistant administrator positions (for instance, Dr. Bernard Goldstein of Rutgers, and Dr. John Moore of the National Institutes of Health); second, some of the blacklisted scientists were conspicuously appointed to advisory committees; third, the consultative role of the National Academy of Sciences was used more often, as were the statutory requirements of the Clean Air Act and the Safe Drinking Water Act; and fourth, a need to "protect" science from politics was articulated by Ruckelshaus first in a speech to the national Academy of Sciences and was later published in a lead article in *Science* (Ruckelshaus 1983).

Thus the decision to separate perceived scientific and managerial aspects of decision making can be seen as part of the political process of restoring the EPA. That it focused so heavily on risk assessment may have been due to two factors: first, the need to find an "objective" metric for evaluating toxic chemicals, which were some of the most controversial issues of the Burford–Todhunter years, and, second, its coincidence with a reevaluation of the risk-assessment methodology used in the regulatory and scientific communities. The quantitative risk-assessment method, as applied to carcinogens, reached its first full methodological statement in the late 1970s (see OTA 1987, for a history of risk-assessment methods in regulation at the EPA and other agencies). The Interagency Regulatory Liaison Group's (IRLG's) statement on carcinogen policy was not finished before the 1980 election, and so a consensus on methods for the quantitative assessment of the carcinogenic risk of chemicals was, ironically, the result of another Reagan initiative to devalue the regulatory impetus. After considerable attempts by administration advisers to weaken the general federal policy statement on the science of cancer risk assessment that had been proposed in 1979, finally by 1984 there was a core doctrine regarding risk assessment and chemical carcinogens that embodied the so-called conservative approach to uncertainty, that is, a set of default assumptions that are, in the presence of uncertainty, strongly biased toward the worst case (OTA 1987).

The growing complexity of the regulatory review of toxic chemicals seemed to support delegating certain parts of the process to technicians. One of the major attacks on early versions of the risk-assessment statement was that it was insufficiently scientific. The final version contains a long discussion of molecular biology and is relatively less relevant to policy, a reverse of the first version. Perhaps it was a technical strategy of the critical reviewers to overwhelm the initial version with science. The price of victory may have been the handing over of the risk-assessment process to technicians while policymakers reserved the management decisions for themselves. Since that time, debates over risk-assessment conclusions have been almost dogmatically divided between science and policy. For instance, in the January 1988 debate over the EPA's proposed cancer risk assessment for inorganic arsenic, the Science Advisory Board declined to examine the appropriateness of lowering the cancer risk, based on an agency assertion that arsenic-induced cancer is operable. This issue was declared a risk-management prerogative and, as such, was beyond the purview of science.

The science of chemical carcinogenesis has become complicated with the growth of molecular biology as a field (and likely to grow more complicated; see Reynolds et al. 1987). The public's ability to participate in debates on risk has been even more severely limited because of the lack of access to appropriate scientific expertise even to understand the language of the process. At present, a confusing debate over the nature of evidence and the definitions of uncertainty occupies many of the chemical-specific controversies. It is not hard to understand why political appointees, economists, and lawyers would like to find scientists to deal with these aspects of the process while reserving for themselves the implementation of their decisions.

PROBLEMS WITH THE DIVORCE

Structurally, for the separation to work, the techniques and performance of the two assessment processes must be assumed to allow a division of resources between technical and managerial or policymaking skills. Specifically, it assumes that the resolution of uncertainty is a policy issue that can be handled by policymakers in a vacuum of science. But uncertainty itself is often so difficult to characterize and explain that this division of expertise and responsibility will increasingly result in inadequate decision making.

Uncertainty is unavoidable in the current risk-assessment methodologies. Only some areas of uncertainty can be resolved. Those types of uncertainty that currently cannot be resolved are those that arise from limits on the prevailing biological understanding or from practical limits on the reduction of uncertainty owing to constraints on experimentation or epidemiological research. For instance, in most cases, it is not possible to determine whether extrapolations—for dose calculations—between animals and humans are more accurate (in general or for specific chemicals) when based on relative body weight or on relative surface area. In practice, such cases have been handled by default assumptions or "answers" that represent consensus agreement among scientists. Other areas of uncertainty in specific data sets can be resolved either by accumulating more information or by further analyzing the existing data. An example of the latter is meta-analysis, or combining several data sets, which the EPA used when determining the ambient air standard for lead.

Identifying and resolving issues of scientific uncertainty can pose particularly formidable problems for a regulatory process based on separated assessment and management. Both fraud and mistakes may become more likely. Uncertainty has been misinterpreted and abused in regulatory processes since 1984. New methods for dealing with uncertainty have been adopted from decision theory (e.g., the surveys of members of the Clean Air Science Advisory Committee and the Bayesian approach advocated by Raiffa and others; see Morgan et al. 1985). Although these exercises may be helpful for the sociology of science, they ignore the origins of much of the uncertainty in risk assessment, which are not in differences of opinion but in gathering scientific data. This is best demonstrated in two issues involving uncertainty that are part of many controversies: the statistical representation of risk-assessment data and the methods for dealing with the weight of scientific evidence. Other important issues in this area are the continuing argument over distinctions between initiators and promoters of carcinogenisis; the definition of potency; and the use of safety factors for noncarcinogenic end points and their basis with respect to positive or negative toxicological data: the so-called lowest observable effect level ($+$) or no observable effect level ($-$). These other sources of misinterpretation and faulty policymaking have been discussed elsewhere (Silbergeld 1987b).

Statistical Representation of Risk-Assessment Data

Cancer risk assessment are essentially probability estimates because cancer is a stochastic event: It either occurs or it does not. The risk, or probability, of cancer can be expressed as a statistically definable relationship between an exposure and an outcome, or a dose and an incidence of cancer. Two types of uncertainty arise immediately in the graphical processing of these estimates of risk: first, the appropriate way to extend the dose response relationship beyond the range of actual empirical data and, second, the appropriate way to handle and illustrate variance. The first issue is important and unavoidable because the doses used in experimental bioassays are much higher than those expected in environmental situations, even in extreme conditions. Thus there is always a need to "go beyond the data" in risk assessments. To detect with statistical reliability a change in cancer incidence at rates that are of concern—in the range below one in one thousand, for instance—is practically impossible in whole animal, lifetime studies.

The resolution of this type of uncertainty is derived from the methodological or default assumptions of the cancer risk model; that is, in the absence of specific information to the contrary, it is assumed that the relationship between dose and incidence is linear at the low doses being examined in the risk assessment. Higher doses may evince hockey-stick relationships between dose and incidence (Hoel, Kaplan, and Anderson 1983) or be confounded by toxicity such that actual incidence is unknown. The application of the model to resolving this uncertainty is a mixture of science and policy: It is based on science in that it conforms with those molecular processes we currently understand best as being involved in cellular oncogenicity (Wright 1983). It is based on policy in that it is expedient and arguably conservative; that is, it is unlikely that the risks of cancer will be greater than those described by a line connecting the range of actual data (extrapolated) and the range of regulatory concern.

The second issue of uncertainty in presenting data concerns the handling of variance. Variance is an accepted property of scientific data fundamental to all phenomena of the physical world (Hawking 1988). More immediately, the precision of measurement is limited even under the best possible circumstances in simple systems. Variance is a source of specific as well as systematic uncertainty: It relates to the lack of absolute replicability of real data and to the lack of relative precision of actual sets of data—that is, the expected differences in measured and measurable outcome in replicated measures. In the cancer bioassay and its extrapolation to human prediction, which are not simple systems, there is an additional source of variance. This is in the extrapolation processes that transform the experimental data into predictions of human response rates. These two sets of variance tend to offset each other empirically; that is, failures to observe significant differences in response between doses would tend to lower the estimates of risk (by lowering the slope of the dose response relationship) unless a generous interpretation of variance were used. The variability of human response is likely to result in significant responses even at low doses in susceptible or multiply exposed persons, so that the lack of precision in the experimental data is to a certain extent "canceled" by the expected variability in the human experience.

The major unmeasurable and uncontrollable sources of variation (in individual metabolism, genetics, and behavior) tend to make predictions of actual human health effects highly variable. Protective public-health policy therefore incorporates the concept of "worst case," which is a philosophical underpinning for accepting the upper limits of the variation in expected response.

Because of resources, the data used in most cancer risk assessment are only minimally adequate for the purposes of calculation. Rarely are more than three actual data points provided on which to generate a line by regression analysis. Each point may have been repeated only twice, and the variance between duplicates at each dose may exceed that between them. Scientifically, this is a rather appalling situation for applying rational techniques of analyzing variance. But given the cost of generating more data points within one bioassay, it is unlikely that in most cases more data will be made available.

There is considerable dispute about how under these circumstances it is appropriate to present the range of variance in the dose incidence relationship. Three methods are usually discussed: upper bounds (based on 95 percent confidence intervals), maximum likelihood estimates (MLE), or most plausible ranges.[3] These all can describe the variance of any computed regression line. Although the distinction among these expressions of statistical uncertainty is technical and as such exemplifies the realm of "scientific" input into risk assessment, there are two ways in which it challenges the distinction between technical and managerial decision making. First, the decision has important policy implications because of the impact of the method on the outcome (expected incidence, or increase in risk, per dose). Second, the data are too sparse to base a choice on technical grounds.

Given that there is rarely an objective basis to select a method of expressing variance, the political goals of the risk managers tend to take precedence in resolving this area of uncertainty. Crudely put, a regulator interested in supporting a lower limit or standard can choose a method that emphasizes the consideration of uncertainty and provides an apparent scientific basis for restricting the worst-case range to a lower set of values. In the current climate of "postconservative" risk assessment (to quote Yosie 1987), MLE are used more often, as opposed to the formerly heavy use of 95 percent confidence intervals. Maximum likelihood estimates tend to circumscribe a narrower range of possible outcomes, and hence a lower risk associated with a given dose or exposure.

In no case are the value implications of the selection evaluated. Choosing the method of determining variance (and thus the range of estimates of risk) entails a decision concerning the relative value of over- and underestimating the true variance using the limited data at hand. Choosing the method of the 95 percent confidence limit is a decision that an error in calculating variance is not as serious in risk regulation as is actually overexposing human populations. Selecting the alternative, the maximum likelihood estimate or MLE method, places greater value on the experimental data themselves. Interestingly, although this method is frequently advocated by those who decry risk assessment as marred by unresolvable uncertainties, relying on MLEs to describe variance places more weight on both the reliability of the data and the accuracy of the methods to predict human response, because it inherently assumes a greater precision of the variance estimates.

In either choice, the textual description of variance is frequently inaccurate. For policymaking purposes, it is often stated in EPA risk assessments that the upper bound represents the upper limit on possible, rather than likely, risk and that the actual risk (as opposed to the statistically defined risk) may be much lower, in fact, as low as zero. This statement implies a uniform probability space below the upper-bound limit (whether the 95 percent confidence limit or defined by MLE). In fact, this is not the case. The determination that the regression line through the data points has a slope significantly greater than zero indicates that a zero risk for a given dose greater than zero is quite unlikely. Moreover, in many data sets, the minimum likelihood estimate, or lower 95 percent confidence limit, can also be determined. In almost all cases, it is extremely unlikely that all values below the upper bound or maximum limit are equally likely to occur. But the statement that actual risk may be much lower is made frequently and presents the regulated community with an invitation to resist control. Scientific oversight might ensure that a more accurate conceptual presentation of probability and variance be included in the conclusions of risk assessment. This would help avoid misleading policy judgments that use these assessment products.

Weight of Evidence

A major source of uncertainty in risk assessment is the nature of the scientific evidence from which the data are derived. Because scientific evidence comes from experiments conducted at different times in different laboratories, the estimation of the relative value, or reliability, of bits of evidence has always been difficult. Determining the "weight of evidence" or value in this sense denotes the influence or impact that should be assigned to a bit of information. Empirical evaluations of weight are based on replications of results in independently conducted experiments or studies and on the standing of each result or study with respect to other information and to more abstract canons of scientific worth.

In most developed countries, scientific peer review is the established means for assigning relative value to scientific data. This is a value-laden standard of ethical behavior that seems to work when perceived to be shared by both the reviewed and the reviewers. Although incidents of scientific fraud have caused some to doubt the unquestioned validation process of peer review, most scientists still use it as the best method of validation.

Evaluating the weight of evidence entails two aspects of data consideration: the content of the evidence (that is, the type of data) and the quality of the evidence (that is, how good the data are). Structures of data types in carcinogen risk assessment have been proposed by both the IARC and the EPA. These structures enumerate the data content and incorporate a limited attempt at qualification of the data, as "sufficient," "limited," or "inadequate." Though apparently simple, these qualitative discriminators are very difficult to objectify and so rely primarily on the judgment of peer review.

Codes can be used to judge scientific data for the purpose of "weighting" (Cox 1989). In toxicology, the Good Laboratory Practices (GLP) compendium compiled by the Food and Drug Administration (FDA) and adopted by the EPA and the Organization for Economic Cooperation and Development (OECD) represents the most ambitious attempt to codify the "ground rules" for acceptable science in the filed. However, the GLP does not fully deal with the issue. This is because some of the most important scientific information (in terms of informing a scientific hypothesis) is conducted in accord with experimental designs unlike those recognized by the GLP. Blind application of the GLP as a gate for acceptability of data would exclude most basic research from consideration, particularly those studies examining molecular mechanisms of action. Basic research studies are frequently conducted under highly specific designs that are quite different from the GLP's simulation type of practices and standard toxicology screening. The risk assessments resulting from the consideration of the GLP- type of data only would be flimsy indeed.

A thorough consideration of the weight of evidence or the evaluation of data thus remains somewhat nonobjective. It seems to work mainly because of the shared training of scientists participating in the peer review. The epistemological basis for scientific evidence is a more fundamental scientific concern. It is best apprehended in the context of a more formal inquiry into the nature of scientific proof or hypothesis formulation, topics outside the scrutiny of this article or my competence.

Policymakers have often misinterpreted the weight of evidence and formulated egregiously unacceptable methods for handling differences in the weight of evidence in their decision making. To incorporate differences in the weight of evidence into sound decisions, it is good to keep in mind those factors that tend to govern the weight of evidence in science. Usually, the weight of evidence—that is, the preponderance of data—is related simply to the resources that have been allocated to the subject: When more research funds are targeted to answer a specific question, more research is done, and more data become available. At the EPA and elsewhere, policymakers have proposed much more complicated interpretations and schemes for incorporating differences in weight that do not reflect the relatively straightforward source of differences in data. For instance, OECD experts have suggested that differences in the weight of evidence could be used to prioritize existing chemicals for further testing and evaluation (OECD 1986). Although this proposal might serve to identify chemicals with substantial, but still not complete, data to determine risk, it would tend to relegate chemicals with few data to a regulatory black hole from which they could never emerge. In 1986, the EPA proposed that differences in the weight of evidence could be used in the management phase to "discount" the level of risk associated with an exposure or dose according to the assessment phase. This represents a complete misunderstanding of the concept of the weight of evidence, and an assumption that relative lack of evidence was somehow predictive of lower risk. However, in 1984, a panel of scientific experts convened by the EPA expressly recommended against this judgment (Federal Register, November 23, 1984, p. 46299).

For both scientific and regulatory purposes, it is essential to avoid enshrining existing uncertainties and reducing the flow of information by inappropriate handling of this type of uncertainty. The best solution is to reconsider the major source of the problems related to the weight of evidence for chemicals: the time and resources allocated to chemical testing. Simply because of the overwhelming lack of data on almost all existing chemicals (see NAS—NRC 1982), the lack of weight cannot be equated with the lack of risk. The balance is so uneven between even partially tested and totally untested chemicals that no such inference can be supported. To give the lack of evidence more significance is to enshrine the dogma of ignorance: If we have not heard about it, it cannot be dangerous—a policy that underlies the Generally Recognized As Safe (GRAS) list but should be seen only as an interim solution to the crisis in toxicity testing.

Under these circumstances, the weight of evidence is perhaps best used in decision making as a means of determining the finality of regulation. That is, chemicals with less weighty evidence, but still some evidence of risk, may be controlled on an interim basis, with the opportunity for review mandated to a reasonable schedule of expected testing. One advantage of this scaling of finality with uncertainty would be to encourage the reduction of uncertainty. That is, it gives an advantage to the regulated community to provide more informa-

tion to reduce the regulatory burden. If, however, the private sector determines
that the economic interests of the product do not support further testing, then
they will accept the market conditions imposed by interim regulation on a permanent basis.

THE FATAL DIVORCE

The impacts on the policy of separating risk assessment from risk management
were demonstrated early at the ASARCO smelter hearings that Ruckelshaus
held in 1985. At issue in this situation was the regulatory use, under the Clean
Air Act, of the risk assessment for arsenic. An old, uncontrolled copper smelter
in Tacoma, Washington, used low-grade, high-arsenic-content copper ores; it
was operated with considerably less than best available control technology, particularly on fugitive emissions from the plant site. (Because arsenic is a human
carcinogen, the potential uncertainties related to extrapolation across species
were not at issue.) Also, distinct from other environmental controversies, there
was no dispute that human exposure was taking place, although the dose and
extent remain debatable.

Nevertheless, in the face of this relatively simple situation, Ruckelshaus chose
to turn the Tacoma decision-making process into a demonstration project of his
concept of separate assessment–management process by abrogating the relationship between policy and the results of the risk assessment. Ruckelshaus and
his staff threw open the confused debate to nonscientists and, indeed, nonpol-
icymakers. This process was called *risk communication*. It was intended to assure
the public that their wishes in environmental regulation would be elicited and
incorporated into the decision making. Unfortunately, the process mainly served
to baffle the participants and to raise issues of apparent conflict between jobs
and health. It had little to do with resolving the controversy at Tacoma: The
smelter was about to be closed for reasons having nothing to do with possible
environmental regulation but, rather, the price of copper worldwide. Thus the
debate as defined by Ruckelshaus—enforcement or jobs—was deceptive. Moreover, according to many polls, the public has clearly indicated its willingness to
accept economic costs in order to have more environmental protection. Since
that time, the EPA has invested considerable resources in conferences and
consultants on risk communication (Conservation Foundation 1987).

Other issues also demonstrate the practical limitations of the decision to
separate risk assessment from risk management. These are the attempts to defer
risk-assessment decision making to the states and to regulate chemicals differ-
entially according to their effect on health.

The most serious challenge to the current structure of separation arises in
those instances in which the EPA has delegated authority to the states. Most
states do not have the resources to support experts in risk assessment and the
urgings of the EPA to isolate the process and "scientificize" it have further
strained many state agencies. The EPA has proposed that the states assume
responsibility for developing and enforcing ways of controlling releases of hazardous air pollutants, including carcinogens. However, because of different ac-

cess to resources, the states have responded with inconsistent guidelines. Some have implicitly adopted risk-assessment methods by relying on the output of the EPA's Carcinogen Assessment Group (e.g., Maryland); other states have transformed occupational health standards (which were not designed to deal with carcinogens) such as occupational threshold limit values (TLVs) (e.g., Massachusetts); some have proposed risk goals (e.g., California); and still others have mostly avoided the issue of carcinogens (e.g., New York).

The problems that have arisen as these states handle carcinogens and risk assessment were exemplified in the controversy in Connecticut over proposed regulations for dioxins in air. Scientists in the state health authority, relatively unfamiliar with risk assessment, pulled together at short notice an advisory panel relatively unfamiliar with dioxin and risk assessment. They made a fundamental technical error in their proposal when they stated that duration of exposure was a factor in selecting the model for the mechanism of carcinogen action: The number thus calculated was based on noncancer models for risk calculation and was orders of magnitude higher than the EPA's Carcinogen Assessment Group calculations and higher than similar work by New York and California, states that have invested considerable resources in a technically proficient staff.

The second issue concerns the attempt to ascribe quantitative differences in risk to qualitative differences in outcome. Over the past three years, the EPA has attempted to promulgate a set of consensus documents outlining the scientific bases for assessing the risks of mutagens, carcinogens, developmental toxins, and systemic toxins. There are still no objective criteria or models to quantify the risks of outcomes other than cancer. The scientific issues of estimating the risks of these types of outcomes are in many ways more complex than are those associated with cancer risk assessment. No singular molecular theory of action can be proposed for these end points; moreover, most of these effects are nonstochastic. Severity, rather than incidence, varies with exposure (see Friberg, 1986, for a discussion of these differing modalities of risk assessment). In such cases, probability estimates are not pertinent ways of denoting the relationship between dose and response to dose. Thus, comparing risks means comparing both the types of end points (or organ system that is the target of the toxin) and the severity of any effect (Hartung 1986). There is little guidance for this undertaking from current methods of risk assessment, and in a policy setting, structuring decisions around such questions can turn environmental decision making into a battlefield of medical specialties, in which diseases are arrayed against one another in competition for allegedly scarce resources. Risks are not fungible or directly comparable. To attempt to use scientific criteria to hide such political decisions is therefore not tenable.

A CULTURE OF INDIFFERENCE

I have asserted that separating environmental decision making into risk assessment and risk management encourages misinterpretations of uncertainty that discourage management action. Within the culture of a regulatory agency, this justifies a separation of activities that is deleterious to sound operations and a sense of shared institutional authority and responsibility. When the scientists are restricted from access to policymaking processes based on the implications of scientific choices (which are prominent in the resolutions of uncertainty), they can only guess how their choices may affect policy. And policymakers, encouraged to remain ignorant of science, may misinterpret uncertainty in some of the ways described in this chapter.

I have highlighted uncertainty in this discussion for two reasons: It is a critical part of the decision to act, and it crosses the line between the technical assessment and the policymaking management dichotomy. As I noted, the concept of uncertainty, the determination of acceptable levels of uncertainty, and the method for resolving differences regarding uncertainty in environmental issues are value-laden matters. Determining that a sufficient amount of uncertainty has been resolved is critical to deciding whether to initiate action or to call for more research.

> [Many environmental] controversies were often debated in terms of "false positives" and "false negatives." What if a given scientific study turned up no evidence of harm from a pollutant [that] general knowledge held to be hazardous? Could one safely rely on that evidence to allow exposure? What if later studies demonstrated this negative evidence to be false? Or, to take the opposite case, what if positive evidence that the chemical was harmful later turned out to be false? What was the relative risk of taking action based on false negatives in contrast with false positives? The debate was lively because at low levels of concentration one study might turn out one way and later ones another. If one wished to be cautious in environmental strategy, one could argue that such false negatives gave rise to an unwarranted confidence of safety and hence were risky. The social risk from making a mistake by regulating a chemical later found to be safe was far less than the mistake in not regulating one found to be harmful. (Hays 1987, p. 278)

The social value of separating risk assessment and risk management has not been positive. It has not resolved the basic problems with risk assessment as a mode of analysis or decision making in environmental issues. Stratifying the process and emphasizing risk communication as a method of resolving disputes over the content of risk-related decision making has not improved the public's acceptance of the EPA's actions over the past few years, despite the best hopes of Ruckelshaus and his colleagues. It is customary in some circles to ascribe this to the hopeless ignorance of the public, but they are hardly more ignorant than are many of the bureaucrats at the EPA. Decision-making processes are supposed to be accessible to the public from which power derives. If the public is made powerless by some methodology that has little inherent validity, the obligation of the decision makers will be to disregard the methodology and readmit a more holistic entity that includes both assessment and management, science and policy, which much of the public inherently understands and still trusts.

NOTES

1. For reasons that I will not discuss here, concern over the carcinogenic potential of chemicals has occupied the major part of the public discussion on the human health

effects of environmental exposures (EDF 1980). For the purposes of this discussion, it is sufficient to indicate that evidence of carcinogenicity is, by definition, the data derived from the standard mutagenicity and oncogenicity tests and adopted by the EPA, OECD, and other regulatory authorities (for further discussion of this, see NAS–NRC, *Safe Drinking Water and Health,* vol. 6 [1986] and National Toxicology Program, *Fourth Annual Report on Carcinogens* [1986]).

2. The Science Advisory Board is a statutorily mandated body composed of distinguished outside advisers who are to give the EPA administrator expert scientific advice and scientific review of research and policy objectives.

3. These three phrases describe different statistical definitions of the range or variation of the data in a regression line (or curvilinear curve fitting). A 95 percent confidence interval is the range of values that have ninety-five out of one hundred chances of being equally probable. An upper bound is the largest likely value of a range of variable data, and a maximum likelihood estimate is the "most plausible" on the basis of either biology or the weight of evidence.

REFERENCES

Conservation Foundation (1987). *Risk Communication.* Washington, D.C.: Conservation Foundation–World Wildlife Fund.

Cox, G. V. (1989). "Quality Assurance in Risk Models for Regulatory Decision-Making." In *Risk Assessment in Setting National Priorities,* ed. J. Bonin and D. Stevenson, pp. 357–70. New York: Plenum Press.

Environmental Defense Fund (EDF) (1980). *Malignant Neglect.* New York: Knopf.

Environmental Protection Agency (EPA) (1987) *EPA Journal* 13: passim.

Finkel, A. M., and Evans, J. S. (1987). "Evaluating the Benefits of Uncertainty Reduction in Environmental Health Risk Management." *Journal of the Air Pollution Control Federation* 37:1164–71.

Friberg, L. (1986). "Risk Assessment." In *Handbook on the Toxicology of Metals,* ed. L. Friberg and A. M. Berlin, pp. 269–93. Amsterdam: Elsevier.

Hartung, R. (1986). "Ranking the Severity of Toxic Effects." Paper presented at the Twentieth Annual Conference of Trace Substances in Environmental Health, Columbia, Mo., June.

Hawking, S. M. (1988). *A Brief History of Time.* New York: Bantam Books.

Hays, S. (1987). *Beauty, Health, and Permanence: Environmental Politics in the United States 1955–1985.* Cambridge: Cambridge University Press.

Hoel, D. G., Kaplan, N. L., and Anderson, M. W. (1983). "Implications of Nonlinear Kinetics on Risk Estimation in Carcinogenesis." *Science* 219:1032–37.

In re Agent Orange Product Liability Litigation (1984). Preliminary Memorandum and Order on Settlement. September 25, U.S. D.C. E.D. N.Y.

Lash, J. (1984). *Season of Spoils.* New York: Grossman.

Morgan, M. G., Henrion, M., Morris, S. C., and Amaral, D. A. L. (1985). "Uncertainty in Risk Assessment." *Environmental Science and Technology* 19:662–67.

National Academy of Sciences–National Research Council (NAS–NRC) (1982). *Toxicity Testing.* Washington, D.C.: NAS Press.

National Academy of Sciences–National Research Council (NAS–NRC) (1983). *Risk Assessment in the Federal Government: Managing the Process.* Washington, D.C.: NAS Press.

National Academy of Sciences–National Research Council (NAS–NRC) (1986). *Safe Drinking Water and Health,* vol. 6. Washington, D.C.: NAS Press.

National Toxicology Program, Department of Health and Human Services, Public Health Service, National Institute of Environmental Health Sciences (1986). *Fourth Annual Report on Carcinogens*. Washington, D.C.: DHHS.

Nichols, A. L., and Zeckhauser, R. (1985). "The Dangers of Caution." Report no. E–85–11, John F. Kennedy School of Government, Harvard University.

Office of Science and Technology Policy, White House (1984). "Chemical Carcinogenesis: A Review of the Science and Its Associated Principles." *Federal Register* 49:21594–661.

Office of Technology Assessment (OTA) (1987). *Identifying and Regulating Carcinogens*. Washington, D.C.: Government Printing Office.

Organization for Economic Cooperation and Development (OECD) (1986). *Existing Chemicals: Systematic Investigation, Priority Setting, and Chemicals Review*. Paris: OECD.

Perera, F. (1984). "The Genotoxic/Epigenetic Distinction: Relevance to Cancer Policy." *Environmental Research* 34:175–91.

Reynolds, S. H., Stowers, S. J., Patterson, R. M., Maronpot, R. R., Aaronson, S. A., and Anderson, M. W. (1987). "Activated Oncogenes in B6C3F1 Mouse Liver Tumors: Implications for Risk-Assessment." *Science* 237:130–38.

Rosner, D., and Markowitz, G. (1984). "A 'Gift of God'?: The Public Health Controversy over Lead-Additives to Gasoline." *American Journal of Public Health* 75:344–52.

Ruckelshaus, W. (1983). "Risk, Science and Public Policy." *Science* 221:1026–28.

Ruckelshaus, W. (1985). "Risk, Science and Democracy." *Issues in Science and Technology* 1:19–38.

Russell, M., and Gruber, M. (1987). "Risk Assessment in Environmental Policymaking." *Science* 236:286–90.

Silbergeld, E. K. (1985), "Response to Ruckelshaus." *Issues in Science and Technology* 1:18.

Silbergeld, E. K. (1987a). "An Environmentalist Perspective on Risk Assessment." *EPA Journal* 13:34.

Silbergeld, E. K. (1987b). "Five Types of Ambiguity: Scientific Uncertainty in Risk Assessment." *Hazardous Waste and Hazardous Materials* 4:139–50.

Silbergeld, E. K., and Percival, R. V. (1987). "The Organometals: Impact of Accidental Exposures and Experimental Data on Regulatory Policies." In *Neurotoxicants and Neurobiological Function,* ed. H. A. Tilson and S. Sparber, pp. 327–52. New York: Wiley Interscience.

Whittemore, A. S. (1983). "Facts and Values in Risk Analysis for Environmental Toxicants." *Risk Analysis* 3:23–33.

Wright, A. S. (1983). "Principles of Chemical Carcinogenesis." In *Developments in the Science and Practice of Toxicology,* ed. A. W. Hayes, R. C. Schnell, and T. S. Miya, pp. 311–18. Amsterdam: Elsevier.

Yosie, T. F. (1987). "Science and Sociology: The Transition to a Post-Conservative Risk Assessment Era." Address to the annual meeting of the Society for Risk Analysis, November.

6

Understanding Uncertainties in Medical Evidence: Professional and Public Responsibilities

VALERIE MIKÉ

The case of Linda Loerch and her son Peter presented to the Minnesota Supreme Court raises the question of whether legal liability can extend beyond the second generation. During the pregnancy leading to Linda's birth, her mother had taken the synthetic hormone diethylstilbestrol, commonly known as DES. Linda herself has a deformed uterus, and her son Peter, born twelve weeks prematurely, is a quadriplegic afflicted with cerebral palsy. The family is seeking damages for the child's condition from Abbott Laboratories, the manufacturer of the drug taken forty years earlier by his grandmother (MacNeil/Lehrer 1988).[1]

The claims of this lawsuit hinge on the evidence available when the drug was prescribed. The case illustrates, with some new ramifications, the interrelated issues of ethics and evidence surrounding the practice of medicine, a major theme of this chapter.

The DES story first became national news at a time that marked the rise of the new field of bioethics. The Food and Drug Administration (FDA) issued a drug alert in 1971 to all physicians concerning the use of DES by pregnant women, as an association had been found between the occurrence of a rare form of cancer of the vagina in young women and their mothers' exposure to DES. The drug had been prescribed widely since the 1940s for a variety of medical conditions, including the prevention of miscarriages. It is estimated that during this period four to six million individuals, mothers and their offspring, were exposed to DES during the mothers' pregnancy. The full dimensions of the medical disaster, the subject of continued controversy, have yet to be firmly established.

DES daughters are at risk of developing clear cell adenocarcinoma of the vagina, the risk estimated to be one per one thousand by age twenty-four. Ninety percent have a benign vaginal condition called adenosis, and many have other genital abnormalities. They are at higher risk of pregnancy loss and infertility. DES mothers also may be at a higher risk for breast and gynecological cancers, and DES sons may be at an increased risk of genitourinary abnormalities, infertility, and testicular cancer. DES may, as well, have affected fetal brain development, leading to behavioral problems and learning disabilities. And beyond all this, there is the immense psychological trauma, the anxiety and fear affecting so many throughout a lifetime (Apfel and Fisher 1984).

To place this story into proper context for our discussion, it is helpful to recall by way of contrast another well-known episode from that period. Within a year after the FDA's warning against DES, a major medical scandal came to dominate the nation's news media: the Tuskegee Syphilis Experiment. This was a study started in 1932 and sponsored by the U.S. government, in which four hundred black sharecroppers afflicted with syphilis were kept under observation without being given any treatment. They were never told that they had syphilis and were not told that effective treatment was being withheld, even after penicillin became widely available. The experiment continued for over forty years and was stopped only in 1973, eight months after the Associated Press broke the story (Jones 1981).

This was a time of growing general awareness of ethical issues concerning medical experimentation. Although perhaps the most shocking, Tuskegee was by no means an isolated incident in the recent history of medicine. But as these practices came to light, there was a strong reaction from the public, and this in turn had an impact on legislators. Philosophers, lawyers, and medical researchers engaged in dialogue. Stimulated by these issues and questions raised by the rapid development of medical technology, the new field of bioethics came into its own as a major discipline, with institutes and scholarly journals devoted to its study. Patient autonomy and informed consent became accepted concepts. Laws were passed to protect research subjects, and experimentation on humans is now regulated by the federal government. It is probably safe to assume that Tuskegee will not happen again. But the story of DES is another matter.

The attention of the public is constantly drawn to the latest medical advances, such as highly publicized life-sustaining technologies, organ transplants, and other sophisticated new procedures. Ethical analyses reported by the media focus on related spectacular issues: surrogate motherhood, the anencephalic newborn as organ donor, the withholding of nutrition and hydration from a nonterminal patient. All of these are important questions of ethics and policy that need to be resolved. But they are not likely to have a personal impact on the vast majority of the viewing, listening, or reading public. Other issues of great impact for many tend, on the other hand, to be ignored.

There is a shadow side to modern medicine that is not part of the general consciousness. Calling attention to it is important because it concerns everyday medical practice; it concerns the lives of people every time they decide to take a pill or visit their doctor. This is the fact of medical uncertainty. The use of medical technology today far surpasses the boundaries of medical knowledge. Further, a large proportion of health care procedures have not been properly evaluated (Institute of Medicine 1985). But even when they have, the results are frequently ignored by practicing physicians, or there may be controversy over conflicting reports. And in many cases there could be as yet unforeseen long-term consequences. The Hippocratic maxim "help or at least do no harm," for two thousand years the basic principle of medical ethics, is especially relevant in our day. The truly effective remedies that are now available to cure disease are not deliberately withheld from patients. But powerful diagnostic and therapeutic procedures can also be harmful. Current FDA regulations concerning proof of safety and efficacy—although far more stringent than when DES came into use—pertain only to the approval of new drugs and new medical devices. There still are no laws to protect patients from the inappropriate use of procedures and commercially available technologies.

Unlike Tuskegee, the DES tragedy could happen again. It is the aim of this paper to explore some of the related issues.

MEDICAL UNCERTAINTY AND MEDICAL PRACTICE

Scientific Evidence
Evidence in the Story of DES

Using the case-study approach, we can examine the nature and extent of the evidence concerning DES and thus try to gain insight into the relationship among uncertainty, scientific evidence, and medical practice.

The argument may be made that the harmful effects of DES could not be foreseen and that the physicians prescribing the drug were trying to help their patients, to enhance their chances of delivering full-term healthy infants. That may very well be true. But their basis for using the drug was even then open to question. Claims of efficacy are frequently based on clinical trials without proper controls. In a paper on the impact of controlled clinical trials on the practice of medicine, Thomas Chalmers reviewed the DES story (1974). Between 1946 and 1955, thirteen studies were carried out to test the effectiveness of DES in preventing miscarriages. Of the seven studies that reached enthusiastic conclusions, none had adequate controls. The other six trials had concurrent controls, and all their conclusions were negative. Yet between 1960 and 1970, long after these results were published, an estimated 100,000 DES prescriptions were still being written each year for pregnant women (Heinonen 1973).

The largest of the controlled trials was carried out by Dieckmann and associates at the University of Chicago (1953), including over two thousand consecutive patients registered in their prenatal clinics for a two-year period between 1950 and 1952. The treatment assignment method was alternation between DES and placebo (not strict randomization, which was not yet a well-established procedure in those days), and the study was carried out in a double-blind fashion; that is, neither patients nor attending clinical staff knew whether the patient was taking DES or a placebo. The rationale for using DES to prevent and treat pregnancy complications was the theory that a decreased production of progesterone and other steroids by the placenta predisposed to, or caused, these complications and that DES could stimulate the secretion of these steroids.

Data from 1646 patients were included in the analysis. The results showed a slight increase in reproductive problems experienced by the DES group over the group that had taken the placebo, the opposite of what had been expected. The duration of pregnancy was significantly shorter for the DES group, and the incidence of prematurity was higher. The rates of spontaneous abortion, neonatal mortality, and toxemia all were somewhat higher than for the placebo group. The results of the study were presented at the 1953 Annual Meeting of the American Gynecological Society and subsequently published in the *American Journal of Obstetrics and Gynecology*. The paper generated a spirited discussion of what was already a controversial issue. Participants at the meeting included George and Olive Smith of Harvard University, the proponents of the stilbestrol theory. These investigators in turn stated that their claim concerning the beneficial effect of DES applied only to high-risk mothers and that the therapy was not intended for routine use. This statement was reiterated by others, and at the end of the session—as well as in the printed version—the authors offered the full set of data to anyone who wished to study it in greater detail.

In spite of these widely publicized results, no properly controlled study of high-risk mothers was ever carried out, and for another eighteen years DES continued to be prescribed routinely by many physicians. The book on DES by Apfel and Fisher shows the reproduction of an ad (1984, p. 26) for desPlex (a brand name for DES) that appeared in a medical journal in the 1950s: "to prevent ABORTION, MISCARRIAGE and PREMATURE LABOR . . . recommended for routine prophylaxis in ALL pregnancies." Several of the supporting claims in the ad are incorrect, and although there is a 1954 reference, the Dieckmann paper, published in 1953, is not mentioned.

There is still a debate, however, concerning the causal relationship between DES and clear cell adenocarcinoma of the vagina. The establishment of causality from observational studies is an extremely difficult problem. This was recognized by the authors of the book on DES (Apfel and Fisher 1984) and in a work on the adverse effects of medicine that used DES as one of its illustrative cases (Siegler et al. 1987). A critical review of the epidemiologic evidence published by McFarlane and associates (1986) attributes potential flaws to the two retrospective case-control studies on which the original DES–vaginal cancer association was based. The first, carried out by Herbst, Ulfelder, and Poskanzer (1971) found DES exposure in utero for seven of eight patients with clear cell adenocarcinoma of the vagina and for none of thirty-two matched controls. The second study, published by Greenwald and colleagues (1971) found prenatal DES exposure for five of five vaginal cancer patients and for none of eight matched control subjects. One of the issues discussed in the critique is that of susceptibility bias. In both studies the patients (cases) tended to be associated with problem pregnancies, whereas this condition for the controls was less prevalent in one study and could not be assessed in the other. Susceptibility bias here means that the DES treatment might have reflected a preexisting problem that subsequently also resulted in cancer, rather than the DES itself being the cause of the malignancy.

Although it is likely that this particular controversy will never be fully resolved, there is now general consensus that DES should not have been widely prescribed for pregnant women in the first place.

Sources of Evidence

A number of studies have shown that the adoption of a drug and the subsequent prescribing behavior of physicians are to a large extent determined by the pharmaceutical industry (McKinlay 1981). It is estimated that 75 percent of practicing physicians first hear of a drug from industry sources, about half from "detail men," the sales representatives of drug companies, and 25 percent from drug company literature. The pharmaceutical industry spends about 20 percent of the retail price of drugs on advertising. The final source of information before adoption are drug-house sources and colleagues for an estimated 60 percent of physicians. Virtually all doctors use the *Physicians' Desk Reference* (PDR) for information on drugs, but the PDR contains essentially the text of the package inserts provided and paid for by the drug companies. Although the package insert is subject to FDA approval, bias may be unavoidable from these interested sources.

Physicians could, in principle, expect to find unbiased evaluations in the professional literature, but here also, as we have seen for DES, the situation may be confusing. And the lay public has no way of knowing about the poor quality of a large proportion of published studies. This has been reported in a number of review papers, but the critiques are themselves printed in medical journals and thus not readily available. For example, Fletcher and Fletcher (1979) reviewed a random sample of six hundred articles published by three leading medical journals between 1946 and 1976. They observed a deterioration in the quality of clinical research design over this thirty-year period. The proportion of studies in which the researchers had used existing data, instead of collecting the data specifically to answer their research questions, increased from 24 percent in 1946 to 56 percent in 1976. Cohort studies, in which the same group of patients is followed over time, decreased from 59 percent to 34 percent. And sample size has remained an unresolved problem. Several surveys of published clinical trials found that the sample sizes used in the majority of these trials were much too small. They thus had a reasonable chance of detecting an improvement in therapy (or a harmful effect) only for treatment differences so large as to be highly unlikely (Freiman et al. 1978; Miké 1982).

Factors Affecting Medical Practice
The UGDP Story: Example of Complexity

Pressure from drug companies and the desire of patients to receive what had been billed as highly effective in the lay press were considered important factors in the continued use of DES. The great complexity of the situation, including the role of financial interests, is well illustrated by the story of the University Group Diabetes Program (UGDP), a large randomized clinical trial designed to assess the efficacy of oral hypoglycemic agents for adult-onset, non-insulin-dependent diabetes (Chalmers 1974). The trial was stopped ahead of schedule in 1970 because of a significant increase in cardiovascular mortality among patients assigned to the treatment groups. But there was a continued rise in the sale of these drugs for another six to eight years. The results of the trial had met with strong controversy and were challenged by the drug manufacturers. The study was then reviewed by several groups of experts, and the case was finally heard by the Supreme Court of the United States. The Court's decision concurred with the verdict of the earlier reviews that the conclusions of the clinical trial were scientifically valid.

In a further analysis of related documents Chalmers (1982) studied the work of fifty-five authors who had written review articles on the treatment of adult-onset diabetes. Twenty-seven were against the UGDP trial and/or in favor of using the agents; twenty-four were for it and/or against the use of the oral drugs; and four were equivocal. A total of 1157 papers published by these fifty-five authors between 1963 and 1978 were identified and reviewed for acknowledged financial support. Nearly 10 percent of the articles published by anti-UGDP authors acknowledged support by the manufacturers of the oral agents, compared with a little over 2 percent of the authors who had accepted the results of the study as valid. Chalmers also found a statistically significant association between criticisms of the trial in the widely read free medical journals financed by drug companies and advertisements in these journals by the manufacturers of the oral agents. These intrinsic conflicts of interest contribute to promoting the continued use of agents in the face of deleterious evidence.

The Slighting of Evidence

But economic factors do not always figure in such situations. Two examples were given by Chalmers, in which the doctors involved had no direct financial incentive in prescribing the treatment (1974). One of these pertained to bed rest for patients with acute viral hepatitis. The practice of requiring bed rest until the patient's liver function test became normal had been established in the 1940s, but in two carefully designed randomized clinical trials reported in 1955 and 1969, respectively, no therapeutic benefit could be attributed to this practice. Yet in a review of hospital records reported by Chalmers it was found that the majority (56 percent) of physicians were still prescribing bed rest years after these definitive studies.

The second example concerns the prescription of special diets for acute peptic ulcer. The efficacy of the so-called Sippy or related bland diet in speeding the healing process was studied in eight clinical trials, with a negative outcome in all. Yet a chart review in a community hospital, again reported by Chalmers, revealed that fifteen of sixteen physicians had prescribed such a diet, as had twenty of twenty-two physicians in a university hospital. These doctors' lack of awareness of the evidence and their preference for their own old methods of treatment are among the underlying causes of this problem.

The impact of clinical trials on family practice was assessed by McGrady (1982) in a study based on a mail questionnaire sent to a random sample of family physicians. Eight clinical cases were presented in the questionnaire, with four possible modes of treatment for each, and the responding doctors were asked to select those that would be closest to their own choice of treatment. In each case one of the four answers corresponded to the conclusions of controlled clinical trials reported in the medical literature. The average concordance of the physicians' choice of procedure with what had been published as the best sci-entifically validated treatment was 33 percent. Note that in a multiple-choice test of this type, selecting treatments at random would yield an average of 25 percent "correct" answers. This study essentially confirmed the observations of Chalmers (1974).

A factor in this situation that should not be overlooked is many physicians' lack of time for carefully reading the literature and their lack of skill in assessing the reported claims of clinical studies. The concepts underlying statistical methodology are nontrivial and often difficult for nonspecialists to understand. A study carried out by Berwick, Fineberg, and Weinstein at Harvard University (1981) used a multiple-choice questionnaire to test physicians' ability to make inferences from quantitative clinical information. The subjects included medical students, interns and residents, and physicians engaged in research or in full-time practice. There was a lack of consensus in all groups on the meaning of frequently used terms such as *P-value* and *false positive rate* and a lack of familiarity with important principles of statistical inference. There was a general tendency to reach conclusions not supported by the data. And the performance scores of the practicing physicians were significantly lower than those of the other three groups.

The Phenomenon of Geographic Variation

There has been much discussion in recent years concerning the phenomenon of geographic variation in medical practice. Attention has been called to the problem by the extensive studies of John Wennberg, who discovered wide regional variations in rates of surgical procedures. He found, for example, that in different regions of Maine, the chances of a woman's having a hysterectomy by age seventy range from 20 percent to 70 percent, and in Vermont the likelihood of a child's having a tonsillectomy ranges from 8 percent in one region to 70 percent in another. The probability of a male resident of Iowa having a prostatectomy by age eighty-five ranges from 15 percent to 60 percent in different areas. Such variations in the use of medical procedures seem to be unrelated to differences in disease incidence and mortality and to the supply of services and facilities (Wennberg 1984).

Geographic variation had also been observed in the prenatal use of DES before the FDA ban (Heinonen 1973). The estimated usage was higher in the East and Midwest than in the South and West (3.67 and 2.93 versus 1.94 and 1.68 DES prescriptions per one hundred live births). There was also considerable variation among hospitals in different regions.

The Institute of Medicine of the National Academy of Sciences held a conference in 1983 to assess the reasons for the variations in medical practice around the country and to find ways of dealing with the problem. Frederick Robbins, president of the Institute of Medicine, said in his concluding remarks: "No practitioner can get away from the potentially damning evidence that these great variations represent. You can cover it up all you want, but it looks bad, and it looks bad because it is bad. It is not an appropriate way for a profession to behave" (Iglehart 1984, p. 4).

The fact that these geographic variations have no discernible effect on the general health of the population supports the view that much of today's treatment is ineffective and that there is a great deal of overtreatment. It has been suggested that if patients were better informed about the risks as well as the expected benefits of recommended procedures, there would be a substantial decrease in their use, resulting in better care at lower cost (Bunker 1985).

The Treatment of Doctors' Families: Further Insights

In trying to assess the complex factors contributing to the observed variations in medical practice, it is relevant to ask about the medical care received by the doctors themselves and their families. If financial incentives are paramount in leading physicians to prescribe treatment that may be unnecessary, then certainly they will not permit this to happen in the case of their own families. Similarly, while patients may accept needless treatment because they are poorly informed, physicians will have much easier access to the latest scientific information that they can assess before agreeing to their own treatment.

A good example to consider is the use of surgical procedures that are believed to be carried out far too often in the general population. This question was studied by Bunker and Brown (1974), and their report contains some surprising findings. They compared the rate of surgical services for doctors and their spouses with that of several control groups. These were other professional groups living in the same high-income area, for whom it could be assumed that there was an unlimited supply of medical facilities and services. The study included common surgical procedures such as appendectomies, cholecystectomies, and hysterectomies. The investigators found that the rates of these operations for doctors and their spouses were at least as high as for the other professional groups. And when the rates were combined for this well-to-do, highly educated group as a whole, the overall rates were found to be some 25 percent to 30 percent higher than the national average.

This study once again illustrates the complexity of factors affecting medical practice. In addition to the influence of economic factors, there is a lack of awareness or understanding of the scientific evidence and, frequently, an unquestioning faith—on the part of both doctor and patient—in the usefulness of medical procedures. Variations in treatment tend to be governed by local custom, with doctors practicing in the same way as their other colleagues in the community, rather than being guided by personal philosophy or independent scientific evidence. Following the standards of accepted practice is also an adequate safeguard in malpractice litigation. Of course, the reluctance or inability of many physicians to confront the extent of medical uncertainty and to discuss it openly with their patients is itself one of the underlying causes of the present malpractice insurance crisis.

MEDICAL UNCERTAINTY AND RESPONSIBILITY

Responsibility for Acceptable Evidence

Presented below are some thoughts concerning responsibility from the perspective of the uncertainties inherent in present-day medicine. This is a brief sketch aimed at offering suggestions, not a full treatment of abstract concepts. The notion of "acceptable evidence," the underlying theme of this book, takes into account ethical and policy values, criteria for judging when evidence is acceptable. But to arrive at such criteria, we need a better understanding of uncertainties in the evidence and the problems of communicating it. Only given such an understanding can anything like an ethics of evidence have meaning. Elsewhere, I have proposed some guidelines for such an ethics based on the principles of autonomy, beneficence, and justice (Miké 1989b) and have discussed the troubling conflict of values in contemporary American medicine (Miké 1988a). My aim here is to clarify further the problem of dealing with uncertainties.

It is my main thesis that we as a society need to (1) make a serious effort to come to terms with the fact of medical uncertainty, (2) calmly and openly share our awareness of this uncertainty, and (3) all of us, professionals and the lay public, assume greater personal responsibility. I believe that this is the only fully ethical approach for professionals and that the public should be given a fair chance to respond. Certainly there is no other way to achieve fully informed consent. The situation is subtle, its details technical and in many ways poorly defined, and it touches on areas of our lives in which it is difficult to be objective.

A multidisciplinary study concerned with the role of moral and legal responsibility in modern medicine was published by Siegler and associates in 1987, in which they analyzed the concepts of causation, responsibility, liability, compensation, and public policy as they pertain to harmful effects of medical innovation. The DES story is used for illustrative purposes throughout the volume. It is one of the conclusions of the introductory chapter that physicians should be held responsible for both assessing new procedures and recommending or not recommending their use to their patients. "Physicians should remain 'on the spot,' caught, on the one hand, between the general public's demand for progress and, on the other, a specific patient's demand for professionally sound judgment." But the situation is far from simple: "Of course, to what extent physicians are in a position to inform or educate their patients is not always clear, and to what extent patients are in a position to give informed consent is also not always clear" (1987, p. 16).

Medical practitioners need to pay closer attention to what has been scientifically established in their field. They need to demand and fully support well-designed clinical trials. And above all they need to tell their patients what is known and not known about suggested procedures—the risks as well as the benefits.

The public needs to be encouraged to be more critical in asking for evidence of safety and effectiveness, *before* agreeing to a test or treatment. One of the first questions should be, How does the doctor know? The public needs to understand that many procedures have never been properly evaluated, including some of those most frequently performed—hysterectomies, for example, of which there are some 700,000 a year in this country. The public needs to understand that statistical validation techniques such as the controlled clinical trial often require the analysis of many variables and may not yield simple yes-or-no answers. For example, even with several large-scale randomized studies, the questions concerning coronary artery bypass surgery, which is performed on about 200,000 patients in the United States each year, have not been fully resolved. The public must understand that the etiology of most chronic diseases is still unknown and that for most there is as yet no cure. On the other hand, many ailments are self-limited, so that one recovers even without treatment. The public should ask questions and continue asking them. Prevention and healthy life-styles require more attention.

It is the responsibility of medical researchers, including laboratory scientists, clinical investigators, epidemiologists, and biostatisticians, to collaborate closely and to communicate the results of their work—what has been established as well as the uncertainties remaining—in such a way that medical practitioners, lawyers, ministers, ethicists, educators, policymakers, the media, and the consumers of health care can gain insight into the nature and extent of the evidence. All would benefit from an attitude of what Robert Merton called "organized skepticism," by which he meant "the suspension of judgment until the facts are at hand, and the detached scrutiny of beliefs in terms of empirical and logical criteria" and which he considered to be "both a methodologic and an institutional mandate" (1968, p. 614).

The news media are in a key position for effecting change. Clearly no one is helped by medical headlines that raise false hopes or instill needless fear. We need more of the kind of responsible reporting that alerts, informs, and educates. Organizations such as the Scientists' Institute for Public Information in New York are available to assist the media in this challenge. We need to remember, however, that the primary source of material for the media is the scientific community and that very few members of the press have the proper training to evaluate independently the results of scientific research. In addition to the all-too-natural desire for prestige and fame, there is pressure on many scientists to achieve recognition in order to ensure continued financial support of their work. This may lead some biomedical researchers to oversell their findings and thus stimulate premature and excessive use of new technology. In her book on the relationship between science and the press, Dorothy Nelkin (1987) discussed the public relations activities of the scientific and technical community—their going to great lengths to influence the information about technical issues that reaches the public. Responsibility must be shared by all of us.

Moral Philosophers and Statistical Inference
Ethics of the Randomized Clinical Trial

With the development of the field of bioethics, more and more professional philosophers have been undertaking the analysis of ethical issues in biomedical research, including those pertaining to statistical aspects. Physicians and other nonspecialists reading these publications can only assume that the theoretical terminology employed by the philosophers correctly represents the situation. Unfortunately, that is not always the case. The arguments at times derive from a lack of insight into basic statistical procedures and from confusion about underlying concepts of statistical inference. When such distortions are used to support charges of unethical conduct of biomedical research, especially of clinical trials dealing with life-threatening diseases such as cancer, the results can cause anxiety and further confusion and can thus be harmful.

The main concern in these discussions has been patient autonomy and the conflict between the dual role of the physician as healer and as scientific investigator, a conflict between the best interest of the individual patient and the advancement of medical knowledge. As applied to randomized clinical trials (RCTs), the question is the ethics of entering patients into each treatment arm of the trial after preliminary trends in favor of one of the treatments have been observed, but before the original research design calls for termination. We are not claiming here that the problem does not exist or that it does not merit serious attention but, rather, that it is not seen in full context; its discussion by some moral philosophers is oversimplified and misleading. One such example is given below, taken from another paper dealing more extensively with the subject (Miké 1989a).

Throughout his article "Leaving Therapy to Chance," Marquis (1983) judges the ethics of various situations in terms of a criterion he considers self-evident, a maxim he calls the "Therapeutic Obligation": "A physician should not recommend for a patient therapy such that, given present medical knowledge, the hypothesis that the particular therapy is inferior to some other therapy is more probable than the opposite hypothesis" (p. 42).

To the casual reader this maxim may sound quite plausible as a foundation for medical ethics. But closer analysis reveals some problems. "Knowledge," "probable," and "hypothesis" are terms that have special meaning in a technical context. The latter two are key concepts in statistical theory and methodology, and medical knowledge—always derived from empirical evidence—is by its very nature "statistical." Thus it cannot be properly assessed without insight into the complexities of statistical theory.

Marquis goes on to interpret the therapeutic obligation in the RCT setting, and we will call this the second formulation, or "Therapeutic Obligation, Form II": "If a trial has reached a stage such that the hypothesis that therapy A is inferior to therapy B is more probable than its opposite, then a physician should not select therapy A for a patient" (p. 42).

The reference now is no longer to medical knowledge in the abstract but to specific data that have been collected in the course of a clinical trial. What is at issue here is not some vague personal judgment but the result of analyses of the data, yielding a clear-cut criterion to guide the therapy assignment. Proceeding along these lines, Marquis presents the next formulation of his criterion, which we will refer to as "Therapeutic Obligation, Form III": "There is at least a presumptive case that if the probability of Type I error is less than .50, then it is unethical for any physician to enter a patient on the protocol if she knows the data from the trial" (p. 42).

Marquis is not seriously proposing the adoption of this procedure, as doing so would preclude the possibility of meaningful trials. Instead, he is saying that only such an approach would make the trial ethically feasible. His conclusion then follows, that there is currently no way to make these clinical trials ethically acceptable.

The fallacy in his argument is that the technical meaning of Form III of the therapeutic obligation is something quite different from the much more general original formulation. The first two forms involve the probability of given hypotheses, which—unlike the Type I error—is not part of the frequentist theory of statistical inference. To see this distinction, consider the null hypothesis that the one-year survival rate with an experimental drug B is equal to or less than the one-year survival rate with a control drug A. The corresponding alternative hypothesis states that the survival rate with drug B is greater than the survival rate with drug A. Then according to Forms I and II of the therapeutic obligation, the physician should not select drug A if the alternative hypothesis is more probable than the null hypothesis is. In this case the probability of the null hypothesis would be less than .50, as the two hypotheses represent complementary events and their probabilities add to unity.

But this is not what is meant by the probability of Type I error. The latter—also called the significance level of a test, denoted by P or α—is the probability of rejecting the null hypothesis when it is true. In formal hypothesis testing, the significance level is usually set at .05 or .01. The various P-values often reported in the medical literature with results of clinical studies refer to probabilities of obtaining the observed outcomes (or ones more extreme) under the assumption that the null hypothesis is true. The observed P-value fluctuates during an ongoing study, and in one-sided tests, such as our example illustrating the therapeutic obligation, it is always less than .50. The main point is that throughout his paper Marquis uses variations of Form III of the therapeutic obligation to argue against the ethical feasibility of RCTs.

An article by Gifford (1986), entitled "The Conflict Between Randomized Clinical Trials and the Therapeutic Obligation," is based largely on the Marquis paper. Gifford refers to the Therapeutic Obligation and calls it "TO." But his actual wording is rather vague, far removed from the technical specificity that Marquis implied: "Physicians are typically thought to have an obligation to treat a patient in a way which yields the best chance of recovery" (p. 348).

The maxim of therapeutic obligation, sounding plausible and now having its own acronym, appears on its way to becoming an absolute moral imperative, accepted as such in the ongoing philosophical dialogue—a world apart from the messy realities of clinical research, statistical data analysis, and medical decision making.

Research in the Foundations of Inference

Some of these moral philosophers may not realize that in their arguments they are touching on basic issues pertaining to the foundations of statistical inference, issues of long-standing concern to statisticians.

The classical theory of hypothesis testing, which is the one generally used in clinical studies and reported in the medical literature, is based on a frequentist approach to statistical inference. Probability is defined as long-run frequency, obtained conceptually from a long series of identical experiments, and hypothesis testing is guided by two types of error probabilities. Type I errors have already been discussed. The probability of a Type II error, denoted by β, is the probability of accepting the null hypothesis when it is false. It is often expressed in terms of its complement $(1-\beta)$, called the *power of the test,* which is the probability of rejecting the null hypothesis when it is false. This theory of inference, however, does not provide a precise general concept of statistical evidence in an observed sample (Birnbaum 1962).

There are other theories, such as the Bayesian, which was developed in the clinical trial setting by Cornfield (1976). This approach involves the quantification of prior probabilities and has been controversial among statisticians. One component of the Bayesian formulation, the likelihood ratio, has been proposed as a measure of evidence, but except for the simplest and hence unrealistic cases, it does not fully resolve the problem. In acknowledging the difficulties inherent in the Bayesian approach, such as the choice of prior distribution, Cornfield stated: "This ambiguity of priors is often regarded as a weakness in the Bayesian view. More cogently, however, it should be considered a strength, since it provides an explication of what in fact everyone, no matter what his behavior, seems to admit theoretically, the equivocality of statistical conclusions" (1976, p. 415).

Procedures for stopping an experiment—the ethical issue under discussion in the preceding example—can be provided by a number of competing approaches, although the theoretical justification for each may not be acceptable to at least some statisticians. Included among these procedures are techniques of sequential analysis that were developed especially in view of the ethical requirement of terminating the trial as soon as possible. There is much related methodological research in progress, and foundational questions concerning statistical evidence are of continued interest to both statisticians and philosophers of science.

The important point to emphasize here is the need for moral philosophers to work more closely with other scholars studying these problems, to make joint progress in understanding the appropriate ways of using and interpreting statistical methods and assessing the evidence. (Some related issues as they pertain to philosophies of risk assessment are discussed in Chapter 12 in this volume.)

Complexities of the Clinical Research Process

There is another reason that the discussions of some moral philosophers tend to be troublesome: They oversimplify the situation. A hypothetical case generally pertains to a comparison of two treatments in an RCT in which there is a single outcome, say success or failure, associated with each patient in the trial. In contemporary clinical research a typical RCT may involve a large computerized data base containing dozens or perhaps hundreds of items of information on each patient, some or many of which are considered important baseline variables. Statistical analysis will utilize elaborate multivariate techniques, sometimes a variety of methods applied to the same data. Nearly always, numerous subgroups of patients, as defined by the baseline variables (such as age, sex, and stage of disease), and several possible end-point variables (such as time to relapse and survival time) will be analyzed. All these analyses yield P-values, and a number of them are likely to be significant even if there is no "real" effect present in the data, simply because many tests have been carried out; that is precisely the meaning of P-value. End points other than those on which the original design was based may call for halting the trial, such as was the case with the UGDP study (Chalmers 1974). Here the higher cardiovascular mortality among patients on oral hypoglycemic agents was a totally unforeseen result.

Experienced statisticians are aware of all this, and they recognize the importance of participating in multidisciplinary data-monitoring committees that regularly oversee all relevant aspects of the trial. Formal statistical techniques are powerful tools and—when properly used—can provide important guidelines; any decisions, however, must be made in the full context of scientific and ethical considerations.

An example of some moral philosophers' lack of insight into the realities of clinical research is provided by the Marquis (1983) paper cited earlier. Holding the view that RCTs as currently practiced are unethical, Marquis claims that they directly violate the second formulation of Immanuel Kant's Categorial Imperative: "Act so that you treat humanity, whether in your own person or in that of another, always as an end and never as a means only." According to Marquis, this maxim "forbids using people merely as means. Enrolling someone in an RCT involves exactly that" (1983, p. 47). The crucial word here is *only* or *merely*. The fallacy is to imply that in an RCT patients are treated only as means, for that is never the case in a properly planned and conducted clinical trial. It could even be argued that a patient enrolled in a trial at a major research institution has a better chance of receiving quality care than does someone who relies entirely on personalized treatment by a physician in private practice. One cannot determine whether the evidence obtained from RCTs is acceptable—say in the sense of Kant's maxim—without understanding the trials themselves in the full context of medical research and practice.

A Challenge for Statisticians: The Example of RLF

Biostatisticians need to play a more active role in a number of ways. They need to urge better and more extensive statistical evaluation of medical procedures, to find more effective ways of conveying to their colleagues and the public the great complexity and basically ambiguous nature of statistical methodology, and to cooperate in helping interpret the incomplete and in itself ambiguous scientific evidence that is at the root of so many medical controversies. Quite aside from problems of statistical analysis and formal inference, various biases inherent in the data may often preclude the possibility of a definitive conclusion.

Uncertainties in medical evidence may be due in part to the extreme complexity of the underlying mechanism of the disease's action, or at least to a lack of insight into this mechanism. The story of retrolental fibroplasia (RLF)—or retinopathy of prematurity (ROP)—is an intriguing example of the interplay between the search for causes and statistical assessment in a rapidly developing branch of medicine, that of neonatology.

In his book *Retrolental Fibroplasia: A Modern Parable,* William Silverman (1980) recounts the early history of RLF, a form of blindness. It is the story of an epidemic of blindness in newborn infants that began in the early 1940s. More than fifty separate causes were proposed; about half were examined formally, and a few were tested in prospective clinical trials. After over twelve years of the epidemic, supplemental oxygen was clearly identified as the major factor in two randomized clinical trials. The problem had been traced to the routine administration of oxygen to prematurely born infants in order to increase their chances of survival: The appearance of RLF coincided with the coming into common use of incubators. In the first RCT, published in 1954, no surviving infant assigned to low-oxygen therapy—under 40 percent concentration—developed RLF, and the observed higher mortality rate in this group—as compared with the group on high-oxygen therapy—was not statistically significant. This led to the widespread belief that oxygen therapy under 40 percent was safe. The results of the second trial, a large cooperative study that had used 50 percent oxygen concentration as the cutoff point between the two treatment arms, also came out in favor of low-oxygen treatment. The problem seemed to have been solved, although the long delay in carrying out these studies contributed to the final toll of blindness in over ten thousand children.

The incidence of RLF fell sharply as the "under 40 percent is safe" policy was widely adopted. In subsequent years, however, incidence rates of neonatal mortality and brain damage rose in surviving infants. RLF was diagnosed in some premature infants who had never received supplemental oxygen. The development in the late 1960s of techniques for measuring arterial blood oxygen tension in newborn infants raised the hope that monitoring the oxygen status of treated infants would lead to the establishment of a safe dose. During the 1970s the continued improvement in life-support techniques resulted in the survival of smaller and smaller premature infants. But even though transcutaneous oxygen monitoring also became possible, the incidence of RLF continued to rise until in the 1980s it once again reached epidemic proportions. It is now recognized that oxygen is only one of the factors contributing to RLF (Lucey and Dangman 1984) and that continuous oxygen monitoring has only limited value (Bancalari et al. 1987). The problem has been analyzed in depth in works edited by Silverman and Flynn (1985) and Flynn and Phelps (1988), and Silverman has characterized the current state of knowledge in the field by the term *epoché*, or "suspension of judgment" (1986).

A reexamination of the initial studies has revealed a number of weaknesses in the evidence that led to the acceptance of the curtailed-oxygen hypothesis (Silverman and Flynn 1985). That is, the sample size in the first RCT was so small (eighty-five infants) that it had very low statistical power to detect a significant increase in mortality for the low-oxygen group. The fact that no RLF was observed in twenty-eight survivors, again a small number, would be consistent with a hypothesis of even a 10 percent true incidence of RLF. The cooperative study enrolled only infants who were alive at forty-eight hours, so that it yielded little information about the safety of early oxygen restriction. The subsequent virtual disappearance of RLF may have been due to the fact that the infants at highest risk were dying in the first hours of life. And no follow-up studies were made of survivors in the large cooperative study to monitor for long-term neurological status. Relevant also is the further observation that the few clinical studies of other suspected causes of RLF (vitamin E deficiency, exposure to light, transfusion of adult blood, etc.) were much too small to rule out important contributing effects.

In his book on RLF, Silverman lists a series of neonatal procedures that—introduced without proper testing—had led to disaster or "misled into fruitless byways" (1980, p. 85). There is clearly an urgent need for a more adequate assessment of the evidence concerning such new therapies, at each step along the way, by a multidisciplinary team, and with open acknowledgment of the extent of remaining uncertainties.

The Role of Education

It would be hard to exaggerate the relevance of education to the problems we are discussing. The acceptance of uncertainty in medicine and the need to justify medical claims—whether viewed on TV or heard in a doctor's office—should be imparted to the young at an early age. Just as computer literacy is now considered desirable, aiming at "health care literacy" would yield a generation of better-informed citizens, more effective doctors, and a healthier population. Formal training should include concepts of probability and statistics, with application to clinical trials and other health care situations, as part of the basic mathematics and health studies curriculum.

The responsibilities of educators in developing a questioning attitude toward issues of health and disease are especially clear today, given the mounting threat of acquired immunodeficiency syndrome (AIDS). Of increasing concern now is the spread of AIDS in the general heterosexual population, and the extent of that danger is a topic of sometimes heated controversy.

Questions include the prevalence of the human immunodeficiency virus (HIV) in the general population and its transmission rate. Studies of HIV infection are subject to serious biases, especially selection and recall bias, which are compounded by the sensitive nature of the information. There are overlapping subpopulations with differing degrees of risk, and longitudinal data are lacking. There is at this time simply no way to estimate accurately the HIV incidence in the United States (Curran et al. 1988).

In discussions of the risk of HIV infection in one heterosexual encounter with a seropositive partner, the rate of one in one thousand has been quoted repeatedly in the mass media, with some experts claiming that it is higher. But the sample sizes in the reported epidemiological studies have been very small, and there is a great deal of unexplained biological variation (Peterman et al. 1988). There is no meaningful way to extrapolate from the data to reassure individuals whose behavior puts them at risk. Another related point is often ignored. An assumed one in one thousand chance of infection may constitute acceptable risk to the adventurous; the actual risk, however, is the cumulative probability of infection in all encounters. Even if the transmission rate in a single encounter is as low as .001, with one hundred encounters involving an infected partner the risk rises to 10 percent and with one thousand, to 63 percent. If the transmission rate is somewhat higher, say .002, the corresponding risks become 18 percent and 86 percent.

Those who derive their main information from the evening news or from thirty-second TV spots during a game show or sitcom, have no way of even suspecting the complexity of the situation and the extent of medical uncertainty. There is no substitute for understanding, and herein lies the responsibility and perhaps unprecedented challenge for educators (Miké 1988b).

Responsibility in Broad Perspective

To round out our discussion of medical uncertainty and responsibility, it is necessary to step back to gain a broader perspective. The practice of medicine touches on questions that are far beyond the scope of science. To be better understood, current attitudes toward evidence and the use of medical technology—on the part of both professionals and the public—need to be seen in the context of our national life and the problems we are facing as a society. Relevant here are the observations of social scientists.

Medicine and Society

Medical sociologist Renée Fox has characterized American society as deeply religious while also highly secularized and pluralistic (1987). Although there is a serious interest in ultimate values and beliefs, there is no common language or cultural framework in which to grapple with them. In the intellectual realm, current world views of medical science tend to be dominated by the "antimetaphysical metaphysics" of reductionist thought (Fox 1989). But medicine, which deals with the human condition—the fundamental questions of life, illness, suffering, and death—presents a natural focus for society's deeply felt concerns in the quest for meaning. The American pursuit of organ transplantation can be seen as "ritualized optimism," the symbolic expression of our relentless drive for success and progress. In a totally secular ethos, life must be continued by the undaunted repair and replacement of failing bodily parts.

Fox believes that the rise of the field of bioethics, another American phenomenon, also was stimulated by our need to consider ultimate questions. Medicine, by providing a shared symbolic language, enables us to "collectively ponder" these issues. But bioethics itself has developed in a secular mode, in the tradition of analytic philosophy, and has largely missed the religious and social dimensions of health, illness, and medicine. The emphasis has been on the individual, rather than the community, and on rights, rather than responsibilities. The tendency has been to reduce problems to a procedural, legalistic level, leaving the fundamental questions unexamined. That is what we are observing today in public debates in our courts and our legislatures.

The public watches the unending parade of legal battles presented by the media, fought to resolve the conflict of the perceived rights of individuals. In the same spirit, the ultimate resolution of conflict in the doctor–patient relationship is also sought through litigation, so that medical malpractice insurance has become a critical issue. As mentioned earlier, the extension of the concept of liability beyond the second generation was being tested in the DES lawsuit against the drug company in Minnesota.[1] But—given the rapid advances being made in biotechnology—this is no doubt only the beginning of the problem.

Habits of the Heart

An in-depth study of American society was published in 1985 by sociologist Robert Bellah and four colleagues. The title of their book, *Habits of the Heart,* is an expression used by the nineteenth-century French social philosopher Alexis de Tocqueville in referring to the mores—meaning the ideas, opinions, and practices—essential to the American character. It was the purpose of this book to "deepen our understanding of the resources our tradition provides—and fails to provide—for enabling us to think about the kinds of moral problems we are currently facing as Americans" (1985, p. 21).

Synthesizing the results of interviews with over two hundred Americans of various backgrounds, the authors bear witness to a pervading individualism that, for many, has led to an inner emptiness and the belief that in the end each of us is really all alone. Alongside the relentless drive for self-assertion, achievement, and the acquisition of material goods, there is confusion concerning basic concepts such as the meaning of life, success, freedom, and justice. The utilitarian and expressive individualist philosophies prevalent today have given rise to a fragmented intellectual and popular culture, a "culture of separation." But if our society still possesses some coherence,

> it is because there are still operating among us, with whatever difficulties, traditions that tell us about the nature of the world, about the nature of society, and about who we are as people. Primarily biblical and republican, these traditions are... important for many Americans and significant to some degree for almost all. (1985, pp. 281–82)

The authors conclude that we must seek to recover the insights of these traditions, that we must reverse "modernity's tendency to obliterate all previous culture. We need to learn again from the cultural riches of the human species and to reappropriate and revitalize those riches so that they can speak to our condition today" (p. 283).

Central to both of these traditions are the concepts of community and commitment to the common good. The biblical tradition, important to American culture since colonial times, is continued today primarily by Jewish and Christian religious communities. The republican tradition, originating in the city states of Ancient Greece and Rome, was basic to the development of modern Western democracies, including the American republic. A citizen of a republic was presupposed to be motivated by "civic virtue," to seek the attainment of justice and the public good.

Recovery of the values of our past, the authors point out, would entail returning to the original concept of "corporation." It was widely accepted into the late nineteenth century that incorporation was a privilege granted to a private group in exchange for service to the public, rather than a right available to any private enterprise. This idea would radically change the notion of public accountability of a corporation, the notion of its "social responsibility," which is now "often a kind of public relations whipped cream decorating the corporate pudding" (p. 290). Recovering the older meaning of "profession" would lead to "calling" or "vocation," with the implication of public responsibility.

Habits of the Heart presents a vision that

> does not seek to return to the harmony of a "traditional" society, though it is open to learning from the wisdom of such societies. It does not reject the modern criticism of all traditions, but it insists in turn on the criticism of criticism, that human life is lived in the balance of faith and doubt.... Such a vision seeks to combine social concern with ultimate concern in a way that slights the claims of neither. Above all,

such a vision seeks the confirmation or correction of discussion and experiment with our friends, our fellow citizens. (p. 296)

Dealing with Medical Uncertainty

Although Bellah and his associates do not discuss medicine, the problems we have been considering fit very well into their conceptual framework. The overuse and misuse of medical technology can more easily be controlled if corporations and professionals accept as given the highest standards of social responsibility. A properly informed and motivated public will provide the necessary support for adequate funding of medical technology assessment programs. And the savings achieved by using only scientifically validated procedures could provide medical coverage for the over 35 million Americans now without health insurance (Miké 1987). This in itself should be a strong incentive for a community-oriented nation.

A greater consciousness of community would shift attention to the importance of human support, compassion, and caring. The availability of other cultural resources would keep people from turning to medicine to solve human problems that by their very nature are beyond its power. Understanding medical uncertainty is a condition for making progress, but expecting the public to deal with this uncertainty without recourse to alternatives may be unrealistic and naive. How does one exercise responsibility in this broad perspective? This, I think, is the real challenge facing everyone who sees the whole picture.

SUMMARY

This chapter has focused on aspects of uncertainty and evidence in modern medicine, subjects that are rarely prominent in public discussions of the American health care system. In recent decades, medicine has made unprecedented advances toward the control and elimination of disease, and a vigorous biomedical research program offers every reason to expect continued progress. But beyond the truly remarkable achievements, the fact of medical uncertainty is not part of the general consciousness.

The rapid proliferation of medical technology has far surpassed the boundaries of medical knowledge, and many of the procedures in use have not been properly evaluated; they may be ineffective, and they may be harmful. But physicians frequently ignore even definitive results. Furthermore, there is a great deal of geographic variation in medical practice, with a strong tendency toward overtreatment.

We explored some of the factors affecting health care practice, by means of examples from the recent history of medicine. Included were economic incentives, a poor understanding of the scientific evidence, and an undue faith in the effectiveness of medical procedures on the part of both physician and patient. The tendency of doctors to follow accepted practice in their own community, and their reluctance to discuss uncertainties with their patients are other characteristics of the problem.

We then examined the notion of responsibility in the face of medical uncertainty, developing the thesis that society must come to terms with the fact of medical uncertainty and that this entails greater personal responsibility for both professionals and the lay public. In this context we described in more detail the special responsibilities of medical practitioners and the public and of moral philosophers and statisticians. Also crucial are the responsibilities of educators and the media.

Acknowledging the extent of our common ignorance is a good habit for everyone to nurture. There is a frontispiece in William Silverman's book on retrolental fibroplasia (1980) showing Maimonides, the twelfth-century physician and Judaic scholar, with an aphorism from the Talmud: "Teach thy tongue to say I do not know and thou shalt progress." I believe we can do no better than to heed these ancient words of wisdom.

ACKNOWLEDGMENT

This research was supported by funds from the National Endowment for the Humanities and the National Science Foundation.

NOTE

1. The Minnesota Supreme Court handed down its decision on the case (C8–88–2161) on September 14, 1989. Chief Justice P. S. Popovich stated that the justices were evenly divided in opinion, with one justice taking no part: This meant that the decision of the lower court became law and the case was dismissed. It is interesting that J. Coyne, the only woman on the seven-member court, had recused herself from the deliberations. The legal situation nationwide is varied. Some states recognize the principle of preconception tort, so that such cases can get a jury trial, whereas others—like Minnesota—do not (F. A. Bremseth, personal communication). The first suit involving a DES granddaughter who died of vaginal cancer was filed in Maryland on March 12, 1990, against Eli Lilly & Company (*New York Times,* March 13, 1990).

REFERENCES

Apfel, R. J., and Fisher, S. M. (1984). *To Do No Harm: DES and the Dilemmas of Modern Medicine.* New Haven, Conn.: Yale University Press.
Bancalari, E., Flynn, J., Goldberg, R. N., Bawol, R., Cassady, J., Schiffman, J., Feuer, W., Roberts, J., Gillings, D., and Sim, E. (1987). "Influence of Transcutaneous Oxygen Monitoring on the Incidence of Retinopathy of Prematurity." *Pediatrics* 79:663–69.
Bellah, R. N., Madsen, R., Sullivan, W., Swidler, A., and Tipton, S. M. (1985). *Habits of the Heart: Individualism and Commitment in American Life.* Berkeley: University of California Press.
Berwick, D. M., Fineberg, H. V., and Weinstein, M. C. (1981). "When Doctors Meet Numbers." *American Journal of Medicine* 71:991–98.
Birnbaum, A. (1962). "On the Foundations of Statistical Inference." *Journal of the American Statistical Association* 57:269–326 (with discussion).

Bunker, J. P. (1985). "When Doctors Disagree." *New York Review of Books* 32:7–12.

Bunker, J. P., and Brown, B. W. (1974). "The Physician–Patient as an Informed Consumer of Surgical Services." *New England Journal of Medicine* 290:1051–55.

Chalmers, T. C. (1974). "The Impact of Controlled Trials on the Practice of Medicine." *Mount Sinai Journal of Medicine* 41:753–59.

Chalmers, T. C. (1982). "Informed Consent, Clinical Research and the Practice of Medicine." *Transactions of the American Clinical and Climatological Association* 94:204–12.

Cornfield, J. (1976). "Recent Methodological Contributions to Clinical Trials." *American Journal of Epidemiology* 104:408–21.

Curran, J. W., Jaffe, H. W., Hardy, A. M., Morgan, W. M., Selik, R. M., and Dondero, T. J. (1988). "Epidemiology of HIV Infection and AIDS in the United States." *Science* 239:610–16.

Dieckmann, W. J., Davis, M. E., Rynkiewicz, L. M., and Pottinger, R. E. (1953). "Does the Administration of Diethylstilbestrol During Pregnancy Have Therapeutic Value?" *American Journal of Obstetrics and Gynecology* 66:1062–81 (with discussion).

Fletcher, R. H., and Fletcher, S. W. (1979). "Clinical Research in General Medical Journals: A 30-Year Perspective." *New England Journal of Medicine* 301:180–83.

Flynn, J. T., and Phelps, D. L. (eds.) (1988). *Retinopathy of Prematurity: Problem and Challenge.* March of Dimes Birth Defects Foundation, Birth Defects: Original Article Series, vol. 24, no. 1. New York: Liss.

Fox, R. C. (1987). "Medical Uncertainty: An Interview with Renée C. Fox." *Second Opinion* 6:90–105.

Fox, R. C. (1989). *The Sociology of Medicine: A Participant Observer's View.* Englewood Cliffs, N.J.: Prentice-Hall.

Freiman, J. A., Thomas, A. B., Smith, H., and Kuebler, R. R. (1978). "The Importance of Beta, the Type II Error and Sample Size in the Design and Interpretation of the Randomized Control Trial." *New England Journal of Medicine* 299:690–94.

Gifford, F. (1986). "The Conflict Between Randomized Clinical Trials and the Therapeutic Obligation." *Journal of Medicine and Philosophy* 11:347–66.

Greenwald, P., Barlow, J. J., Nasca, P. C., and Burnett, W. S. (1971). "Vaginal Cancer After Maternal Treatment with Synthetic Estrogens." *New England Journal of Medicine* 285:390–92.

Heinonen, O. P. (1973). "Diethylstilbestrol in Pregnancy: Frequency of Exposure and Usage Patterns." *Cancer* 31:573–77.

Herbst, A. L., Ulfelder, H., and Poskanzer, D. C. (1971), "Adenocarcinoma of the Vagina: Association of Maternal Stilbestrol Therapy with Tumor Appearance in Young Women." *New England Journal of Medicine* 284:878–81.

Iglehart, J. K. (1984). "From the Editor." *Health Affairs* 3(2):4–5.

Institute of Medicine (1985). *Assessing Medical Technologies.* Washington, D.C.: National Academy Press.

Jones, J. H. (1981). *Bad Blood: The Tuskegee Syphilis Experiment.* New York: Free Press.

Lucey, J. F., and Dangman, B. (1984). "A Reexamination of the Role of Oxygen in Retrolental Fibroplasia." *Pediatrics* 73:82–96.

MacNeil/Lehrer NewsHour (1988). "DES Update." PBS Television, February 29.

Marquis, D. (1983). "Leaving Therapy to Chance." *Hastings Center Report* 13(4):40–47.

McFarlane, M. J., Feinstein, A. R., and Horwitz, R. I. (1986). "Diethylstilbestrol and Clear Cell Vaginal Carcinoma: Reappraisal of the Epidemiologic Evidence." *American Journal of Medicine* 81:855–63.

McGrady, G. A. (1982). "The Controlled Clinical Trial and Decision Making in Family Practice." *Journal of Family Practice* 14:739–44.

McKinlay, J. B. (1981). "From 'Promising Report' to 'Standard Procedure': Seven Stages in the Career of a Medical Innovation." *Milbank Memorial Fund Quarterly* 59:374–411.

Merton, R. K. (1968). *Social Theory and Social Structure*. 2nd ed. New York: Free Press.

Miké, V. (1982). "Clinical Studies in Cancer: A Historical Perspective." In *Statistics in Medical Research: Methods and Issues, with Applications in Cancer Research*, ed. V. Miké and K. E. Stanley, pp. 111–55. New York: Wiley.

Miké, V. (1987). "Saving Money and Improving Health by Evaluating Medical Technology." *Technology Review* 90 (April):22–25.

Miké, V. (1988a). "American Medicine Today: Values in Conflict." *Bulletin of Science, Technology & Society* 8:374–77.

Miké, V. (1988b). "Modern Medicine: Perspectives on Harnessing Its Vast Emerging Power." *Technology in Society* 10:327–38.

Miké, V. (1989a). "Philosophers Assess Randomized Clinical Trials: The Need for Dialogue." *Controlled Clinical Trials* 10:244–53.

Miké, V. (1989b). "Toward an Ethics of Evidence—And Beyond: Observations on Technology and Illness." *Research in Philosophy and Technology* 9:101–13.

Nelkin, D. (1987). *Selling Science: How the Press Covers Science and Technology*. New York: Freeman.

Peterman, T. A., Stoneburner, R. L., Allen, J. R., Jaffe, H. W., and Curran, J. W. (1988). "Risk of Human Immunodeficiency Virus Transmission from Heterosexual Adults with Transfusion-associated Infections." *Journal of the American Medical Association* 259:55–58.

Siegler, M., Toulmin, S., Zimring, F. E., and Schaffner, K. F. (eds.) (1987). *Medical Innovations and Bad Outcomes: Legal, Social, and Ethical Responses*. Ann Arbor, Mich.: Health Administration Press.

Silverman, W. A. (1980). *Retrolental Fibroplasia: A Modern Parable*. New York: Grune & Stratton.

Silverman, W. A. (1986). "Epoché in Retinopathy of Prematurity." *Archives of Disease in Childhood* 61:522–25.

Silverman, W. A., and Flynn, J. T. (eds.) (1985). *Retinopathy of Prematurity*. Boston: Blackwell Scientific Publications.

Wennberg, J. E. (1984). "Dealing with Medical Practice Variations: A Proposal for Action." *Health Affairs* 3(2):6–32.

7

Evidential, Ethical, and Policy Disputes: Admissible Evidence in Radioactive Waste Management

E. WILLIAM COLGLAZIER

A sustained and definitive radioactive waste management policy has been a elusive goal for our nation since the beginning of the nuclear age. An atmosphere of contentiousness and mistrust among the interested parties, fed by a long history of policy reversals, delays, false starts, legal and jurisdictional wrangles, and scientific overconfidence and played out against the background of public concern with nuclear power and weapons issues generally, has dogged society's attempts to come to grips with the radioactive waste–management issue. The policy conflicts have become so intense and intractable that Congress has been forced to deal with the issue periodically. The year 1982 was one watershed year for congressional action on high-level nuclear waste, and 1987 proved to be another.

This chapter will examine ethical and value issues in radioactive waste management (RWM), with a special emphasis on disputes about scientific evidence. Controversies over evidence have been particularly important because of the many scientific uncertainties and problems inherent in trying to ensure that nuclear waste in a geological repository will harm neither people nor the environment for the thousands of years that the waste will remain hazardous. This requirement of guaranteeing adequate safety over millennia is an unprecedented undertaking for our regulatory and scientific institutions. The first section of the chapter will provide a brief historical overview of the national policy disputes in radioactive waste management, and the second section will discuss some of the key value issues that have been at the heart of the controversies.[1]

Our approach is to delineate key policy issues and to separate the value components of each into three categories: procedural, distributional, and evidential. Key stakeholders—Congress, federal agencies, the nuclear industry, utilities, environmental groups, state governments, Native American tribes, local communities—take particular policy positions justified in part on the basis of procedural, distributional, and evidential values. Procedural values refer to who should make what decision for whom and by what process. Distributional values concern what is a fair allocation of costs, benefits, and risks to the affected parties and to society as a whole. Evidential values refer to what counts as evidence, for example, what type and degree of scientific evidence is sufficient and admissible in making a particular societal decision, especially in the face of large scientific unknowns and significant social and scientific debate.[2] Categories of "value concerns" thus include fairness and appropriateness of process, outcomes, and evidence.

HISTORICAL OVERVIEW OF THE POLICY DISPUTES

A brief summary of RWM policy debates in the 1980s—focusing primarily on the United States' attempts to manage and dispose of intensely radioactive spent fuel from civilian reactors and reprocessed high-level waste from nuclear weapons production—provides a vivid picture of the contentiousness of this issue (Colglazier and Langum 1988). (Both types of waste will be referred to as high-level waste, or HLW.) The safe management of these wastes is imperative regardless of the future of nuclear power.

The Nuclear Waste Policy Act of 1982

Many congressional committees, federal agencies, special-interest groups, and a presidential commission were involved in the several years of struggle and debate that led to the comprehensive Nuclear Waste Policy Act (NWPA) of 1982 (Public Law 97–425). The act attempted to create a stable institutional framework for making decisions and reducing technical uncertainties during the two or more decades required for a permanent disposal solution. The NWPA addressed many issues, some of which we shall discuss next.

Permanent Disposal Versus Long-Term Interim Storage

The overarching purpose of the NWPA was "to establish a schedule for the siting, construction, and operation of repositories" and "to establish the federal responsibility and a definite federal policy for disposal." The act set a deadline of January 31, 1998, for the initial operation of the first geological repository and a very tight schedule for the intermediate steps in achieving this goal.

Some observers in 1982 believed that the schedule would be impossible to achieve. Congress adopted it in order to force the federal government to get on with the job. Nuclear power proponents felt that demonstrating a permanent solution to the public was essential to removing a dark cloud obscuring the future of nuclear power. Many environmentalists agreed that the United States should not leave this problem to future generations but that if a permanent solution could not be found expeditiously, the use of nuclear power should be halted. The NWPA ratified into law the development of a permanent geological repository as the primary purpose of the federal nuclear waste–management program.

The act also stated that before the president could select one site for the submission of a construction license application to the Nuclear Regulatory Commission (NRC), the Department of Energy (DOE) would have to characterize at repository depth three sites in at least two different rock types. This requirement was to ensure a sufficient diversity of sites to make a reasonable comparative assessment. One consequence of the tight schedule for the first repository was that the DOE concentrated on the sites and rock types then under investigation: tuff on the Nevada test site, basalt on the Hanford reservation, and salt sites in Utah, Texas, Louisiana, and Mississippi.

The NWPA also required the DOE to examine sites for a second repository that could be ready by the year 2003. It did not authorize the second repository—leaving that decision to a later Congress—but the capacity of the first repository was restricted until a second repository opened. This provision for a second repository was adopted for several reasons: reducing perceived regional inequities, as many of the original proposed repository sites were in the western United States, whereas reactors were primarily in the East; minimizing transportation distances; and investigating additional rock types such as granite.

Some influential members of Congress opposed the emphasis on the repository program. They favored, instead, developing long-term interim storage in an engineered facility above the ground, called monitored retrievable storage (MRS), for several reasons: The technology was readily available; the facility could be safely sited almost anywhere; the country would not have to decide now about the ability of a mined repository to isolate waste for thousands of years; the delay would permit development of new technologies for permanent disposal; and the future reprocessing of spent fuel could be easily accommodated. Some of the proponents of an MRS also came from states that were prime candidates for a permanent repository. The permanent repository program, however, received the highest priority in the act; the MRS was not authorized. The NPWA did require the DOE to present to Congress a study of the need for and feasibility of the MRS as well as two specific MRS designs and three possible MRS sites. In general, the MRS was viewed as a backstop in case the repository schedule slipped too much.

Sharing Power Among the Federal Government, States, and Tribes

The NWPA stated that the president's recommendation of a site to the NRC, thereby initiating the licensing process, would be effective unless the host state or Native American tribe disapproved. Congress would have ninety days to override state or tribal disapproval by a joint resolution; otherwise the veto would stand. This mechanism, whose exact provisions were in dispute up to the last hours before the act was passed, was intended both to assure states and tribes that their concerns would be seriously considered and to minimize arbitrary or capricious action by any level of government. The act also prescribed a detailed partnership approach among federal, state, and tribal governments, requiring timely consultation via binding written agreements, grants to states and tribes that are notified as having potentially acceptable sites in order for them to conduct their own technical reviews and to hire consultants, and financial and technical assistance to states and tribes to mitigate the impacts of repository development. Extensive opportunities for public participation were also required.

Governor Richard Bryan of Nevada stated the philosophy behind this approach:

> The framers of the NWPA realized that, in order for any state ever to be able to accept a repository, a situation must be created whereby the leaders and citizens in that state are able to see and believe that the site selected was the product of an impeccable, scientifically objective screening process. No amount of compensation or federal "incentives" can ever substitute for safety and technical suitability in the site selection effort. (Bryan 1987, p. 33)

Away-from-Reactor (AFR) Storage Facilities

The act determined that nuclear power owners and operators have the principal responsibility for the interim storage of their spent fuel. Utilities, whose current interim storage facilities were nearing capacity in some cases, wanted federal AFRs because they viewed the government as being responsible for the delays in developing a permanent repository. Some states were adamantly opposed to AFRs without a realistic plan for permanent disposal because, as one governor observed, "There is a basic law of political physics, often overlooked . . . waste stays where it is first put" (Riley 1982, p. x). Utilities dropped their demands for an AFR during negotiations over the legislation in exchange for a mandated schedule for permanent repository development. However, the act required the federal government only to begin accepting the utilities' spent fuel when the repository was ready, although the utilities would have preferred a specified date for federal acceptance, as the repository might be delayed.

Financing and Management

The act required that the costs of the federal repository development program for commercial HLW be paid by the generators and borne by the beneficiaries through a one mill per kilowatt-hour fee on nuclear-generated electricity. Industry had originally favored funding by general governmental appropriations. The act created the Office of Civilian Radioactive Waste Management, headed by a political appointee, in order to give the program more visibility and bureaucratic focus.

The NWPA's complex procedures strove to set out in advance, as far as practicable, a clear and public set of ground rules for the decision-making process. As Governor Richard Riley of South Carolina stated in 1982,

> Of highest importance, today, is not only what is to be done, but also how we decide it is to be done. A process of decision making must be established that will allow us to have confidence in the results of that process. There will be remaining uncertainties no matter what the decisions are. Only confidence in the process which leads to those decisions will enable us, as a society, to live with those remaining uncertainties. (Riley 1982, p. xi)

Conflicts and Issues from 1982 to 1987

The respite from conflict did not last long after passage of the NWPA. Lawsuits and acrimony over the regulatory framework and the siting process quickly ensued.

Regulations

In 1985, the U.S. Environmental Protection Agency (EPA) finally issued its generally applicable standards for the management and disposal of spent nuclear fuel, high-level waste, and transuranic waste (EPA 1985). The standards limit projected releases of radioactivity to the accessible environment for ten thousand years after disposal. The EPA projected that compliance with the containment requirements would cause no more than one thousand premature cancer deaths over the entire ten thousand–year period. The EPA also estimated that this level of residual risk would be comparable to the risks that future generations would have faced from the uranium ore used to create the wastes if the ore had never been mined. The containment requirements apply to accidental disruptions of a disposal system as well as to any expected releases. The release requirements are stated probabilistically as follows: Cumulative releases of radionuclides should have a likelihood of less than one chance in ten of exceeding certain limits specified in the regulations and a likelihood of less than one chance in one thousand of exceeding ten times these same limits. Performance assessments using analytical models and various release scenarios will be required to demonstrate compliance with these release limits. The containment requirements call for a "reasonable expectation" that their various quantitative tests will be met.

The standards also include individual protection requirements that limit annual exposures from the disposal facility to members of the public, assuming undisturbed performance of the repository system. Also included are groundwater protection requirements that limit radioactivity concentrations in water withdrawn from "a special source of groundwater" near a disposal system for one thousand years after disposal, again assuming undisturbed performance. The EPA's assurance requirements mandate the use of both manufactured and natural barriers to isolate the wastes, avoiding sites where there is a reasonable expectation of exploration for scarce or easily accessible resources, and the use of sites where the removal of most of the wastes is not precluded for a reasonable period of time after disposal.

On July 17, 1987, the U.S. Court of Appeals for the First Circuit Court in Boston ruled, in a suit filed by environmental groups and affected states, that portions of the standards addressing groundwater migration and the protection of individuals from radiation exposure were in conflict with the Safe Drinking Water Act (SDWA). Consequently, the court remanded the relevant part of the standards to the EPA to be rewritten. Two years later, it was still not clear what changes the EPA would propose in order to comply with the court order or what the impact on DOE would be.

The NRC's regulations for geological disposal were promulgated in 1983 before publication of the EPA's generally applicable standards (NRC 1983).

With minor modifications, the NRC's regulations were determined to comply with the 1985 EPA standards, but they may have to be revised if the EPA standards are significantly altered. Among the NRC's technical requirements are the following: complete containment of HLW within the waste packages for a period between three hundred and one thousand years; a release rate from the engineered barrier system following the containment period that does not exceed 1 in 100,000 parts per year of the radionuclide inventory present after one thousand years; and a groundwater travel time along the fastest path of travel from the disturbed zone to the accessible environment that is at least one thousand years. The 1983 NRC regulations also require development of site characterization plans before sinking exploratory shafts and the characterization at repository depth of three sites in at least two rock types before the selection of one site. In addition, the regulations require a repository design that enables any or all of the emplaced waste to be retrieved on a reasonable schedule for a period up to fifty years after the initial waste emplacement.

First Repository

The NWPA established a process for selecting sites for the first HLW geological repository. The first steps were identifying potentially acceptable sites and developing general guidelines. The guidelines that were adopted require that site selection decisions be based on multiple objectives, dealing with health and safety, the environment, socioeconomics, and cost (DOE 1984b). The guidelines are divided into two categories: (1) preclosure, which refers to the period of construction, operation, and sealing, and (2) postclosure, which refers to the period of up to 100,000 years after sealing. Critics still complained that these requirements were too general and too easily met, that a site is acceptable under the guidelines unless there are data that would clearly disqualify it. Several lawsuits were filed by states and environmental groups.

The NWPA required the DOE to use the guidelines to nominate five sites as suitable for further investigation, to prepare environmental assessments, and then to select at least three sites for characterization as candidate sites for the first repository. During the characterization, exploratory shafts would be constructed for underground testing at repository depth to determine the suitability of the site for meeting the EPA's and the NRC's regulations. When the site characterization was completed and the environmental impact statement was prepared, the president would submit one site to the NRC licensing process.

The DOE selected five sites—at Hanford, Washington; Yucca Mountain, Nevada; Davis Canyon, Utah; Richton Dome, Mississippi; and Deaf Smith, Texas—from which the final three would be chosen for characterization. In December 1984, the DOE issued draft environmental assessments for each, including a comparative evaluation of the five and the tentative selection of Hanford, Yucca Mountain, and Deaf Smith (DOE 1984a). Because the methodology of selection came under intense scrutiny and criticism, the DOE requested the Board on Radioactive Waste Management of the National Academy of Sciences to review the selection methodology. In April 1985, the board reported that "the methodology of comparative assessment is unsatisfactory, inadequate, undocumented, and biased and should be reconsidered" (BRWM 1985, p. 1).

The DOE revised its selection methodology and requested that the board conduct an independent review of the revised methodology during its implementation by the DOE. The methodology that the DOE used was an application of multiattribute utility analysis (MUA). In its final letter report in April 1986, the board commended the DOE for the high quality of its application of the methodology. The board went on to state: "While recognizing that there is no single, generally accepted procedure for integrating technical, economic, environmental, socioeconomic, and health and safety issues for ranking sites, the Board believes that the multiattribute utility method used by DOE is a satisfactory and appropriate decision-aiding tool" (p. 2). The board was "disappointed that DOE did not follow the recommendation (made twice in writing) that independent experts be brought into the assessment process itself as well as into the review of the process" (p. 2). The board was "concerned that DOE's use of its own technical experts to assess performance by this subjective method may mask the degree of real uncertainty associated with post-closure issues" (p. 2).

On May 28, 1986, President Ronald Reagan announced the selection of the three sites favored in the original DOE report. He also stated that the DOE's investigation of sites for a second repository in the East would be indefinitely deferred. The secretary of energy provided the supporting analysis for the decision as well as a background document describing the application of the decision-aiding methodology (DOE 1986a, 1986b).

The application of multiattribute utility analysis yielded the following ranking of the first-round sites, in order of preference: Yucca Mountain, Richton Dome, Deaf Smith, Davis Canyon, and Hanford. Following the DOE's recommendation, the president chose the first-, third-, and fifth-ranked sites. In its final report the DOE tried to explain the reasons that its decision was preferable to selecting the three sites ranked highest. In the DOE's view, the methodology was capable of "providing only a partial and approximate accounting of the many factors important to the site recommendation decision." The department indicated that it gave considerable weight to having maximum rock diversity in the set selected. The DOE also emphasized that all the sites scored exceedingly well on postclosure factors. On the preclosure factors, the DOE stated that preliminary repository and transportation costs controlled the rankings. By aggregating the preclosure performance objectives, excluding the cost—the least important factor according to the guidelines—and stressing rock diversity, the DOE justified its choice.

The affected states, several decision analysts who had been consultants for the DOE and the board, and a congressional committee strongly criticized the DOE's choice of sites. One respected analyst, who was part of the board's review, stated: "The conclusions drawn in the Recommendation Report are based on selective and misleading use of the analysis described in the Methodology Report ... [where] I find a convincing analysis that clearly rejects the Hanford site" (von Winterfeldt 1986, p. 1). A congressional committee staff report claimed that the DOE officials systematically deleted passages of the methodology report critical of the Hanford site and distorted the methodology in the recommendation report to justify its decision (Staff to Reps. Weaver and Markey 1986).

The three states selected for characterization had, for many years, been very critical of the DOE's HLW program. All believed that their sites should have been disqualified by a proper application of the guidelines; all complained about the DOE's approach to consultation; and all had brought lawsuits against the department. Washington State attempted to negotiate a consultation and co-operation agreement with the DOE but eventually withdrew from the negotiations.

Governor Booth Gardner of Washington, who had taken the most neutral position of any of the affected governors, stated that he would support a repository in his state if it could be shown that the site was technically acceptable and the best of the sites under consideration. He strongly challenged the selection of Hanford in May 1986, because the site ranked lowest among the five sites on both the postclosure and preclosure factors of the MUA. He stated that the citizens of his state would "overwhelmingly demand that such a decision be fought in every possible way until it is overturned" (Gardner 1986).

Both Texas and Nevada had taken even more adversarial stances toward the DOE during the site selection (Bryan 1987; Frishman 1986). Like Washington, they raised numerous technical problems with their sites and berated the DOE's optimism in the interpretation of the data. Nevada had particular reason to be concerned, as the DOE ranked the Yucca Mountain site the highest of the five sites. The special advantages of the Nevada site included the repository's location above the water table. The Deaf Smith site had the potential problem of having to drill through a major aquifer, the Ogallala, to reach the repository horizon. Texas was especially concerned about the potential for contaminating the aquifer and about the difficulty in ensuring retrievability in salt for fifty years.

The NRC made detailed comments on the environmental assessments for the five sites, stating that the assessments were deficient in three areas: "(1) not identifying the range of uncertainties associated with the existing limited data base, (2) not identifying the range of alternative interpretations and assumptions that can be reasonably supported by existing data, and (3) not incorporating a reasonable range of uncertainties and alternative interpretations into evaluations and conclusions" (Noe 1986, p. 3). These comments indicated that the DOE would have a more difficult time justifying its technical views during the licensing process, when there would likely be major disputes over the validity of models and the magnitude of uncertainties.

Second Repository

In January 1986, the DOE selected twelve areas in seven states—Maine, New Hampshire, Virginia, North Carolina, Georgia, Wisconsin, and Minnesota—for consideration for the second repository, and areas in ten other states—New York, New Jersey, Connecticut, Pennsylvania, Massachusetts, Rhode Island, Vermont, Maryland, South Carolina, and Michigan—were eliminated. The DOE then conducted public hearings in early 1986 in the seven states, during which the department received intense public opposition and criticism.

Legislators from the seven affected states enthusiastically greeted the president's decision in May 1986 to suspend indefinitely all site work for the second-round repository. The DOE based its decision on a reassessment of need and stated that because the volume of spent fuel was growing more slowly than was anticipated several years earlier, the first repository would be adequate for the foreseeable future. The secretary stated that the DOE would reassess the second repository in the mid-1990s but that spending money now for site-specific activities would be premature and a waste of money. The department planned to continue technical studies related to crystalline rock but would cancel all site-specific work.

Some members of Congress, particularly from the West, criticized the DOE's decision, stating that it was politically motivated. A bipartisan group of thirteen key members warned the secretary of energy that his decision "could destroy the delicate balance and might ultimately lead to an erosion of" the NWPA. Congress eventually reduced funding for all of the DOE's HLW program and imposed a moratorium on site-characterization activities.

In 1987, the DOE determined that its deferral of the second-round program was illegal without action by Congress, and so the department prepared to restart the program. The DOE's decisions regarding the second-round repository—first deferral and then resumption—increased support for those who said that Congress would have to intervene and to amend the NWPA.

MRS and Transportation

Monitored retrievable storage (MRS) in engineered facilities at or near the surface was not authorized by the NWPA, but the DOE was required to study its need and feasibility. The act gave the potential host state certain rights of consultation with the DOE and the right to issue a notice of disapproval for an MRS facility that only Congress could override.

In April 1985, the DOE produced a draft version of its study (DOE 1985). The agency proposed a facility that had significant differences from the MRS in the NWPA. Beginning in 1996, spent fuel from eastern reactors would be shipped by truck and rail to the MRS for consolidation and repackaging, using special equipment within hot cells. It would then be placed in temporary storage in dry concrete silos at the site. When the repository in the West was ready, the consolidated fuel would be shipped from the MRS in very large metal casks on unit trains. This proposed MRS was no longer a backup or alternative to the repository; it was an integral part of the overall waste-management system. The DOE proposed three sites, all in Tennessee, with the favored site near Oak Ridge.

The DOE argued that the MRS was not absolutely necessary but that it provided significant advantages. In order to alleviate concerns that the MRS might become a "de facto" permanent storage facility, the DOE proposed that the MRS not accept spent fuel until the repository received a construction license from the NRC. The DOE also proposed a cap on interim storage capacity at the MRS.

The DOE appeared to attach great symbolic importance to building in the near term a major facility like the MRS, in order to show that the United States was solving the nuclear waste problem. The DOE also appeared to feel a moral commitment to begin taking utility-spent fuel by 1998, which would still depend on the permanent repository's receiving a construction license. Some supporters of the MRS wanted the federal government to accept spent fuel even if the repository were delayed. They were willing to live with the constraints on the MRS proposed by the DOE in order to get the facility authorized and felt that a future Congress would likely remove these restrictions if the repository ran into problems.

The DOE provided $1.4 million to Tennessee for the state to conduct its own independent evaluation of the MRS. Part of these funds went to the affected local communities for their evaluations. The city of Oak Ridge and Roane County set up the Clinch River Task Force, which produced a report stating that the MRS could be safely built and operated and that the community would accept the facility provided that certain conditions were met (Clinch River Task Force 1985). The DOE attempted to accommodate some of these conditions in its revised proposal to Congress.

The Tennessee governor's Safe Growth Team sponsored studies of the economic impacts of the MRS and the need for, feasibility of, and siting of the MRS. The "need and feasibility" study compared waste-management systems with and without the MRS and concluded that the system with the MRS would cost about $2 billion more than a no-MRS system would, and would not necessarily reduce transportation impacts. The risks of either system were estimated in the Tennessee study to be low and comparable. The study also concluded that the preference of one system over another was a policy judgment about which reasonable people could disagree (Colglazier 1986; Energy, Environment, and Resources Center 1985).

Concerning siting, the Tennessee study concluded that a major reason for the DOE's preference for the Oak Ridge site was that the community would be favorably inclined. A separate study on economic impacts concluded that negative public perceptions of the MRS could affect industrial location decisions and tourism in the region (Center for Business and Economic Research 1985). The state also conducted a number of public meetings to solicit citizens' views, which were largely negative except for those of the Oak Ridge community.

In January 1986, Governor Lamar Alexander of Tennessee announced his decision to oppose the MRS. He concluded that nuclear waste was Tennessee's problem, too, that the MRS could be operated safely, but that it was unnecessary and therefore a waste of money. Moreover, he believed that Oak Ridge was exactly the wrong place for it because he thought that it would turn away many more jobs that it could ever attract. The MRS would, he felt, create a negative image for the Oak Ridge–Knoxville corridor, which otherwise had a bright future for attracting high-technology companies (Alexander 1986).

The potential "stigma" effect associated with the MRS appeared to carry significant weight in the governor's decision. He believed that negative perceptions of the MRS could cause real economic damage that could not be easily offset. The nuclear industry and the DOE believed that there would be no negative economic effect on a region hosting the MRS, and they disputed the inferences from the polling data on which the governor's conclusions were based.

The governor did comment that a site in a poor county, such as Morgan County which is not far from Oak Ridge, might have been more acceptable because there were no jobs to turn away.

The DOE was finally able to deliver its MRS proposal to Congress in March 1987 after it won a lawsuit brought by the state (DOE 1987). The new governor of Tennessee attempted to deliver a notice of disapproval to Congress, but the legislative parliamentarians ruled that his timing was wrong. The MRS proposal was then caught up in the general debate over whether or not the NWPA should be amended.

Socioeconomic Impacts

The NWPA directed that the DOE give grants to the affected states to assess the potential economic, social, public health and safety, and environmental impacts of a repository. These assessments would be part of the information used by the DOE in providing financial assistance to mitigate these effects on the state selected for repository development. The DOE's own assessments estimated the socioeconomic impacts as if the proposed facilities were like any other industrial activity. The states, however, have been concerned with the special damage that might be caused because of the public's strongly negative perceptions concerning nuclear waste.

Whether or not there will be significant economic fallout from the stigma effect associated with nuclear waste has become a major issue for the states. These perceived effects were probably the most important reason for the Tennessee governor's negative decision on the MRS. Like Tennessee, the state of Nevada has been worried about the potential impact on tourism, conventions, industrial location decisions, and migration (Mountain West 1989). Texas was concerned about the stigma on agriculture in the Deaf Smith region.

Social scientists have convincingly demonstrated that the public does have negative perceptions of nuclear waste (Slovic 1987), and several surveys indicate that many people claim that their future behavior will be affected by nuclear waste siting. Whether or not people would actually behave as they state on the surveys is open to question. But perceptions can have real economic impacts. Negative economic effects might be substantial, and excessive public rhetoric about a stigma effect could help create a self-fulfilling prophecy.

Waste Fund

The groups most concerned with the waste fund have been the nuclear utilities and organizations representing rate-payer interests. These groups have been appalled by the staggering projected life-cycle cost of the DOE's civilian HLW program, estimated to be over $30 billion and still rising in 1987, which suggested that the one mill per kilowatt-hour fee on nuclear-generated electricity might have to be increased. (The department's revised mission plan in 1987 changed the scheduled opening of the first-round repository from 1998 to 2003. In 1990, DOE slipped the opening date to 2010.) They also were concerned about the size of the financial contribution of the DOE's defense programs to the repository effort, which was under negotiation within the DOE following the president's determination that a defense HLW would be buried in the civilian repository.

It appeared to these groups that defense programs were not offering to pay their fair share and that it would be almost a conflict of interest for the DOE to negotiate with itself.

1987 Amendments

The cancellation of the second repository program was seen as a move to help the Republicans in the 1986 election, which outraged several powerful western senators and members of Congress who had been defenders of the DOE Office of Civilian Radioactive Waste Management. Even though the DOE had reasonable arguments for deferring the second repository, the action was interpreted as an illegal voiding of the compromises struck in the NWPA. The DOE action made these western members of Congress look impotent to their constituents, and it created the impression that the federal government could flout the law and continue using the West as a "dumping ground" for whatever the politically powerful East did not want. The NWPA, with its two repositories, had at least created the appearance of regional equity and scientific credibility in siting.

The inclusion of Hanford in the three sites selected for characterization for the first repository was also interpreted by some as a political decision, as that site was rated last among the five candidates under consideration. Although there were rational arguments to support the DOE's inclusion of Hanford, the perception was otherwise. As a consequence of these two decisions, many members of Congress came to believe that the DOE was making political rather than technical decisions. So even though the cancellation of the second repository reduced the intense outcry from the East, the DOE program lost much of its credibility and some of its main supporters.

The fragile consensus represented by the NWPA unraveled completely when both Representative Morris Udall (Arizona) and Senator Bennett Johnston (Louisiana), who had been key figures in passage of the act, called for its amendment. The nuclear industry had reluctantly come to the same conclusion. But it was far from obvious what course of action could reestablish a political consensus. The environmental community, the first-round states, and Tennessee wanted a moratorium on the DOE program while a high-level presidential commission studied the issue. This course appealed to many politicians because it would have postponed decisions and DOE site-specific activities until after the 1988 election. If no legislative proposal amending the NWPA could be agreed upon, then it was likely that the DOE program would continue to spend money but be prevented from engaging in site-specific activities while awaiting a reassessment by a new administration.

Johnston's Proposal

In July 1987 Senator Bennett Johnston unveiled his proposal for amending the NWPA. His approach contained several novel and politically appealing elements. First, the DOE would select only one site for characterization for the first repository. Work would be deferred at the other two sites unless problems were uncovered during the characterization at the preferred site. Second, his bill would fully authorize an MRS (leaving the siting to the DOE) and remove the constraints on the MRS that the DOE had promised to Tennessee. In Johnston's proposal, the MRS would not be restricted in storage capacity, and its acceptance of spent fuel would not be tied to progress on the repository. And last, the second repository program would be canceled.

Sequential characterization for the first repository and cancellation of the second repository would result in considerable savings to the program, as the projected cost of characterizing a site had risen from less than $100 million in the 1982 estimates to $2 billion or more in the 1987 estimates. Johnston's bill would use these savings to fund a significant compensation program for both the repository host state and the MRS host state. The repository host state would receive $50 million annually following selection and $100 million annually upon operation. The MRS state would receive $25 million annually following selection and $50 million annually upon operation. Although critics called this a bribe, the amounts were large enough to attract attention and indicated how eager some in Congress were for a solution to the siting problem. Johnston cleverly attached his amendment to two legislative vehicles—the budget reconciliation and the omnibus appropriation—that Congress would necessarily have to address before recessing in December. His amendment prevailed in the Senate with only minor modification.

Johnston's bill had enormous political appeal to all the states except Nevada and Tennessee, as well as to fiscal conservatives because of the significant budgetary savings. It appealed to many utilities, as the MRS would be able to begin accepting spent fuel by 1998 no matter what happened to the repository. Representatives of Washington and Texas were still concerned, as their states were not completely "off the hook." But Johnston's bill was opposed by many environmentalists, as it abandoned what appeared to be a more objective process that compared sites and it seemed to offer a bribe to overcome health and safety concerns. It was certainly a more risky approach, as it placed "all the eggs in one basket" by selecting a single repository site before a detailed characterization to determine technical suitability.

Congressional Debate

Although no bill amending the NWPA passed the House, Johnston's approach was a required item for discussion at the conference on the budget reconciliation between the House and Senate in December. The conferees knew that if agreement could not be reached, Johnston's amendment would have to be voted on, up or down, in the House, and it would likely pass because of its appeal to the second-round states, Washington, and Texas.

The House conferees rejected the MRS and surprised the Senate by offering to select Nevada explicitly as the site to be characterized for the repository. Johnston was unwilling to give up the MRS and refused to compromise. The House conferees, however, were able to persuade the Speaker of the House to allow them to offer their amendment as a substitute for Johnston's in case there was a vote before the House, which greatly increased their negotiating leverage. The then Speaker Jim Wright and the Majority Leader Thomas Foley, who hailed from Texas and Washington, respectively, also let it be known that they wanted a bill. Many Representatives felt that they were compromising their principles in voting for the amendment, and a Washington State member stated that the plan was an outrage but that he had to protect his constituents.

Ultimately, both sides compromised in order to get a bill that would provide political relief to so many legislators. The legislation's *raison d'être* was the selection of Nevada. The conferees agreed to authorize the MRS, but only with its schedule tied even more tightly to the repository. The House conferees added a number of other far-reaching provisions to the final bill. The president was directed to appoint a special negotiator charged with seeking a state or a Native American tribe willing to host a permanent repository or MRS. Congress would have to approve and enact legislation to implement any agreement reached by the negotiator and a state or tribe. Congress also reduced the compensation package in the original Johnston bill. The repository host state or tribe now is eligible to receive $10 million per year upon selection and $20 million per year upon operation; the MRS host state is eligible to receive $5 million per year upon selection and $10 million per year upon operation. These amounts are one-fifth of Johnston's proposal. Moreover, in order to receive the funds, the host state or tribe must forgo its right of disapproval and other federal impact mitigation assistance, but not technical assistance. The DOE was also required to report within a year on its estimate of the impacts of siting the permanent repository in Nevada, including which impacts should be paid by the state.

Program Redirection

The Nuclear Waste Amendments of 1987 have had a profound effect on the DOE's civilian radioactive waste–management program. The most striking provision—the congressional directive that only the Yucca Mountain site in Nevada be characterized—entails some technical risk. If Yucca Mountain turns out to be unacceptable or unlicensable after characterization, the DOE is required to return to Congress for direction. Site-specific activities at Hanford and Deaf Smith, as well as at the second-round sites, have been completely terminated. The process of making a comparative evaluation of sites in diverse rock types and settings was voided, with the financial savings estimated to be many billions of dollars.

Congress authorized the MRS but imposed a number of constraints. First, a three-member commission was created to report to Congress in November 1989 on the need for the MRS. Siting the MRS was left to the DOE's judgment, but the DOE was forbidden from recommending a site until after it had recommended to the president a repository site for development, which follows the characterization process. The construction of the MRS could not begin until after the repository receives an NRC construction license, which is a considerable delay from the DOE's original schedule. The capacity of the MRS was capped, and no MRS could be located in Nevada. The host state could disapprove its selection, but that disapproval was subject to override by Congress. (In November 1989, the MRS Review Commission recommended two MRSs but placed severe restrictions on both. One MRS would be for the emergency removal of spent fuel if there were another accident like Three Mile Island. The other MRS would be limited to five thousand metric tons and would be paid for by the utilities that use it. Neither MRS would be linked to the repository. The commission offered little insight into how these facilities might be sited, given the expected political opposition.)

In essence, Congress reaffirmed in 1987 its position that the repository program is the highest federal priority. The MRS was authorized, but the delay and tight linkage to the repository schedule implied that the MRS would not be a substitute for the repository. If these provisions remain unchanged, progress on the MRS will be hostage to progress on the repository, which requires resolution of the uncertainties surrounding the ability of Yucca Mountain to meet the NRC's licensing requirements. Utilities will be unable to count on government promises to begin accepting spent fuel within this century.

VALUE COMPONENTS OF POLICY ISSUES

Stakeholders' values—expressed as special interests and general principles—arise in discussions about five key policy issues concerning HLW disposal: (1) the need for a repository, (2) intergovernmental sharing of power, (3) siting, (4) safety, and (5) impacts.

Need for a Repository

The core policy dispute over high-level radioactive waste concerns the primary goal for the federal program over the next twenty years: development of permanent disposal in a geological repository versus long-term monitored storage in an engineered facility at or near the surface. In other words, is a repository really needed now? The principal value concerns in the debate have been mainly distributional, pertaining to the perceived costs and benefits of either course of action to current and future generations and to various stakeholder groups. Stakeholders have used arguments based on evidential concerns as well. The procedural aspects of this issue have been less prominent because all the stakeholders have assumed that Congress would ultimately decide.

Congress attempted to resolve the issue in 1982 by directing that the development of a repository be the highest priority of the federal government. With its proposal in 1985 for an integral MRS, the Department of Energy attempted to combine features of both approaches. Although Congress ratified again the preeminence of the repository program in 1987 and placed significant constraints on the MRS by linking its schedule to that of the repository, it did authorize an MRS for the first time. If the repository effort in Nevada should falter, which is quite possible, long-term monitored retrievable storage could emerge in the 1990s as the focus of a revised U.S. HLW management effort. The debate over the final goal of the federal program is far from over.

The distributional value issues originate in concerns about the future of nuclear power and the welfare of those utilities that have relied on nuclear power and about protection of people and the environment now and in the future. Pronuclear stakeholder groups believe that society's interests are best served by promoting and using an energy source that they perceive to be cost effective, environmentally benign, and secure; and solving the nuclear waste problem appears to them to be a key element in convincing the public to adopt their point of view. Many of these groups feel that the federal government, because of its past promotion of nuclear power, has a special responsibility to take spent fuel, in a timely manner, from the utilities.

In advocating a course of action that they believe to be fair to society and to nuclear utilities, the pronuclear stakeholder groups have generally believed that developing permanent disposal is necessary to convince the public that the nuclear waste problem can be solved. A tight schedule is favored to force the federal government to demonstrate a solution and to remove the liability of spent-fuel management from the utilities.

Those pronuclear members of Congress who in 1982 favored the unconstrained MRS, however, felt that interim storage by the federal government was the only option that could be relied on to remove spent fuel from the utilities, because political problems might stymie the discovery and licensing of a permanent repository. Regardless of which policy option is favored, the pronuclear stakeholders seek a solution to the nuclear waste problem in order to be fair to nuclear utilities, as promotion by the federal government involved the utilities with nuclear power in the first place, and to be fair to society by facilitating an energy source that for them has many positive attributes.

Many of the environmental stakeholder groups see nuclear power and nuclear waste as a special threat to people and the environment. Although some of these groups have attempted to use the nuclear waste issue as a means to build public opposition to nuclear power, many have focused on finding a solution to the nuclear waste problem over the next twenty years, in order to fulfill this generation's responsibility to protect future generations. The majority of this group has united with the pronuclear group to favor development of permanent disposal as the highest priority. Although a substantial segment of the pronuclear group advocates increasing the role of long-term interim storage in national planning, the environmental groups are adamantly opposed to any type of MRS or AFR (away-from-reactor) storage facility.

In the 1970s the environmentalists, in conjunction with the EPA, helped stop the effort by the Atomic Energy Commission to develop long-term storage, then called retrievable surface storage facilities (RSSF). They believe that this generation is responsible for solving the nuclear waste problem and should not leave it as a legacy to future generations. They view perpetual care as unfair to future generations and possibly unsafe because of uncertainties in maintaining long-term institutional control (an evidential argument). They also feel that if a permanent solution cannot be found soon, nuclear power should not be used. The RSSF, AFR, and MRS are seen by environmentalists as counterproductive, as they might encourage society once again to put off finding a permanent solution. Interim storage options also are regarded by environmentalists as an attempt by the nuclear utilities to be bailed out of a problem caused by poor planning, and a course of action that would inevitably lead to more reliance on nuclear power without having a permanent remedy for the waste problem.

The responsibility of this generation to future generations has been used on both sides of the policy debate. Some proponents of long-term engineered storage contend that this generation should not make irreversible decisions that may be costly to correct in the future. Relying on evidential judgments, they argue that new technological developments may occur within the next century that could change our view of how to handle nuclear waste, that we currently know how to store waste safely in monitored engineered facilities where if anything goes wrong, it can be corrected and repaired, and that there is no reason to rush prematurely into developing a means of permanent disposal that contains many scientific uncertainties and could lead to costly and potentially dangerous remedial action in the future. They feel that we should be more humble about our ability to project technological developments and human behavior over the next ten thousand years. All stakeholder groups appear to feel strongly about the importance of this generation's fulfilling its responsibility to future generations, but they do not agree on how to turn this belief into policy.

Intergovernmental Sharing of Power

Procedural values were preeminent in the NWPA's approach, not only because of having to allocate undesirable "goods" (i.e., the host for a HLW repository), but also because of evidential uncertainties. As Governor Riley stated, "A process of decision-making must be established that will allow us to have confidence in the results" (1982, p. xi). In 1982 Congress established procedural rules for sharing power among the affected governmental entities. Nevertheless, there have been conflicts over making trade-offs between the procedural and schedule requirements of the NWPA, that is, between fair process and efficient implementation. The states have felt that the federal agencies, particularly the DOE, have not lived up to the spirit of "consultation and cooperation." In their view, the DOE has generally chosen in its discretionary actions to try to meet milestones rather than to slow down the program in order to accommodate more consultation with the states. The federal agencies, on the other hand, have felt that they had a mandate to move forward expeditiously, pointing to the rigid timetable. To the DOE, the states have sought to delay in order to throw obstacles in the way of the federal program. The DOE perceives that it has sincerely tried to accommodate the states' concerns, but being fair to society as a whole has required moving forward in a timely manner. This distributional concern of being fair to the country overall has been given greater weight by the DOE than have procedural equity concerns emphasized by the affected states.

The degree to which the affected states and Native American tribes have felt that they have not been treated equitably on procedural issues is indicated by the inability of the states and the DOE to reach acceptable "consultation and cooperation" agreements and by the numerous lawsuits brought against the agency. Moreover, Nevada has interpreted the 1987 amendments as manifestly unfair to the state on procedural (as well as evidential and distributional) grounds. Nevada still retains its right of disapproval as set forth in the NWPA, but it is universally assumed that based on what happened in 1987, Congress would override Nevada's veto. On the local level, those communities that have favored the repository or the MRS have complained of mistreatment by their parent states in terms of process; that is, they have felt that their views have been ignored. In the 1987 amendments, Congress increased the role of local government, particularly in decision making regarding the use of impact aid.

Siting

The NWPA repository site selection process might have worked with some different decisions by the DOE. One can speculate that preserving the second-round program might have enabled the department to move forward with the characterization of the three first-round sites. A decision to select the three top-ranked sites from the multiattribute utility analysis, thereby eliminating Hanford, might have increased the DOE's credibility and the perception that the department was being objective rather than political in its decisions. But these speculations ignore the large political furor and discomfort for elected officials caused by investigating potential repository sites in a number of states, as well as the large financial cost to the country of multiple-site characterization and two repositories.

In making politically difficult siting decisions, political leaders have always had two basic options: (1) to make the choice internally and impose it (e.g., to overpower a weak constituency by sheer political force) or (2) to set up a more open selection process that might be perceived as objective, scientifically credible, and procedurally fair. In the first case, the top political leadership had better be sure that it has sufficient power to impose its will on any opposition. In the second case, the host area, even though opposed, might be willing to accept its fate. Evidential and procedural guarantees are required in this approach in order to gain consent for an unpleasant distributional decision. This philosophy is similar to that expressed in a 1981 Utah newspaper editorial:

> Neither Utah nor any other state can properly refuse to bear the nuclear waste burden once it [the repository site] has been established to the best of human conditions. However, the honor of making such sacrifice for time without end must confer on the luckless lamb the satisfaction of knowing first-hand that the duty couldn't have been just as well assigned elsewhere. (*Salt Lake City Tribune* 1981)

At the time of the NWPA's passage, an objective and comparative site selection process with the involvement of all parties appeared necessary for both technical and political reasons. A site selected for political reasons seemed doomed to failure in an era of new federalism and participatory democracy; the de facto power of a state seemed sufficient to obstruct the federal effort even if it lacked the legal power to do so. Thus, evidential and procedural concerns with regard to site selection were given great weight in the NWPA. Multiple-site characterization was required for each of two repositories, and the DOE used evidential methodologies like multiattribute utility analysis to select sites for characterization.

The DOE's selection of sites for the first repository and its elimination of the second repository were cited by critics as proof that it was being political rather than objective in its decisions. This perception helped lead to the stalemate eventually broken by the 1987 amendments. The department lacked sufficient credibility with its critics, and credibility is essential to implementing the siting approach set forth in the NWPA.

The three types of value concerns—procedural, distributional, and evidential—have been frequently linked in proposals for reaching consensus. For example, Governor Gardner of Washington commented that he would support the repository at Hanford if it could be shown that the site was technically acceptable and the best of the sites under consideration. (Perhaps he was confident that Hanford would not meet this standard.) In his proposal, the evidential and distributional concerns were clearly linked. Moreover, in 1986 he simultaneously offered a procedural approach—a national mediation effort—to end the stalemate. Nevertheless, in 1987, the Washington state congressional delegation overwhelmingly favored the selection of Nevada by legislative designation. As one congressman from Washington rationalized, he had to compromise his principles because he had to protect his interests.

With the 1987 amendments, Congress de-emphasized some of the evidential and procedural requirements for siting a repository set forth in the NWPA. The legislators' distributional concerns, particularly the impact on their careers and states, outweighed the arguments for conducting a broad siting process that investigated sites in a number of states. There was too much political discomfort and friction from the siting approach in the NWPA.

Safety

Stakeholders' concerns are for a fair evidential process and an appropriate standard of proof to determine that a repository will be acceptably safe for thousands of years. Congress emphasized the importance of independent scientific review in the NWPA and the 1987 amendments. Licensing by an independent regulatory agency, independent review by the host state, oversight by a new technical review panel, and oversight by the Board on Radioactive Waste Management of the National Academy of Sciences all are part of the formal and informal mechanisms for reviewing the technical aspects of the DOE's repository effort. Questions of How sure is sure enough? What can we really prove? and What weight of evidence is sufficient? will dominate the debate because the science of repository "safety" has many uncertainties.

There are significant technical risks for the DOE program arising out of the congressional action in 1987. The Nevada site may prove to be unacceptable and/or unlicensable. A 1987 unreviewed study by a physical scientist who works for the DOE in Las Vegas concluded that if a certain conceptual model of the hydrology at Yucca Mountain is substantiated, serious consideration should be given to abandoning the site and declaring it unsuitable for the permanent disposal of HLW (Szymanski 1987). The critical question for him is whether the geological record contains evidence of large-scale fluctuations of the water table, which could result in more rapid transport of radionuclides to the accessible environment. This study, delivered to the DOE two days after the amendments passed Congress, was widely publicized by the governor of Nevada in early 1988. The state's nuclear waste office stated that the study "provides a credible theory, with supporting data, that is consistent with the state's observations and is in direct conflict with DOE's simplified model of the existing geologic and hydrologic setting of the Yucca Mountain area" (Nuclear Waste Project Office 1988, p. 1). The point is not that this study will stand up but that proving with "reasonable expectation" that the EPA's and the NRC's containment requirements and other licensing conditions will be met is an enormous and unprecedented evidential challenge (BRWM, 1990).

Even if more sites were simultaneously characterized, significant problems could occur in the licensing process. The United States has adopted a set of licensing criteria that require considerable certainty about geohydrological conditions at a site and release predictions based on modeling with a hypothetical set of long-term disruptive scenarios. The time frame for resolving these uncertainties with reasonable scientific assurance may be much longer than that allowed in the licensing process, and it may be impossible. As is often the case with science, in the short run, the more that is known may actually increase rather than decrease the uncertainties. It is very hard to speculate about a site until the conditions are explored at repository depth, and there will always be credible scientists who will interpret the remaining uncertainties in ways different from that of the DOE. The standard of proof required by the NRC will be the key, and the NRC is probably as unsure as the DOE is of what that will be in practice. An alternative approach would be to overengineer the manufactured barriers in the repository, as Sweden may do. The way to live with the remaining uncertainties may be to rely on very conservative engineering.

Impacts

The debate over the impacts of the repository program include issues such as who should pay for the costs of the program, how the impacts should be calculated fairly, and what fair compensation is for negative impacts. The NWPA determined that the cost of the repository program should be paid by the beneficiaries of nuclear-generated electricity through a one mill per kilowatt-hour fee. Everyone agreed on this general approach for fairly allocating the costs. Concerning defense wastes that will go into the civilian repository, however, there was considerable debate. Rate-payer groups and the utilities believed that the DOE's defense program was not offering to pay its fair share. The debate was over the methodology used to calculate "fair share," an evidential dispute.

Another evidential dispute over distributional impacts concerns the potential stigma on the host state. The affected states are concerned that public apprehensions about nuclear waste may result in real and significant negative impacts, including lost business and jobs for the local region. They are concerned that talking about a potential stigma effect may be a self-fulling prophecy. The social science methodologies for assessing such a stigma have received attention in the Nevada socioeconomic impact assessment studies (Mountain West 1989). The nuclear industry and the DOE ridicule the claim of a stigma effect and dispute that believable evidence can ever be found to substantiate it. They assert that the Nevada test site, for example, appears to have had no such impact on Las Vegas.

A final important controversy about impacts is the use of incentives and compensation. Congress created special payments for the host state, provided that the state forgoes some of its rights to object, and it created a new procedural mechanism through the office of the special negotiator. What the special negotiator is able to accomplish, either in pacifying Nevada or in attracting other states to volunteer for an MRS or a repository, is problematical. Nevada appears unalterably opposed to the repository at Yucca Mountain, and no state appears willing to host an MRS. A negotiated approach for handling this contentious distributional issue would have to overcome the perception that a bribe is being offered in exchange for taking unacceptable risks; any negotiated solution would probably require additional evidential guarantees of safety.

FUTURE PROSPECTS

The nuclear waste issue may heat up again politically on the national level in the 1990s, especially if it appears that the characterization and licensing process for Yucca Mountain is not reaching closure. In that case, there will be pressure from the utilities and the nuclear industry to have an MRS that can accept spent fuel regardless of the outcome of a repository. Many environmentalists will claim that we can no longer be confident that the nuclear waste disposal problem can be solved. Congress may be forced to deal with this thorny issue once again and hold another round of debate over long-term storage versus permanent disposal.

ACKNOWLEDGMENT

The material in this chapter is based on a research project supported by the National Science Foundation under grant no. RII–8419118. The project study team (with listed disciplines) includes David L. Dungan (religious studies) and Mary R. English (planning and sociology) of the University of Tennessee, Sheldon J. Reaven (philosophy of science) of the State University of New York at Stony Brook, John J. Stucker (political science) of Carter Goble Associates, and myself. This project was also supported by the Waste Management Research and Education Institute at the University of Tennessee. Any opinions, findings, and conclusions or recommendations expressed in this paper are mine and/or those of the study team and do not necessarily reflect the views of the National Science Foundation. The complete report of the project will be published in a book by the study team.

NOTES

1. For additional details and references on radioactive waste–policy issues, see Colglazier and Langum (1988), from which some of the material in this chapter has been taken.
2. Sheldon Reaven is the originator of the term "evidential value issues."

REFERENCES

Alexander, L., Governor of Tennessee (1986). News Release. January 21.

Board on Radioactive Waste Management (BRWM), National Research Council (1985). Letter to Ben Rusche, April 26. Reprinted in DOE 1986a.

Board on Radioactive Waste Management (BRWM), National Research Council (1986). Letter to Ben Rusche, April 10. Reprinted in DOE 1986a.

Board on Radioactive Waste Management (BRWM), National Research Council (1990). *Rethinking High Level Radioactive Waste Disposal*. Washington, D.C.: National Academy Press.

Bryan, R. H. (1987). "The Politics and Promises of Nuclear Waste Disposal: the View from Nevada." *Environment* 29:14–38.

Center for Business and Economic Research (1985). *An Economic Analysis of Monitored Retrievable Storage Site for Tennessee*. Knoxville: University of Tennessee Press.

Clinch River MRS Task Force (1985). *Recommendations on the Proposed Monitored Retrievable Storage Facility*. Roane County–Oak Ridge, Tenn.: Clinch River MRS Task Force.

Colglazier, E. W. (1986). "The MRS: An Assessment of Its Need and Feasibility." *Forum for Applied Research and Public Policy* 1:25–37.

Colglazier, E. W., and Langum, R. B. (1988). "Policy Conflicts in the Process for Siting Nuclear Waste Repositories." *Annual Review of Energy* 13:317–57.

Department of Energy (DOE) (1984a). *Draft Environmental Assessment, Yucca Mountain Site, Nevada Research and Development Area, Nevada*. DOE/RW–0012. Washington, D.C.: National Technical Information Service.

Department of Energy (DOE) (1984b). "General Guidelines for the Recommendation of Sites for the Nuclear Waste Repositories." *Code of Federal Regulations*. Title 10, Part 960. Washington, D.C.: Government Printing Office.

Department of Energy (DOE) (1985). *The Need for and Feasibility of Monitored Retrievable Storage—A Preliminary Analysis*. DOE/RW–0022. Washington, D.C.: Department of Energy.

Department of Energy (DOE) (1986a). *A Multiattribute Utility Analysis of Sites Nominated for Characterization for the First Radioactive-Waste Repository—A Decision-aiding Methodology*. DOE/RW–0074. Washington, D.C.: Department of Energy.

Department of Energy (DOE) (1986b). *Recommendation by the Secretary of Energy of Candidate Sites for Site Characterization for the First Radioactive-Waste Repository*. DOE/S–0048. Washington, D.C.: National Technical Information Service.

Department of Energy (DOE) (1987). *Monitored Retrievable Storage Submission to Congress*. DOE/RW–0035–1. Washington, D.C.: National Technical Information Service.

Energy, Environment, and Resources Center (1985). *Evaluation of the Need, Feasibility, and Siting of the MRS in Tennessee*. Knoxville: University of Tennessee Press.

Environmental Protection Agency (EPA) (1985). "Environmental Standards for the Disposal of Spent Nuclear Fuel, High-Level, and Transuranic Radioactive Wastes." *Code of Federal Regulations*. Title 40, Part 191. Washington, D.C.: Government Printing Office.

Frishman, S. (1986). "High-Level Waste Repository Siting: A State Perspective." *Tennessee Law Review* 53:527–40.

Gardner, B., Governor of Washington (1986). Testimony Before the Subcommittee on Energy Research and Production of the House Committee on Science and Technology. July 22.

Mountain West (1989). "An Interim Report: Nevada Socioeconomic Studies." Report for the Nevada Nuclear Waste Project Office. Carson City: State of Nevada.

Noe, C. (1987). *Oversight* 5:3.

Nuclear Regulatory Commission (NRC) (1983). "Disposal of High Level Radioactive Waste in Geologic Repositories." *Code of Federal Regulations.* Title 10, Part 60. Washington, D.C.: Government Printing Office.

Nuclear Waste Project Office (1988). *Yucca Mountain May Be Unsuitable for Isolation of High-Level Nuclear Wastes.* Carson City: State of Nevada.

Riley, R. (1982). "Nuclear Waste and Governance." In *The Politics of Nuclear Waste,* ed. E. W. Colglazier, pp. vii–xi. Elmsford, N.Y.: Pergamon Press.

Salt Lake City Tribune (1981). Editorial. April 24.

Slovic, P. (1987). "Forecasting the Adverse Effects of a Nuclear Waste Repository." In *Waste Management '87 Volume 2,* ed. R. G. Post, pp. 91–94. Tucson: University of Arizona Press.

Staff to Representatives James Weaver and Edward Markey (1986). *Preliminary Results of Staff Investigation into DOE's Selection of Three Sites for Characterization as the Nation's First Repository for High-Level Radioactive Waste.* October 20.

Szymanski, J. S. (1987). "Conceptual Considerations of the Death Valley Groundwater System with Special Emphasis on the Adequacy of This System to Accommodate the High-Level Nuclear Waste Repository." Manuscript.

von Winterfeldt, D. (1986). Letter to Ben Rusche, July 22.

8

Expert Claims and Social Decisions: Science, Politics, and Responsibility

RACHELLE D. HOLLANDER

Concern for relationships among ethics, values, policy, and science and engineering is prominent in modern society. The existence of a program called Ethics and Values Studies in an agency of the U.S. government, the National Science Foundation, provides some evidence of this (Hollander 1987a, 1987b; Hollander and Steneck 1990). The bills introduced in the U.S. Congress to support bio(medical) ethics centers through the National Institutes of Health also provide evidence (U.S. Senate 1988). New initiatives support research and related activities in areas of biomedical ethics in the National Center for Nursing Research and the Office of Human Genome Research in the National Institutes of Health. In July 1988, the Board of Radioactive Waste Management of the National Research Council devoted one day of a four-day retreat to considering the ethical and value aspects of that issue (BRWM 1988).

In this chapter I shall attempt to show why such issues occupy particular attention now. My thesis is that a new acknowledgment of our collective moral responsibility is needed because of the political and social context in which science now operates. This context requires more sophisticated scientific and ethical analysis, as well as scientists, engineers, policymakers, interested scholars, and others working together to determine not just acceptable risk but also acceptable evidence.

AN ENEMY OF THE PEOPLE

To provide perspective on these matters, we should note that interactions of science, technology, and society have raised these kinds of problems for a long time. A play by Henrik Ibsen, *An Enemy of the People*, written in 1882, raises all these concerns.

An Enemy of the People is a story about the possibility of contamination in the water supply that feeds a town's new mineral baths. The baths attract the summer visitors that have rejuvenated the community. A Dr. Thomas Stockmann has investigated and discovered the problem; he has documented it, and he is delighted to have made the discovery. He, after all, had warned the town fathers about the problem when they designed the water supply, and they did not listen. Now he presents the truth as he sees it—and he sees it in the worst possible light—to his brother Peter, the mayor, who had organized the efforts to construct the baths. The mayor is much less concerned and convinced than is the doctor about the threats to people's health—typhoid, gastric troubles— and is much more concerned about the costs of reworking the water supply. Mayor Peter Stockmann does not believe the threat can be as bad as all that, and he believes that the doctor's report should be kept confidential. When the mayor tells representatives of the press and small trade and business interests that the gentlemen who own the baths do not intend to pay the bill for the expensive remodeling that will be necessary and that the tax assessment will have to rise to accommodate it, the town turns against the doctor. The newspaper, which had been about to publish the report, suppresses it. In the end, the doctor is dismissed from his position; his sons are harassed at school; and he and his family are nearly run out of town. But he decides to stay—out of spite and conviction—and carry on his campaign for truth, morality, and justice. He will start by educating his sons and street urchins and guttersnipes, at home. The play ends with the doctor's statement that "the strongest man in the world is he who stands most alone" (Ibsen 1882).

Notice that the doctor believes that there is a problem based on results from scientific tests that he had done by sending away water samples for examination. He believes that it is his moral responsibility to do this and, given the results, that it is his moral responsibility to alert everybody. He believes—that is, one way to interpret his responses is that he believes—that having the "facts," his moral responsibility is simple and clear-cut. It follows from the facts, and everybody else's responsibility is similarly clear-cut.

But what are the facts? What are the risks? What are the nature and extent of the contamination of the water supply and their implications for the health of the visitors to the baths and the community residents? This list of questions can be expanded considerably. Given the state of scientific understanding at the time, the answers were even less clear than they might be now.

Then there are other kinds of facts, or empirical concerns. What social and economic risks do the townspeople face? What alternatives are there to shutting down the baths and completely overhauling the water supply? There are normative concerns. What should the visitors to the baths be told? Who should pay for monitoring the water supply and making changes in it? Finally, how should the empirical and normative concerns be treated in a risk assessment? In risk management? Do certain problems or kinds of problems need to be considered independently of the risk-assessment and risk-management approaches? Who should decide, and how should the decisions be made?

Both the mayor and Dr. Stockmann must recognize the limits of scientific understanding as they can be applied to their problem. This does not mean that the problem does not exist or that one answer is as good as any other. But it does mean that the problem is not a scientific one and that science cannot solve it. When factions are at war, science is just another weapon. The risks to the townspeople and visitors are medical, social and economic, in different proportions; the solutions are similar, requiring consideration and negotiation. And given the "marketing" of the town's baths, any negotiations would be very delicate. Times have not changed much, at least since 1882.

COLLECTIVE RESPONSIBILITY

An ethical resolution of the problem in Ibsen's play requires the exercise of collective moral responsibility. John Ladd identifies some major philosophical dimensions of moral responsibility in an article called "Collective and Individual Moral Responsibility in Engineering: Some Questions" (Ladd 1982). Ladd points to several qualities that he feels distinguish moral from other kinds of responsibilities.

One quality is that moral responsibility is forward looking. Unlike legal responsibility, focusing on who or what to blame when something goes wrong, moral responsibility is the exercise of moral foresight, the attempt to prevent things from going wrong in the first place.

A second quality is that moral responsibility is nonexclusive. Unlike many task or job responsibilities, it is not the case that just because someone is morally responsible for something, "other persons are not responsible" (p. 9). Moral responsibilities, on and off the job, are not exclusively the jobholder's. In fact they mostly are and can be shared. They may vary in degrees, says Ladd, "only to the extent that one person is better able to do something about [them] than others" (p. 9).

What are the policy and practice implications of this position? Individuals and collectivities have distinct and distinctive capabilities and functions that enable them to exercise, and to help others exercise, moral foresight. Of course, these responsibilities are not exclusively theirs nor can they be exercised in isolation and without assistance. Persons need to help one another draw up policies and practices that will assist professionals, along with everybody else, to act in a morally responsible manner. To quote from Ladd:

> If many people can be responsible for the same thing, then there can be such a thing as . . . collective responsibility. . . . thus, the whole family is responsible for seeing that the baby does not get hurt. The whole community is responsible for the health and safety of its citizens. And all the engineers, as well as others, working on a project are responsible for its safety. (1982, p. 9)

In the end, Dr. Stockmann's tragedy, despite his good intentions, is his inability to help the other people in his community exercise their moral responsibilities. The tragedy for the others is that they are not able to help him exercise his. If today's professionals are to behave responsibly, they will need help from others to do so, just as others will need help from them. Only in this way, will efforts toward collective moral responsibility be successful.

An Enemy of the People shows that these problems have a long history; today's societies are only now becoming more aware of them. This example— and others in this volume—illustrate concerns in two main and interrelated areas in ethics, policy, science, and technology: professional responsibility and the appropriate uses of science in decision making. One such question is about ethical norms in science, what they are, what they should be, and how they can be bolstered and violated. Another category is how decisions should be made: what processes, outcomes, and evidence should be used in deliberating about and arriving at conclusions about what we should do. Value-laden choices must be made to answer either kind of question.

THE SCIENTIFIC STATE

In an interesting article in the journal *Science, Technology, & Human Values* in the winter of 1986, Jurgen Schmandt and James Everett Katz call attention to what they believe is a significant political change in the industrial or postindustrial world, that from an administrative state to a scientific state. Schmandt and Katz claim that there are three ways in which science plays a role in this new political world. One way is as a product, where technology usually comes in. Schmandt and Katz are thinking of promotion and control of science in the service of innovation. A second use of science is as evidence: information and the interpretation of the results of scientific research in policymaking. Finally there is scientific method: analytical, experimental, and empirical techniques (Schmandt and Katz 1986, pp. 43–46).

People value using science and engineering in these ways today even more than they did in Dr. Stockmann's time. Scientists and engineers, science and engineering are unavoidably and intrinsically part of these processes. These are human activities, part of decision-making processes, and as such are inevitably called to human account. People deliberate about whether the processes, outcomes, and justifications, including scientific justifications, for the choices made are themselves justified. There is no turning back.

This is a dynamic process. Forces including science, engineering, and technology reshape our world, and vice versa. Both the activities and their results, being human made, can and will be judged by standards that also are human made: Is this responsible? Good? Desirable? How should decisions be made? How can the process or its results be improved? How can those harmed be compensated? Should they be? And as human interventions become pervasive, mammoth, and more sophisticated, so do the scale and the requirements of human accountability. Scientists and engineers discover themselves "on the spot."

William Lowrance captures the tragic aspects of this human endeavor, this human condition, in his book *Modern Science and Human Values:* "I take tragedy to mean the deliberate confrontation of deeply important but nearly irresolvable life issues. Tragedy begins in our knowing of causalities, in our intervening in particular cases, and in our technical enlargement of interventional possibilities" (1985, p. 17). Humans are tragically aware of what has been done, what can be done, and what the consequences of human commitments are. With awareness of themselves as causal agents comes awareness of individual and collective responsibility.

An example from a project the National Science Foundation supported several years ago can help explain these points, in particular, that the scale and sophistication of scientific and technical intervention require, as well, increasing sophistication in making value judgments. The project examined equity issues in the management of radioactive wastes, and it resulted in a delineation that has been commonly accepted and often used. It separates the ethical issues into those of locus, legacy, and labor. All three raise questions. Locus pertains to the ethical issues created in locating facilities that produce or handle radioactive wastes. A nearby group of people is or may be harmed to benefit another, perhaps larger group of people, and those harmed may well be different from those benefited. Legacy raises similar ethical questions concerning future generations. Such questions are complicated, as future generations can have no voice in current choices. Labor brings similar and different questions. A particular group may be harmed or placed at risk in order to benefit another. But the group itself may benefit or perceive a benefit—jobs—and so may consent (Kasperson 1983, pp. x–xii).

Decisions regarding locus, legacy, and labor lead to many perplexing moral puzzles and value questions. They arise from both philosophical and scientific difficulties and illustrate why increasingly sophisticated scientific and ethical analysis are necessary and interdependent. They help us understand how scientific and value choices are intertwined. The result is that value-free science, in these contexts, becomes impossible, and a process of participation, negotiation, and compromise becomes necessary if important human values are to be protected. An examination of these problems also helps demonstrate the need for a better understanding of professional moral responsibility. Much is at stake here because of the complicated, contentious, and fragile world in which we all live.

The first example that I shall examine is of issues raised by questions concerning our responsibilities to future generations. The analysis indicates the limits of cost–benefit or risk–benefit approaches that do not distinguish among different kinds of values. It also shows why values need to be considered when determining acceptable evidence even within cost–benefit or risk–benefit approaches. The second illustrates that scientific approaches to acceptable evidence contain value choices. And the third example reminds us that agreement on values cannot ensure that our scientific assumptions and procedures will provide correct answers to our questions about whether we are achieving the ends that we have agreed are those we want or will accept.

The Identity Problem

People can try to dismiss the philosophical puzzle that we shall describe next, viewing it as a fanciful kind of semantics, the kind of problem that only a philosopher could love. It is all that, but it is more than that, because it addresses the question of what we owe to future generations.

The problem is now known as the *identity problem,* which Derek Parfit described in 1983 (Parfit 1983, p. 167). Much simplified, the problem is as follows: If who the individuals in future generations are—that is, their specific identities—depends on the energy policy (along with other) choices that we make, than those persons cannot blame their ancestors for having made their lives worse than they might otherwise be, for it is only because their ancestors made the choices that they did that they now exist at all. Those choices were influential in determining that they, and not other individuals, are alive.

People who are not philosophers—and even some people who are philosophers—are made very unhappy by this kind of problem. It is a linguistic puzzle, viewed with suspicion as a trick. On the other hand, to persons who take ethics seriously, it is important to try to overcome this problem. Surely, they say, moral intuitions tell us that future generations are owed a better world than the current one. Making things better for posterity has been a powerful moral idea and ideal. But can it be justified without appealing to the happiness or interests of particular future individuals and apparently falling prey to the identity problem?

In the same book in which Derek Parfit describes the identity problem, Douglas MacLean (1983, p. 190–93) attempts to overcome it by pointing out that the moral justification for the position that posterity matters may not depend on anyone's happiness—current or future. Rather, he believes, it depends on our recognition that in moral accountings, experiences (in the sense of how we feel or how things make us feel) are not all that is important. Posterity happens to matter to us because it adds meaning to our lives.

MacLean gives other examples of things that people value similarly: friends, reputations, cultural heritages. He does not identify our natural or environmental heritage, but I think he could (Hargrove 1989). These objects matter not (or certainly not solely) because they make us happy, he asserts, but because they are necessary for the kinds of experiences that make our lives meaningful or worthwhile.

MacLean states that these kinds of experiences are different from those that are stimulated by fine cars, nice clothes, and tasty food. Posterity—along with friends, heritage, self-respect, and other objects or phenomena valued at least in part for themselves and not just for the pleasurable sensations or results they bring—exemplifies moral values. Moral values are not the kinds of things that it is appropriate to measure just in economic terms or to trade for equivalent other experiences. In fact, it is difficult to decide what an equivalent other experience would be. Energy policies, like other policies, have to be made with attention to this distinction if they are to preserve important moral values.

I think that MacLean is right to distinguish kinds of values, but he has not discovered what makes the difference. In my opinion, one kind of value derives from the personal satisfaction we receive from the object; another comes from our relationship with or to the object. This is, I think, what MacLean is trying to say when he talks about "meaning" in our lives. These two types of values cannot be exchanged for each other equivalently. Part of the difference is in our relationship with the kinds of objects or phenomena that we value at least in part for themselves. It is our understanding of the significance of our relationship to posterity—like friends and children, community and environment—that makes the difference.

Working through this philosophical puzzle is important, then, because it demonstrates a moral limit on cost–benefit or risk–benefit analysis. When considering energy policy, it will not suffice simply to weigh everything on the same scale. Even if the current generations need not weigh future individuals' happiness against their own in making energy policy decisions, a moral concern for posterity and cultural and environmental heritages will have to be accounted for if human lives are to remain meaningful. It will damage the human community if bureaucratic technocracy rides roughshod over this distinction.

Elizabeth Anderson (1988) makes somewhat the same point:

> Workers commonly find risks more or less acceptable depending on the social relations in which they are imposed and managed.... Saving lives, pursuing truth, creating art—these are intrinsically valuable accomplishments which can make risk-taking really worthwhile.... Workers with families ... see their choices as attempts to discharge their responsibilities to their families. (pp. 60–61)

Thus, acceptable risk does not turn on a paycheck; rather, it turns on a personal moral calculus.

The attempt to reduce calculation of risk to a single scale, Anderson notes, finds expression in a political calculus, that "the market provides an avenue for the expression of discontent through exit, not voice" (p. 56). Critics of this position object that it improperly "ignores the way this lack of voice influences the acceptability of the risks" (p. 57). As Nelkin and Brown's empirical research—which resulted in the book *Workers at Risk* (1984, pp. 178–83)—tells us, acceptability is a function of trust, control, voice, and the perception that the end is worthwhile. Slovic and Covello make similar points in this volume. Furthermore, Anderson is concerned that relying on the market mechanism of exit, even when fair, informed, and voluntary, limits the political discussion to concerns for risk reduction and compensation. It leaves out the possibility for people "at risk" to establish their own agenda for discussion in which they might have substantial control over how risks would be managed.

Calling on Nelkin and Brown's empirical work and MacLean's normative work, Anderson's essay—on the limits of market norms as a basis for cost–benefit or risk–benefit analysis—provides evidence of the need for both in grappling with issues posed in the scientific state. It demonstrates the merit of philosophical analysis in thinking about these problems.

Using a Scientific Test

Turning from philosophical to scientific puzzles shows us another way in which scientific and value choices are intertwined. Examining a little more closely just one aspect of the issues raised by equity and exposure to risk reveals how increasingly sophisticated science requires increasingly sophisticated social and ethical analysis. How can we determine that people have been harmed or are at risk of harm? A scientific test is used.

The decision to use a scientific test at all sometimes has moral dimensions, of course. That is, it is a choice made by human beings that may help or harm them or their surroundings. Once the choice is made, the decisions about which test to use must also have moral dimensions. As pointed out by M. J. Panner and N. A. Christakis in "The Limits of Science in On-the-Job Drug Screening," the tests inevitably result in false positives or false negatives, even if only to a small degree (1986, p. 8). What are the grounds for deciding to use a test that will indicate that some people are affected who in fact are not or a test that will not identify everyone who is affected?

This same difficulty arises more generally in tests for chemical toxicity. It is a statistical rule that guarding against false positives increases the likelihood of false negatives. Two papers by Carl Cranor (1988, 1990) point out that for relatively rare diseases, when sample populations are small, using a statistical rule that guards against false positives increases the likelihood of false negatives. It is a statistical fact that guarding against false positives in these circumstances decreases the statistical "power" of the study. This in turn decreases confidence in the reliability of results that show no increase in the disease being tested (see the chapter in this volume by Deborah Mayo for further explanation of these points).

Thus, says Cranor, "good" science (in the sense of science that does not predict a toxic effect when there is none) is purchased at the cost of overlooking some toxicity. This kind of "good" science is also more likely to predict no toxic effect when there is one. And it also may conflict with "good" regulation, which may purchase the security of preventing a greater incidence of disease at the cost of banning or slowing the development of harmless chemicals (the chapter by Ellen Silbergeld discusses these points).

Once science enters the policymaking arena, ostensibly scientific decisions—those concerning the use of a statistical rule, for instance—are not (or are not just) scientific. They are social, indeed moral, decisions. That is, a decision, in the case of the workplace, to set test standards to allow more false negatives and to avoid false positives could be made on the grounds that it preserves equality of opportunity and worker choice (Marshall 1988). In effect, those values are being ranked above worker protection, as workers themselves tend to do. On the other hand, in the traditional doctor–patient setting, test standards might well allow more false positives, because there is no conflict between the doctor's and the patient's goals. Both rate protection and knowing at the top. Testing for environmental toxicity might be another circumstance in which a risk-averse society would opt for more false positives.

These decisions are not easily generalized, however. Suppose in the first case that the workplace disease is particularly gruesome or deadly. Suppose in the second that there are no further tests to rule out the false positives and no treatments for the purported disease or that the only treatments are themselves dangerous. Suppose in the case of environmental testing that a chemical offered a way to overcome a very serious pest problem in confined areas. The particular circumstances may require careful consideration in arriving at a final decision, which will inevitably mean moral choices, as Cranor notes as well (1990).

Given these differing priorities, all scientific tests cannot be equally "good." The goodness of a scientific test depends on personal and social values (Cranor 1990). In this circumstance, questions about the acceptability of evidence are as important, perhaps more important, than are questions about the acceptability of risk.

All of these examples help illustrate Schmandt and Katz's claim that the United States has moved from an administrative to a scientific state. They help show that science and values are inextricably linked in contemporary society.

They may also help show why professionals need to develop new understandings of their moral responsibilities and why not just they need to do this. Indeed, the process requires the involvement of many different kinds of scholars and organizations.

One thing is sure. "Good" science is not enough for professionals to fulfill the requirements for moral responsibility. The use of a scientific test itself and the setting of standards for a test are themselves moral choices. Even the questions of whether to use a test and what standards to use are not scientific questions, as they do not have scientific answers. Scientific tasks often seem bound up with nonscience; scientists cannot separate their moral responsibilities "as scientists" from their responsibilities as persons and citizens.

On the other hand, if these are not scientific questions and do not have scientific answers, then scientists "as scientists" cannot and should not be expected to answer them, particularly by themselves. The responsibility for answering them is social, that is, shared. This is particularly true when the social investment in finding scientific answers is, for intrinsic and extrinsic reasons, so limited that finding definitive answers is unlikely. Nonetheless, scientists—as well as other scholars in the social sciences and humanities who study science as a phenomenon—have important roles in deliberating these questions. Their knowledge is critical to understanding what science can accomplish and what its limits are. Coming to this understanding is particularly relevant to a scientific state. Often, scientists or other scholars are the only ones who will examine these limits and explain them so as to enable both society and science to use that information in making decisions about the future. (As an example, see the discussions by Ronald Giere and Kristin Shrader-Frechette in this volume.)

Questionable Scientific Assumptions

One more example elucidates this point further. Roger Kasperson describes a study that examines whether workers who accept more on-the-job risks are compensated proportionally better (Kasperson 1986). Economists have long believed that this is so, that the market will provide this additional compensation, thereby enhancing social justice. Many major studies have purported to find this is so. However, once they drop several crucial but unwarranted assumptions that had been made in these studies, Graham, Shakow, and Cyr (1983) come up with results contrary to this. In particular, rather than assuming that the work force is homogeneous, they identify conditions that distinguish two kinds of jobs and two kinds of employees. In this way the economists' belief is justified for one group—unionized, with high wages and security, in large firms, primarily white prime-aged males—but not for another with the opposite characteristics. In this case, even if everyone agrees on the value issues, that is, agrees that extra compensation for extra risk is a good way to mitigate social inequities, questions remain about whether the goal is being accomplished and about whether the economists' studies show what they purport to show.

These difficulties are to be expected even when science is "good" science. If we cannot accept these difficulties, science cannot be used in social contexts. But the necessity of accepting these difficulties implies that in many cases, science will not answer societal questions about what should be done. Science may help, but there will be times when it will not.

Sheldon Reaven worked with E. William Colglazier on a study of ethical and policy issues in radioactive waste management reported on in this volume. Reaven makes a related point in a draft manuscript entitled "How Sure Is Sure Enough?" (1988) when he states:

> The NWPA [National Waste Policy Act] sets up a *scientific trap* in which the public is encouraged to demand or expect what is close to certainty (from virtually "flawless science"), and in which scientists and engineers are encouraged to believe or pretend that they can supply it. But reasonable assurance cannot be *demonstrated* because it simultaneously presupposes reasonable doubt. (p. 17)

Sheila Sen Jasanoff also makes this point in Chapter 2 in this book, arguing that political needs for accountability in the United States push regulators toward finding a "scientifically correct" answer even when there is none. This attempt then is doomed to political challenge.

PROFESSIONAL MORAL RESPONSIBILITY

There is a two-headed hydra here, and it may be useful to recognize this duality in approaching questions about the moral responsibility of professionals. On the one hand, as the false negatives–false positives examples show, parties outside a particular profession must help establish standards for acceptable risk and acceptable evidence; they must help set "scientific" standards. On the other, as the risk compensation example reveals, careful scientific—or what Deborah Mayo calls metascientific—considerations of epistemic issues for science, about what we are entitled to say we know, are essential. Kristin Shrader-Frechette's notion of "scientific proceduralism," as a process subjecting ostensibly scientific analysis to criticism from many interested parties, requires both these kinds of considerations. This process occurs over time, and so these answers should be expected to change.

Recognizing this duality is important to the new understanding of professional moral responsibility that needs to be developed. Scientists and engineers and their professional groups need to establish, implement, and evaluate their principles and practices with this duality in mind if they are to behave in a morally responsible manner. Doing so is a distinctively professional moral responsibility because it arises with professional capabilities and functions, those recognizing the epistemic and moral limits of science.

Of course, for such efforts to be successful, individuals or groups must initiate them (George 1987); that is one of the reasons for this book and one of the reasons for whistleblowers like Dr. Stockmann (Glazer and Glazer 1987). Because professional organizations occupy positions in which they can voice these concerns, they also have a responsibility to try to do so. That is why the meeting of the Board on Radioactive Waste Management, in which an attempt was made to examine these issues, provides a hopeful sign.

Without these kinds of discussions and debates, the details that must be considered if recommendations and guidelines are to be developed will not be. In an essay entitled "Ethics, Science and Moral Philosophy," Marcus Singer points out that applied ethics (i.e., the careful consideration of ends, principles, procedures, etc. and the development of a consensus and agreement on codes and policies of conduct, etc.) is an appropriate and necessary task for branches of science (1981). This chapter and others in this book make it clear that the task cannot be an "exclusive" one. It requires participation from analysts, critics, and interested parties willing to challenge and question the assumptions, methods, and conclusions that scientists and engineers may bring to this endeavor in order to assess their epistemic and moral limits.

Identifying and explaining these connections will become more and more important if the scientific state is not to evolve into a totalitarian one. This evolution can happen for several reasons: the perception that ordinary people are not competent or knowledgeable enough to make these decisions, the desires of interested individuals and groups to preserve their dominion and advantages, and the lack of awareness by those who should know better—scientists and engineers, for example—of the moral dimensions of ostensibly scientific decisions.

To recognize the tragicomic aspects of these phenomena, consider the recent proliferation of proposals to test Americans for drugs (Holden 1987, 1988; Marshall 1988). In this debate, it may be most apparent that social, legal, and ethical analysis lags significantly behind scientific or technical analysis and the political analysis of when to create or jump on a bandwagon and prey on fears of the "drug menace." Here niceties of concerns for values such as equality of opportunity, worker protection, and personal privacy and freedom disappear, as proposals to build more jails and test every worker and bureaucrat and every teacher and teenager in the country proliferate. The debate over testing for AIDS adds another complication, and elements of tragedy, to the discussion. Sober heads may prevail, but these proposals demonstrate the powerful tug for Americans of the idea that there are technological fixes for social and societal—even family—problems.

The lure of this idea can be seen in a letter to the editor of *Chemical and Engineering News* (January 26, 1987). H. Peter Lehmann, professor of pathology at the Louisiana State Medical Center, is discussing drug-testing programs. From the letter, he appears certainly to be an ethical gentleman:

> There are, unfortunately, cases where, through ignorance and carelessness, an individual's rights may be placed in jeopardy. . . . [This] can be avoided only if drug tests are performed in accredited laboratories that are directed and staffed by qualified and certified clinical laboratory personnel. . . . Clinical chemists, pathologists, and medical technologists have expertise in the complete drug testing procedure, from specimen collection to validation of the test results, including all the security procedures necessary to ensure that an individual's rights are maintained. . . . Now is the time to cooperate rather than to divide. . . . Analytical chemists, clinical chemists, pathologists, and others, by combining their skills and expertise, should be able to provide the community with drug testing programs that are reliable and secure, and will perform the desired wishes of our society without endangering the rights of any member of that society.

This sounds good, and it is a comfortable position for chemists to put themselves in. But it ignores the possibility that the desired wishes of society might conflict with human rights. It limits the possibilities that chemists will interfere with individual rights to technical problems of ignorance, carelessness, blunders, and lack of reliability or security. These are important sources of interference, it is true. But they are probably not the most important sources. Scientists and engineers cannot limit their professional, moral responsibilities in this fashion.

Why not? There are two kinds of reasons. One is that moral responsibility is never fulfilled when an individual says, "I'm just doing my job, and I am doing it well." Moral responsibility is getting the right job done, not just getting the job done right.

Another kind of reason grows out of the nature of the scientific job. As noted earlier, Lehmann and other scientists and engineers need to recognize the moral dimensions of ostensibly scientific decisions. The development and application of a scientific test, in a social context, mean choices that cannot be characterized as scientific. In this case—and in most cases in the complicated society in which we live—jobs are fraught with uncertainties—scientific uncertainties, quasi-scientific uncertainties, social uncertainties—all of which pertain to the justice and moral worth of the jobs and their outcomes. Individual chemists and their professional societies cannot defend their decisions to administer tests with the simple statement "This is what society wants, and we are doing it in a manner consistent with 'good' science." Neither can the organizations in which they work. They have to add that they cannot guarantee their samples. They probably have to add that their lab personnel are more or less reliable given various exigencies (as just noted: ignorance, blunders, whether it is Monday or Tuesday, etc.) They have to add, "The larger the population is that we test, the larger the number of wrong answers that we will give you. The wider we cast our nets, the more false positives we will catch." Given this outcome, the question of safeguards for those unjustly accused moves high up on the agenda. Getting the job done right cannot be isolated from value-laden decisions, even if it is under the exclusive control of the individual laboratory scientist or technician.

On the other hand, politicians or employers who make regulations insisting on random workplace testing, as well as the workers, judges, and voters who accept it, are also responsible. All must recognize the damage to the community's moral values and trust to which these initiatives can lead. All must recognize that monitoring technologies may be an example of hidden, value-threatening hazards, as Roger and Jeanne Kasperson define them. These impersonal "control technologies"—like computer monitoring for keystrokes on the job for example—can be justified in the name of economic efficiency or law enforcement, although it is not clear that they do contribute to these ends. However, they may interfere with a culture that values privacy, diversity, and individual liberties and life-styles. They may discourage individual initiative and responsibility. They can also encourage the false credibility of "scientific" findings and "technological fixes" and reinforce views that favor isolating and punishing ostensibly aberrant persons. Without adequate procedures for notification and appeal, many persons can find their livelihoods and reputations compromised. Unless we develop political processes that enable us to confront and discuss these kinds of potential losses, democracy may suffer greatly.

A brief digression may be appropriate here, to raise an additional point about a notion of moral responsibility that Ladd does not make. Another aspect of moral responsibility that relates to the quality of nonexclusivity that Ladd does identify is the commitment to make right what has gone awry, independent of who is to blame legally (or even morally, for there can be moral blame even if it is not part of the notion of moral responsibility). This is also a nonexclusive notion. One person's having the moral responsibility to make right what has gone wrong does not preclude others having it too. And individuals will need to cooperate if they are to succeed in repairing damage.

One of the alluring but woefully misleading aspects of the search for technological fixes is that such fixes are expected to preclude the need to correct mistakes. Ronald Giere identifies the problems in this approach in Chapter 9 in this volume. And some scientists and engineers are beginning to realize the illusory nature of this search (Whipple 1988). We all need to realize how illusory it is. If we do, we may be on the path beyond the false certainty and isolated rectitude that the nineteenth-century Dr. Stockmann shares with the twentieth-century chemist Professor Lehmann.

ACKNOWLEDGMENT

Dr. Hollander is the coordinator of Ethics and Values Studies, National Science Foundation, Washington, D.C. 20550. The views in this chapter are her own and do not represent those of the foundation. The chapter is based on and draws from an earlier essay entitled "Toward a Model of Professional Responsibility" that appeared in *Research in Philosophy and Technology* 9 (1989).

REFERENCES

Anderson, E. (1988). "Values, Risks, and Market Norms." *Philosophy and Public Affairs* 17:54–65.
Board on Radioactive Waste Management (BRWM), National Research Council (NRC) (1988). *Agenda Book*. Session 2: "Moral and Value Issues." Summer Retreat Meeting, Santa Barbara, Calif., July 11–15.
Cranor, C. F. (1988). "Some Public Policy Problems with Risk Assessment: How Defensible Is the Use of the 95% Rule in Epidemiology." Paper prepared as part of the University of California Riverside Carcinogen Risk Assessment Project, Toxic Substance Research and Teaching Program.
Cranor, C. F. (1990). "Some Moral Issues in Risk Assessment." *Ethics* (forthcoming).
George, K. P. (1987). "Individual Ethics and the Social Goals of Agriculture." *Agriculture and Human Values* 4:100–4.
Glazer, M. P., and Glazer, P. M. (1987). "Whistleblowers Who Risked It All." *Recreation News* 5:1, 10–12.
Graham, J., Shakow, D. and Cyr, C. (1983). "Risk Compensation—In Theory and in Practice." *Environment* 25:14–20, 33–42.

Hargrove, E. (1989). *The Foundations of Environmental Ethics.* Englewood Cliffs, N.J.: Prentice-Hall.

Holden, C. (1987). "Doctors Square Off on Employee Drug Testing." *Science* 238: 744–45.

Holden, C. (1988). "NIH Scientists Balk at Random Drug Tests." *Science* 239:724.

Hollander, R. (1987a). "Commentary: In a New Mode—Ethics and Values Studies at the NSF." *Science, Technology, & Human Values* 12:59–61.

Hollander, R. (1987b). "Ethics and Values Studies at the National Science Foundation." *Professional Relations Bulletin* 40:2–3.

Hollander, R., and Steneck, N. (1990). "Science- and Engineering-related Ethics and Values Studies: Characteristics of an Emerging Field of Research." *Science, Technology, & Human Values* 15:84–104.

Ibsen, H. (1882/1970). *An Enemy of the People.* In *Ibsen, Four Major Plays.* Vol. 2, trans. Rolf Fjelde, pp. 115–222. New York: New American Library.

Kasperson, R. E. (ed.) (1983). *Equity Issues in Radioactive Waste Management.* Cambridge, Mass.: Oelgeschlager, Gunn & Hain.

Kasperson, R. E. (1986). "Scientific Evidence, Ethics, and the Management of Occupational Hazards." Paper presented at the 1986 Annual Meeting of the American Association for the Advancement of Science, Philadelphia, May 27.

Ladd, J. (1982). "Collective and Individual Moral Responsibility in Engineering: Some Questions." *IEEE Technology and Society,* June, pp. 3–10.

Lehman, H. P. (1987). "Reliable Drug Testing." Letters. *Chemical and Engineering News,* January 26, p. 2.

Lowrance, W. (1985). *Modern Science and Human Values.* New York: Oxford University Press.

MacLean, D. (1983). "A Moral Requirement for Energy Policies." In *Energy and the Future,* ed. D. MacLean, pp. 180–97. Totowa, N.J.: Rowman and Littlefield.

Marshall, E. (1988). "Testing Urine for Drugs." *Science* 241:150–52.

Nelkin, D., and Brown, M. S. (1984). *Workers at Risk, Voices from the Workplace.* Chicago: University of Chicago Press.

Panner, M. J., and Christakis, N. A. (1986). "The Limits of Science in On-the-Job Drug Screening." *Hastings Center Report,* December, pp. 7–12.

Parfit, D. (1983). "Energy Policy and the Further Future: The Social Discount Rate." In *Energy and the Future,* ed. D. MacLean, pp. 31–37. Totowa, N.J.: Rowman and Littlefield.

Reaven, S. J. (1988). "How Sure Is Sure Enough: How and Why Stakeholders Differ on Repository Scientific Issues." Manuscript.

Schmandt, J., and Katz, J. E. (1986). "The Scientific State: A Theory with Hypotheses." *Science, Technology, & Human Values* 11:40–52.

Singer, M. G. (1981). "Ethics, Science and Moral Philosophy." In *New Directions in Ethics: The Challenge of Applied Ethics,* ed. J. P. DeMarco and R. M. Fox, pp. 282–98. New York: Routledge & Kegan Paul.

U.S. Congress. Senate. Committee on Labor and Human Resources. *National Research Institutes Reauthorization Act of 1988.* 100th Cong., 2d sess. May 26. S. Bill S.2222. S. Rept. 100–363.

Whipple, C. (1988). "How Uncertain Is Too Uncertain? Setting Reasonable Expectations for Risk Assessment." Paper presented at the Board on Radioactive Waste Management Summer Seminar, Santa Barbara, Calif.: July 11–15.

III

PHILOSOPHY AND
SCIENTIFIC EVIDENCE

The contributors to Parts I and II showed how values infuse both the development of evidence and the uses to which evidence is put in individual, group, and societal decision making. They demonstrated why understanding this relationship between science and values is necessary to make responsible decisions regarding risk and to understand disagreements and controversies regarding risk assessment and management. The thesis was that such an understanding is possible, that we can become better at recognizing what counts as acceptable evidence of risk. Work from philosophers and methodologists of science are a needed part of any program that emphasizes the acceptability of the evidence of risk.

The contributors to Part III illustrate two such philosophical roles. One is the task of understanding evidence of risk. A major difficulty in judging the acceptability of risks is that the evidence is based on probabilistic reasoning about the effects caused by various substances and practices. A central task of philosophers of science is to provide and examine models of probabilistic reasoning, causal inquiry, and decision making under conditions of uncertainty. Clearly, then, such work is relevant to understanding and evaluating risk evidence. The chapters by Ronald Giere and Kenneth Schaffner describe and analyze models of decision making and probabilistic causality, respectively. Deborah Mayo's chapter sets out "metastatistical" rules for better understanding tests of statistical hypotheses about increased risks.

Philosophical challenges to the rationality and objectivity of science form the basis for a second role for the philosophy of science. This role stems from the fact that views of the nature and role of evidence in risk analysis—whether such views are those of policymakers, risk assessors, sociologists, or others— are closely tied to general philosophical views of the nature of scientific knowledge and the rationality of hypothesis appraisal in science. The arguments that have caused some philosophers to question the traditional foundations of science form the basis for a growing tendency to question the reliability of scientific evidence as it enters both individual risk judgments and expert risk assessments. The philosophical challenges to traditional foundations of science are often traced to Thomas Kuhn's 1962 work *The Structure of Scientific Revolutions.* Kuhn questions the traditional view of science as embodying impartial algorithms for appraising claims objectively. As Kuhn and others stress, looking at actual historical cases reveals that scientific debates are not adjudicated simply by empirical rules; "extrascientific" values enter as well.

The problem now is whether to abandon the view that science is characterized by rational methods or to seek a new account of scientific rationality. An extreme version of the first approach is Paul Feyerabend's anarchy, in which "anything goes" in science. Here, beliefs about the world are not restrained by "the facts of the matter" at all. The response in a great many fields has increasingly opened science to suspicion and mistrust, a theme running through recent works relating science and technology to society. But what about the second approach to the problem, to seek new models of scientific rationality? Any solution that these new models provide for the problem of scientific rationality will be relevant to the problem of the reliability of scientific evidence in risk assessment. This takes us to the second key role for philosophy of science.

Because science plays so important a role in risk assessment, it is not surprising that philosophical disagreements about the validity of science form the basis of many disagreements about the validity of risk assessments. Nevertheless, there is typically no explicit recognition of these philosophical underpinnings in risk research or in the science studies work on risk. The chapters in Part III attempt to identify these philosophical presuppositions, in order to explore the premises on which the major opposing views rest and to question their conclusions. The contributions by Shrader-Frechette and Mayo deal with charges of relativism in risk perceptions and risk assessment, respectively. The source of these charges, they argue, is the philosophical presupposition stemming from the "old" or positivist image of science. According to this image, scientific objectivity requires a fact–value dichotomy, and for evidence to have scientific validity, it must be value neutral. This explains the desire that Silbergeld discusses for risk assessments divorced from value considerations. It also explains why the recognition by Kuhn and others that such value neutrality is impossible has led many to infer that risk assessment (if not risk itself) is largely a matter of subjective values. This explains why many have tended to downplay the role of scientific evidence in adjudicating disputes about risk assessments. If scientific evidence is largely a matter of subjective values, how can it help adjudicate disputes?

Having laid bare these presuppositions, authors Schrader-Frechette and Mayo can begin to question them. Why hold such a value-neutral view of scientific rationality and objectivity? By raising such questions and providing alternative views of objectivity and rationality, their chapters provide a philosophical basis for avoiding charges of relativism and irrationality in risk judgments.

RISK AND MODELS OF DECISION AND CAUSALITY

Because statistical knowledge is so important to risk assessment, philosophical work on statistical foundations and rational decision making offers insights into the roles of evidence in risk management. In "Knowledge, Values, and Technological Decisions: A Decision Theoretic Approach," Ronald Giere shows that a standard decision model can offer a clear and simple framework for both analyzing a controversy and judging its outcome. His focus is on just one decision ingredient: scientific knowledge, particularly the statistical knowledge of low-level radiation. The limit of this knowledge is typically taken to imply that the driving force in the nuclear power controversy, and policy controversies generally, is a conflict over policy values—the presumption being that knowledge gaps are necessarily filled with subjective policy judgments. The implication is that the controversy would be resolved if we could resolve the underlying value conflict. Giere's main conclusion denies that implication and brings out the various uncertainties in the knowledge that block a rational decision (e.g., the difficulties of extrapolation from animals to humans, from large doses to small ones, etc.). Understanding these uncertainties explains the current controversy and why scientific evidence fails to force either side to concede. What is needed are strategies for dealing with the uncertainty itself. Giere proposes a strategy in which decisions are made sequentially, so that information gained at earlier stages can be used to improve decisions at later stages. He suggests how a carefully planned use of a *sequential decision* strategy could sufficiently reduce uncertainty so as to make possible a fairly rational resolution of the issue—or at least reveal where various value conflicts are causing continued controversy.

In determining fair compensation for harm associated with a risk, it is necessary to ascertain the likely extent to which it has been caused by a given hazard exposure. However, as Kenneth Schaffner points out in "Causing Harm: Epidemiological and Physiological Concepts of Causation," there are two very different ways that statistical and causal evidence might be used to assign liability for harm. He proposes that this distinction in how evidence is used corresponds to two different concepts of causation—that of epidemiological causation (involving, for example, risk factors) and in more usual scientific causation (involving, for example, physical mechanisms).

To develop his thesis, Schaffner considers the philosophical concept of causation, a probabilistic notion, which is implied by the clinical methods by which epidemiologists reach causal claims. The notions of epidemiological causation appear very much like, if not identical to, what philosophers of science (e.g., Good, Reichenbach, Salmon, and Suppes) have termed "probabilistic causation." However, causal accounts that concentrate only on the formal machinery of probabilistic causation lack criteria for distinguishing causal connections from coincidences, associations based on subtle biases, and those with an unidentified common cause. Schaffner introduces a different approach to probabilistic causation that he maintains is clearer and more coherent than previous philosophical accounts are. He compares this new approach with the causal accounts of Ronald Giere and Nancy Cartwright and then uses it to find the essential difference between the epidemiological notion of cause and that based on physical mechanisms. He discovers this difference not in the probabilistic aspects of epidemiology but, rather, in the lack of homogeneity in the actual populations involved. Schaffner's approach helps us make sense of some of the epidemiologists' criteria for assessing causal claims and helps sharpen our analysis of inference from epidemiological studies.

TOWARD A POSTPOSITIVIST PHILOSOPHY OF RISK ASSESSMENT

Where the authors in Part II tried to show that separating scientific risk assessment from policy values is self-defeating in practice, the chapters by Kristin Shrader-Frechette and Deborah G. Mayo discuss the philosophical views underlying such separatist attempts. In "Reductionist Approaches to Risk" Shrader-Frechette argues that two of the major accounts of societal risk—the cultural relativist and the naive positivist positions—err by presupposing highly doubtful reductionist positions. Cultural relativists (e.g., Douglas and Wildavsky) view risk as a collective construct and thereby reduce risk evaluation to sociology. Naive positivists (e.g., Starr and Whipple), on the other hand, reduce risk characterization to objective natural science viewed as value free. Shrader-Frechette contends that the cultural relativist and the naive positivist positions, respectively, overemphasize and underemphasize the value judgments in risk analysis. Both views also lead, though for different reasons, to discrediting public attitudes toward risk. That is, the cultural relativist discredits public attitudes by viewing them as the outgrowth of biased ethical and social backgrounds, whereas the naive positivist discredits public attitudes by claiming they lack the technical, scientific expertise that characterizes rational risk assessment. Shrader-Frechette's defense of the rationality of public attitudes toward risk complements the empirical work discussed by Slovic and Covello in Part I, which offers a more rational means to communicate risk and model public perceptions of risk. Shrader-Frechette favors a middle position, "scientific proceduralism," that is based on a notion of objectivity in risk assessment as being able to withstand criticism. Contrary to the two reductionist positions, her view recognizes that no risk assessments are free of values, but she also maintains that some are more warranted than others are. Her position is "procedural" because it relies on democratic procedures—on criticism by affected publics as well as by scientists. By enabling policymakers and the public to grasp the consequences of their risk decisions, such criticism promotes public attitudes toward risk that are rational and ethically defensible.

The final chapter by Deborah Mayo shares with the previous one a view of objectivity based not on value-free algorithms but on rational criticism. It shares with Giere's paper the denial that risk-assessment conflicts largely reflect policy conflicts. In "Sociological Versus Metascientific Views of Risk Assessment," Mayo focuses on a key question in risk-assessment literature, namely, the question of separability: Can (and should) risk assessment be separated from the policy values of risk management? As seen in Part II of this volume, attempts to improve risk assessment through such separation have been challenged both in principle and in regard to the form they have actually taken. Criticisms of separability and the corresponding endorsements of nonseparatism, however, have very different meanings. These nonseparatist views, Mayo suggests, may be divided into two broad camps, the sociological view and the metastatistical view, each corresponding to an underlying philosophy of scientific rationality. The difference between the two may be found in what each finds to be problematic about any attempt to separate assessment and management. Whereas the sociological view argues against separatist attempts on the grounds that they give too small a role to societal (and other nonscientific) values, the metascientific view does so on the grounds that they give too small a role to scientific and methodological understanding. Mayo calls our attention to the problem that has resulted because nonseparatists of the metascientific stripe are lumped together with nonseparatists of the sociological stripe.

Mayo shows how challenges to attempts to separate risk assessment from policy relate to a more general philosophical challenge to the "old image" of scientific rationality and outlines the argument that leads the sociological view to reject separability. This argument, however, is based on a premise that she denies: that the inseparability of science and policy in risk assessment entails the inability to adjudicate objectively among competing risk assessments. The implications of the metascientific view, which also rejects separability, are contrasted with the sociological view and with two other positions.

Mayo's conclusion is that what matters most is not whether or not a view espouses separatism but whether it adheres to the old image of science or, alternatively, sets the stage for a new or postpositive image, in which scientific scrutiny—neither algorithmic nor value free—can nevertheless appraise objectively the adequacy of risk assessments. By "appraising objectively" Mayo means determining (at least approximately) what the data do and do not say about the actual extent of a given risk. Mayo illustrates these issues and arguments by considering a controversial ruling during the Gorsuch EPA regarding the significance of the risks associated with formaldehyde. Only with a critical understanding of the uncertainties involved, she stresses, is it possible to determine what is and is not warranted by the evidence and thereby to determine the standard of protectiveness applied in reaching such risk assessments. Only in this way is it possible to hold risk policymakers accountable.

Using the philosophy of science for these tasks in the field of risk analysis may be viewed as "applied philosophy of science." As such, the chapters in Part III demonstrate a new trend in recent philosophy of science: the emphasis on being relevant to the actual practice of science. According to this view, positions in philosophy of science should be judged not on a priori grounds but on how well they fit with and illuminate scientific practice. On the one hand, applied philosophy of science serves the preceding roles in solving problems in risk analysis. On the other hand, these applications to risk provide test cases for the empirical appraisal of various philosophical models of causality, statistical inference, and scientific rationality. We hope that these chapters encourage the use of applied philosophy of science in the service of these mutually beneficial ends.

9

Knowledge, Values, and Technological Decisions: A Decision Theoretic Approach

RONALD N. GIERE

Before World War II, most decisions involving the introduction of new technologies were made primarily by individuals or corporations, with only minimal interference from government, usually in the form of regulations. Since the war, however, the increased complexity of modern technologies and their impact on society as a whole have tended to force the focus of decision making toward the federal government, although this power is still usually exercised in the form of regulation rather than outright control. Given the huge social consequences of many such decisions, it seems proper that the decision-making process be moved further into the public arena. Yet one may wonder whether the society has the resources and mechanisms for dealing with these issues. Thus, the nature of such controversies, and the possible means for their resolution, has itself become an object of intense interest.

One may approach this subject from at least as many directions as there are academic specialties. Many approaches are primarily empirical in that they attempt to determine the social and political mechanisms that are currently operative in the generation and resolution of controversies over new technologies (Nelkin 1979). Such studies usually do not attempt to determine whether the social mechanisms actually operating are effective mechanisms in the sense that they tend to produce decisions that in fact result in the originally desired out comes.

The approach of this chapter is much more theoretical. It begins with a standard model of decision making and then analyzes the nature of technological decisions in terms of the postulated model. The advantage of such an approach is that it provides a clear and simple framework for both analyzing a controversy and judging its outcome. The disadvantage is that it tells us little about the actual social and political processes in the decision. Eventually we would like an account that incorporates both theoretical and empirical viewpoints.[1]

Regarding the proposed model, there are several ingredients in any decision. This chapter concentrates on one of these ingredients: scientific knowledge, particularly statistical knowledge of the type associated with studies of low-level

environmental hazards. There is no presumption, however, that statistical knowledge, or scientific knowledge generally, is the most important ingredient in any
decision. Indeed, it is an immediate consequence of the model itself that even
perfect scientific knowledge is not sufficient to produce a decision. Some representation of values is also required.

It is widely believed that the driving force in technological controversies is
a conflict over values. The implication is that such controversies would be easily
resolved if we could find some way to resolve the underlying value conflict. The
main conclusion of this chapter denies any such implication. Even if there were
perfect unanimity on all value issues, there might be sufficient uncertainty in
the scientific knowledge so as to make an effective decision virtually impossible.
In such cases any attempt to force a decision through political or other means
would likely lead to consequences that few participants in the controversy would
desire. What is needed are strategies for dealing with the uncertainty itself, for
example, ways of making minimal decisions while attempting to obtain scientific
information that would significantly reduce the major uncertainties. Such information might or might not be sufficient to overcome value differences. If it is
not, resolving the value conflict would then remain a major problem in resolving
the controversy.

In order to keep the discussion from becoming too abstract, I shall focus on
the example of nuclear power as a source of electricity. There is now a growing
consensus in the United States that decisions to invest in nuclear power were a
very costly mistake. During the 1980s only about twenty-five new plants were
completed, and more than twice that many previously planned units were canceled. And no new orders were placed. The accidents at Three Mile Island and
Chernobyl dramatized the dangers of nuclear power, but the immediate cause
of the demise of the nuclear power industry in the United States has been
economic. Indeed, by 1988 at least one public utility company had filed for
bankruptcy, something that had not happened since the 1930s. Experts estimate
the eventual cost of abandoned plants as high as $1 billion (Wald 1988). It would
be rash to claim that attention to decision theoretic details could have prevented
this mistake. But the model can at least help us better understand where we
went wrong.

A DECISION THEORETIC FRAMEWORK

Decision theory, like systems theory and operations research and like the idea
of nuclear power, came into being during World War II. Cost–benefit analysis
is among its more recent offspring. For most of this chapter, however, the
framework of classical decision theory will be sufficient.[2]

Decision theory is often presented as if it applied only to individual agents.
This type of presentation has been reinforced by Kenneth Arrow's well-known
work on the difficulties of amalgamating several different preference rankings
(that is, value rankings) into a single group ranking (Arrow 1951). Yet Arrow's
work does not show that it is impossible for a corporate entity or other collective
to have values; it shows only that these values cannot be derived from the values

of individuals in what appears to be a simple, straightforward way. In regard to the formal structure of classical decision theory, there are no restrictions on the type of agent making the decision.

For our purpose, the generality of the framework is necessary if we are to speak meaningfully of a decision without specifying the type of agent (or agency) making the decision. A more empirically oriented investigation would attempt to link the elements of the framework to various social roles, groups, and the like.

In classical decision theory, a decision problem has only a few basic components related to a set of options (that is, possible courses of action) and a set of possible consequences of the options. The rule (or decision strategy) to follow when selecting an option depends on the available knowledge concerning which consequences are likely to result from which options. Thus there is a clear separation between the decision problem itself and the role of scientific knowledge in reaching a decision. This makes classical decision theory an ideal vehicle for investigating the role of scientific knowledge in technological decisions.

Options

Any decision problem must include a set of options of which one, and only one, is to be selected. We are thus reminded at the outset that an action can be judged relative only to other, contrary, actions that might have been chosen. An adequate formulation of the problem must specify in a positive way what the contrary options might be. Moreover, for any moderately complex problem the requirement that the options be both exhaustive and mutually exclusive means that each option will in fact be a whole set of actions.

In our example we are invited to consider alternative programs for generating electricity. Indeed, given the obvious connections between electricity and other forms of energy, it is difficult to avoid formulating the options as comprehensive alternative energy programs. For example, one might begin specifying two options, as follows:

1. Make all new power plants nuclear; phase in breeders as rapidly as possible; improve safety standards for nuclear plants; and devote minimal resources to developing other energy sources.
2. Phase out nuclear and oil generators in favor of coal-powered systems; improve safety standards for remaining nuclear plants; improve emission controls for coal plants; and step up research on solar and other forms of power generation.

Formulating an appropriate set of options is clearly no easy task.

Standard decision theories cannot provide a complete account of decision making. They require only that one have some set of mutually exclusive options regarded as exhaustive. Thus decision theory itself might sanction a choice that is clearly suboptimal. Some additional principles for selecting the options themselves thus are needed. One suggestion is to expand or refine the set of options until further changes make no important differences in the resulting decision.

Implementing this rule requires attention to other elements of the overall decision problem.

Consequences

Decisions have consequences in the real world. Part of any decision problem, therefore, is an exhaustive set of mutually exclusive consequences that might result from choosing any of the options. As in the case of options, the requirement that the consequences considered in a decision problem be mutually exclusive means that these consequences are really complex sets of circumstances.

In the nuclear power debate, relevant consequences have included such things as the availability of electrical power, the cost to consumers, the risks to workers, and the risks to the general public. Breaking down these general categories would lead one to consider various possible accidents that might occur in a nuclear plant as well as such things as the possible consequences of increasing the carbon dioxide content of the atmosphere by 10 percent. A specification of one such set might begin as follows: Ample reserve capacity, average cost of ten cents per kilowatt hour, average risk to workers of one death per gigawatt year, and average risk to general public of two deaths per gigawatt year.[3]

Here again standard decision theories permit using consequence sets for which all options might be obviously suboptimal. On the other hand, it is impossible to consider all logically possible consequences. We need a principle to guide us in selecting an appropriate consequence set. And again a suggested rule is to expand and refine the consequence set as long as the new set might lead to a different decision. In general, the consequences will be irrelevant to making the decision if they have the same value and the same probability for every option. This leads us to consider these further components of the decision problem.

Outcomes and Values

In the simplest decision theories, the options and consequences are represented in a two-dimensional matrix, as shown in Figure 9.1. The elements of the matrix are thus couples consisting of one option and one possible consequence. It is these couples, called *outcomes,* that are the bearers of value. According to standard decision theories, it is not options as such, or consequences alone, that we value—but outcomes, that is, achieving a particular consequence having chosen a specified course of action.

The minimum representation of value required by decision theory is a rank ordering of the outcomes according to their relative desirability. One might object that even this minimal requirement is far too stringent in demanding a degree of comparability among different values that is totally unrealistic. More specifically, if applied at the level of social decision making, it requires that the values of different individuals and groups be reduced to a single ordering. Moreover, it requires an amalgamation of different kinds of values, for example, economic, legal, and moral. Surely these cannot all be reduced to a common denominator such as dollars.

Consequences

	Con_1	Con_2	Con_3
$Option_1$	V_{11}	V_{12}	V_{13}
$Option_2$	V_{21}	V_{22}	V_{23}
$Option_3$	V_{31}	V_{32}	V_{33}

Options

Figure 9.1   A general decision matrix (V_{12} represents the value of the outcome consisting of Option$_1$ with Consequence$_2$.)

Of course it is difficult to specify how the value rankings of different individuals may be combined in a fair and democratic manner, as Arrow's work demonstrates. But this does not show that it cannot be done. Indeed, it is done all the time. The political process does in fact yield social decisions, for example, through the actions of the federal government. One may object that this process is not sufficiently fair or democratic but not that it is impossible.

The situation is similar for the combination of different kinds of values. Individuals make decisions to which different kinds of values, such as economic or moral values, are relevant. Thus different kinds of values are in fact compared and weighed. We may wonder how this is done, but we cannot reasonably conclude that it is impossible. And if it is possible for individuals, there is no reason that it should not be possible, even if more difficult, for groups.

Although these issues are as interesting as they are difficult, they need not be pursued further here. The purpose of this chapter is to investigate the role of scientific knowledge in technological decisions. For this purpose it is sufficient that it be possible in principle to represent the relevant values compatible with the decision theoretic framework of the discussion. It is surely of interest to know the extent to which scientific knowledge can be useful in resolving certain types of controversy even in the absence of any value conflicts.

Decision Strategies and Scientific Knowledge

Imagine now that we have the decision problem adequately formulated with the options, possible consequences, and even values all in place. That is still not enough to make a decision. We also need a rule, a decision strategy, that tells us how to use the information in the matrix to reach a decision. The appropriate rule, however, depends on our knowledge concerning which consequences will, or are likely to, follow which actions. Here it is instructive to begin with the case of perfect knowledge and then consider various possible deviations from perfection.

Perfect Knowledge: Deterministic Case

Suppose there were a deterministic connection between our actions and their consequences so that any option chosen would lead necessarily to exactly one possible consequence. Moreover, suppose that we knew all these connections beyond any reasonable doubt. In this case there can be no dispute over the appropriate decision strategy. One simply compares the values of all outcomes that can be achieved by means of some option or other and chooses the option that leads to the highest valued outcome. Thus once we have added perfect knowledge to the decision problem, the correct decision depends only on our evaluation of the possible outcomes.

Looking at this ideal case in another way, we have a proof, in decision theoretic terms, of the claim that knowledge alone cannot determine the best course of action in any situation. Lacking a relative evaluation of the outcomes, a decision matrix with even perfect knowledge of deterministic relationships is incapable of yielding an optimal choice. A value input is also essential.

Perfect Knowledge: Stochastic Case

If the world were genuinely stochastic, then even perfect knowledge of the empirical laws connecting actions and consequences would yield only a distribution of probabilities over all consequences for each option. Although the appropriate decision strategy may not seem obvious, the rule almost universally advocated by decision theorists is to choose the option with the greatest *expected value*. The expected value is really a weighted average of values, with the probabilities of consequences supplying the weights.

Using the expected value rule requires more than a mere ranking of the outcomes by value. The values must also be measured on a scale that allows meaningful comparisons of differences between the values assigned to any two different outcomes. But this requirement may limit the practical usefulness of a decision theoretic approach to technological decision making, as at the moment no one knows how to construct value measures suitable to this type of case. But this need not detract from the usefulness of the model as a conceptual aid in understanding the role of scientific knowledge in technological decision making.

A more interesting issue is raised by an apparent conflict between the expected value rule and everyday intuitions. Most people would react quite differently to the possibility of two deaths per year with a probability of one-half than to the possibility of ten thousand deaths with probability 1/10,000. Yet both have the same expected value of one death per year. Despite this obvious conflict with most people's intuitions, studies of alternative energy sources use the numbers of deaths and the incidence of disease as a means of comparison. Moreover, the use of probabilistically weighted averages has become common, particularly in discussions of possible accidents involving nuclear plants. If this practice conflicts with intuition, it is not because of any defect in the expected value rule but because simple numbers of deaths, injuries, and so forth are not an adequate measure of our real values in these matters. We need better measures of the value of life and health. Without them, the comparisons now commonly made can be dangerously misleading.[4]

Imperfect Knowledge

From a formal standpoint, there is no difference between perfect knowledge of a stochastic system and imperfect knowledge of a system that might be either deterministic or stochastic. In either case all one knows are the probabilities of the various possible consequences conditional on the available options. And from a practical standpoint, there is no reason that one should not treat a system that, based on all available data, appears to be a stochastic system, as actually being such a system. The primary difference is methodological. If one's knowledge of systems that produce the various possible outcomes is only partial, one might hope, by means of further research, to uncover sets of variables that either determine the consequences or reveal the ultimate probabilities for which there are no further "hidden" variables.

Uncertainty

The empirical data may be so meager that one cannot even provide meaningful estimates of the probabilities of consequences relative to the options. In this situation there seems to be no general decision strategy that one can apply. There are, however, several special cases worth considering.

Complete Dominance. When comparing two options, sometimes every outcome associated with the first option has a higher value than does every outcome associated with the second. In this very special circumstance, the first option is clearly preferable to the second, as no matter what the true empirical laws are (whether deterministic or stochastic), the first option must yield a greater value (or expected value) than the second will. Similarly, it is possible, though correspondingly much less likely, that one option will completely dominate all the others. Such an option, if one exists, is clearly the preferred choice, as no possible further information could make any other option more desirable.[5]

This case is interesting because it demonstrates the possibility of arriving at a decision on the basis of values alone. No empirical information beyond that required to formulate the decision problem is required.

Sufficiency. Decision theory and economics have been dominated by the strategy of maximization. There is, however, a less ambitious strategy that does not necessitate our finding the "best" option but merely one that is "good enough."[6] In somewhat more precise terms, a sufficient option is one all of whose outcomes have an "acceptable" value. Thus no matter which possible consequence of a sufficient option in fact occurs, the resulting outcome will be judged acceptable. Of course this strategy requires that one be able to say what level of value is acceptable, and that is no easy task. But it is surely easier than attempting to define a uniquely best option lacking any meaningful empirical data concerning the possible consequences.

Caution. Lacking even a sufficient option, it is still possible to define an option that is best if one wishes to exercise a maximum of caution. Similarly, we can specify which action is best if one is willing to throw caution to the winds.

The best cautious option is determined by the strategy known as *minimax* (minimize your maximum possible loss) or, equivalently, *maximin* (maximize your minimum possible gain). To find the minimax option for any matrix, one simply notes the value of the lowest-valued outcome for each option. The desired option is then that option for which the lowest-valued outcome is the highest. The minimax strategy thus provides the best guaranteed floor for the decision problem. It sets a lower value limit on the result of the decision. But obtaining this guarantee may entail forgoing any possibility of achieving some very highly valued outcomes.

For those who prefer keeping their eyes on the ceiling and ignoring the floor, there is also a strategy to follow—*maximax*. To follow this strategy one simply looks for the highest-valued outcome in the entire matrix and chooses the corresponding option. This strategy guarantees one a crack at the best outcome, but of course the floor may be very far below. Indeed, it seems to be a feature of real life that the options with the highest possible payoffs tend also to be those with the lowest possible payoffs.

It is difficult to argue that caution is always the best strategy. On the other hand, when the issue involves the lives, health, and well-being of so many people, it seems extremely difficult to justify any other strategy—unless, of course, there is at least partial knowledge to relieve the uncertainty. Let us look, then, at the kinds of knowledge that might be available.

THEORETICAL SCIENTIFIC KNOWLEDGE

The idea of using controlled nuclear fission as a source of energy to generate electricity emerged from developments in nuclear physics together with the project to build a fission bomb. This required highly abstract knowledge of such things as nuclear reactions, reaction rates, half-lives, mean free paths, and capture cross sections. Though highly theoretical, the reliability of this knowledge is now beyond reasonable doubt. Moreover, such knowledge is essential to both the design of reactor cores and the analysis of possible accidents involving the core.

It is worth noting that atomic and nuclear systems are essentially stochastic systems and that the underlying theory, quantum theory, is a probabilistic theory. Moreover, physicists regard these probabilities as being objective features of the world on a par with atomic weights and electrical charge. Thus, being expressed in terms of probabilities does not necessarily lower the status of a body of scientific knowledge. Some of our most highly regarded knowledge is of this type.

If it were not for theoretical nuclear physics, we would never have had the option of utilizing nuclear power. On the other hand, many other types of knowledge are necessary for the conversion of nuclear energy into a safe and reliable source of electricity. The significance of this obvious fact has not been generally appreciated. Perhaps this is because the early champions of nuclear energy were mostly physicists who took nuclear physics as their paradigm of scientific knowledge. Within such a paradigm it is easy to overlook the diversity of knowledge required to build a nuclear power plant and to underestimate the difficulty of obtaining it (Ford 1982).

ENGINEERING KNOWLEDGE

There has been relatively little written about the general nature of engineering knowledge. This is regrettable but not so surprising, given an intellectual tradition that has always valued theoretical knowledge above all other forms. Lacking recognized categories, therefore, we are free to devise our own. The example at hand suggests two broad categories: knowledge required for the design of power plants and knowledge required for their actual operation.

Design

Questions about reactor design have always been central to discussions of nuclear power. Construction permits have been issued on the basis of design specifications, and operating licenses on the grounds that the design had been carried out according to specifications. For many in the field, questions about reactor safety have been principally questions about reactor design. This was certainly true of the first major study of reactor safety, the Rasmussen report of 1974 (AEC 1974). This study focused on the design characteristics of boiling water reactors (BWRs) and pressurized water reactors (PWRs) which include most commercial reactors now operational in the United States. The Rasmussen group early concluded that the main danger from nuclear plants to workers and the general public arises through damage to the reactor core itself and subsequently to a possible breaching of the containment structures. Their study focused on the ways the core could be damaged, especially loss-of-coolant accidents (LOCAs), and ways that there could be a subsequent containment failure. Because empirical data on such accidents were then quite limited, the Rasmussen study relied on investigations of individual components and their role in the overall design.

The Rasmussen group concluded that among the total U.S. population of roughly 200 million people, we could expect 1 death and 2 injuries per 2500 reactor years. Accidents with a significant amount of core melt but with no injuries or fatalities would occur once in 20,000 to 200,000 reactor years. These were incredibly low numbers, especially if one considers the roughly 50,000 deaths and 1.5 million injuries suffered yearly in auto accidents alone.

The accident at Three Mile Island was of the type that, according to the Rasmussen group, should have occurred only once in 20,000 to 200,000 reactor years. Yet it occurred after less than 500 reactor years. So here the report was off by a factor of between 40 and 400. Of course this is only one case, but the disparity is great enough to make it significant even in a purely statistical sense. Where might the Rasmussen group have gone wrong? One possibility, of course, is that they misjudged the probability of various possible sequences of events involving the reactor itself. But this could be judged only by an expert in reactor design. More likely the disparity was due to things not evident from the design, namely, features of the construction, maintenance, and operation of nuclear plants.

Construction, Maintenance, and Operation

There are manuals covering the construction, maintenance, and operation of power plants, both nuclear and conventional, but much of what is really known of these activities must be learned on the job. Moreover, there seems to be little in the way of statistical knowledge about what works and what does not. Thus any estimates of reliability that depend on assumptions about these matters can only be rough guesses. Here a few examples may be helpful.

At the time of the Three Mile Island incident, construction was shut down on a reactor being built on the Ohio River near Madison, Indiana. Testimony given by a former worker to environmental groups revealed that the cement in various containment buildings incorporated "honeycombs" that compromised their strength. The containment at Three Mile Island was preserved partly because the containment buildings were built extra strong owing to its proximity to the Harrisburg airport. Who really knows how many existing plants might contain improperly laid cement? The same applies to plumbing and electrical work that cannot be subjected to routine inspection.

The accident at Three Mile Island was apparently triggered by a man working on the feedwater system. The subsequent shutdown would have been routine except that two valves controlling water to the emergency pumps had been closed during maintenance several weeks earlier. This mistake was discovered too late to prevent the subsequent damage. Finally, as the Kemeny report emphasized, the accident was compounded by operator errors. Pumps forming part of the emergency core cooling system (ECCS) were manually shut off by operators who misunderstood the unusual behavior of some of their instruments. There was confusion all the way up the chain of command.

In sum, there are many engineering factors relevant to the safety of nuclear plants about which there is little systematic information. Estimates of the probability of various events involving such factors can only be guesses based on the intuitions of experienced personnel. Such judgments are important, but it is doubtful that even trained operators can distinguish probabilities of 1/1000 from those of 1/10,000 or 1/100,000. Yet it is just such distinctions that are at issue.

Beyond Engineering Knowledge

The Rasmussen report did not stop at estimating the probability of various types of accidents involving power plants. It went on to estimate the effects of these possible accidents, in particular, the health effects on workers and the general public. Clearly, such conclusions require knowledge that goes well beyond the purview of nuclear engineering. For example, in discussing the effects of a containment failure in the surrounding population, one must make some assumptions about the prevailing weather conditions at the time. How far radioactive materials will spread, and how fast, is highly dependent on the weather. But the weather, or course, is a highly variable and only partially predictable phenomenon. The use of average values when dealing with a highly variable condition can be extremely misleading.

Finally, of course, there is the question of the effect of radiation on people unfortunate enough to be exposed. There is also the parallel issue of the health effects of fossil fuel emissions. Curiously, the former have in many respects been more thoroughly studied than the latter have. In any case, enough has been written about this question to merit further consideration here.

BIOMEDICAL AND EPIDEMIOLOGICAL KNOWLEDGE

Knowledge of the effects of low-level environmental hazards is generally very difficult to obtain. Our knowledge of the effects of radiation on humans has several sources: experimental studies on animals; studies of humans exposed to various doses of radiation, especially the Japanese survivors of Hiroshima and Nagasaki; and studies of humans exposed to x-rays and other forms of diagnostic and therapeutic radiation.

The typical experimental or epidemiological study is well suited to determining causality, and from such studies we know that radiation has serious health effects, particularly cancers such as leukemia, lung cancer, and cancer of the thyroid. But to estimate the health effects of various possible nuclear accidents, we need to know the response rate of the population at risk to various levels of exposure, particularly low levels. This is far more difficult to determine than is the simple fact that a causal relationship does exist. Let us look first at the types of studies used to determine causality and then consider the special problem of determining the dose–response relationship for low-level doses.

Simple Causal Hypotheses

There is still considerable controversy, in both philosophical and scientific circles, about the nature of causal hypotheses. Because this is no place to engage in such controversies, I shall simply employ an account that seems particularly well suited to the issues at hand.[7]

Causality is best thought of as operating in individual systems, for example, individual human bodies. The internal, dynamic structure of any individual system may be either deterministic or stochastic. If it is deterministic, then the hypothesis—for example, that a specified dose of radiation would produce leukemia in a particular individual—means that given the radiation as input, the deterministic causal connections would lead to the development of leukemia in that system and, moreover, that the system as it is without that input would not, by itself, produce the same result. For a stochastic system we say that the radiation is a causal factor for the production of leukemia if it raises the probability of that result above what it would otherwise have been in that particular person.

Although causality operates at the individual level, it is impossible at present to study it at that level in systems as complex as a human body. The best we can do is to study populations of similar systems. For a given population we can say that a specified dose of radiation is a causal factor for leukemia in that population if there are some individuals in the population for whom the radiation is either a causal determinant or a stochastic factor for leukemia. Thus, when dealing at the population level it generally matters little whether we imagine the individuals to be deterministic or stochastic. In either case we say that there would (or very probably would) be more cases of leukemia if every member were so exposed than if none were. In investigating a causal hypothesis, we are trying to discover whether or not this counterfactual statement is true of the population in question.

Only rarely, as in a census, does anyone even attempt to examine the whole population of interest. The individuals studied generally constitute only a sample from the population of interest. Thus one automatically faces the standard problems in obtaining a sample that is at least very likely to be representative of the population in the respects under investigation. In addition to questions about sampling are important issues of experimental design.

The strongest evidence of the existence of a genuine causal factor is provided by *experimental* studies.[8] Ideally one begins with a large sample, divides it randomly in two, and then administers the causal agent to one group while keeping the other uncontaminated. If the incidence of the effect in the two groups is significantly different, that is good evidence of a causal connection. There have been many experiments of this type in which laboratory animals, such as rats, were subjected to varying amounts of radiation. The results show quite conclusively that radiation causes death, leukemia, and a host of other effects—in rats.

There are two reasons that these studies are of only limited usefulness in determining the effects of low-level radiation on humans. First, of course, rats are not humans. The population studied is not the population of interest. And although one can often be confident that the cause would produce similar effects in humans, it is very difficult to be very precise about the rate. But it is the rate we need to know. Second, in order to have any chance of observing a statistically significant difference in the incidence of the effect, fairly large doses must be used. With low doses the incidence of the effect is so low that one would need literally thousands of subjects in order to get enough cases for statistical significance. Moreover, the difficulties of maintaining such large numbers of animals introduce the worry that the effect is due to some other factor involved in running the experiment. So not only do we face the difficulty of inferring from rats to humans, but we also must extrapolate from large doses to small ones. The resulting margin of error is so great as to make the results relatively useless for the type of decision at issue.

Although it is disturbing to think in these terms, the atom bomb attacks on Japan created the conditions for a large-scale experimental study of the effects of radiation on humans. It is easy to find other Japanese cities whose populations were similar to that of the affected populations, thus providing an ideal control group. In fact, however, a systematic study of the survivors was not begun until 1950. Because it is very hard to reconstruct information about the people who moved or died in the intervening five years, it is highly unlikely that any sample selected in 1950 could be representative of the original survivors. Indeed, one would expect that on the average, the population remaining in 1950 would be more robust than the original was. These studies are also complicated by the fact that we have only very rough estimates of just how much radiation any person actually received. Thus although these studies provide much useful information about the long-term effects of radiation exposure, they fail to yield precise information about the response rate to specified doses.

A *prospective* study of workers at a government nuclear facility could provide evidence of greater relevance to these questions. The fundamental weakness of prospective, as opposed to experimental, studies is that there is no deliberate random assignment of subjects into experimental and control groups. Rather, the subjects select themselves into the "experimental" situation, in this case by becoming workers at a particular facility. On the other hand, it is difficult to imagine any reason that people who decided to seek employment at a particular plant should be any more or less susceptible to the effects of radiation than would others living in the same region. In general, then, a prospective study could be very relevant to this sort of case because it focuses on the effects in a sample of ordinary people who have been subjected to fairly well documented low levels of radiation for long periods of time.[9]

The Dose–Response Curve

There are few direct data on the long-term response of humans to low levels of radiation. Claims about low-level effects are based primarily on extrapolations from known responses to higher levels. The validity of these extrapolations clearly depends on the nature of the relationship between dose levels and response rate. There are three recognized general types of dose–response relationship, represented by three differently shaped curves in a dose–response diagram, as shown in Figure 9.2.

When the biological effects of radiation were first studied early in this century, it was widely believed that there was a threshold dose below which there was no effect at all. Curve b in Figure 9.2 represents a dose response curve with a threshold. For a single stochastic individual, doses below the threshold would produce no increase in the normal probability for any of the effects observed at high doses. If one supposes that the threshold for different individuals varies, then one also must assume that the most sensitive individual in a given population has a threshold above the zero dose level. Otherwise there would be no threshold for the population as a whole. Similarly, for deterministic individuals, the population threshold is the lowest dose that would trigger a response in any member. Response rates for higher doses then reflect the percentage of the population in which members have an individual threshold between the minimum threshold and the given higher dose.

Studies since World War II have failed to reveal the assumed threshold. Thus, by the 1970s it became common practice to assume a *linear* dose response relation, as shown by curve c in Figure 9.2. Not that anyone believes the response really to be linear. Rather, it has typically been assumed that the curve is less than linear, so that the linear assumption represents a "conservative" extrapolation.

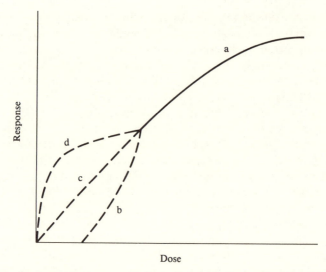

Figure 9.2 A dose response curve: (a) the relationship for moderate to high doses, (b) the low-dose extrapolation with a threshold, (c) the linear extrapolation, and (d) a supralinear relationship.

Recent evidence suggests that the linear assumption may not be conservative at all.[10] The real relationship instead may be *supralinear*, as illustrated by curve d in Figure 9.2. One hypothesis suggested to explain recent experiments is that biological damage (explained biochemically) is a function of the dose rate as well as the dose itself. Moreover, for a particular dose, the damage, according to this model, is greater for lower dose rates than for high rates. If this suggestion is confirmed, the potential damage from low-level radiation may be much greater (perhaps by a factor of 10 or 100) than most authorities have believed.

The Official Conclusions

The official view of the health hazards of radiation is that of the National Academy of Sciences–National Research Council, Advisory Committee on the Biological Effects of Ionizing Radiation (BEIR). By 1980, this committee had issued three reports: BEIR I, 1972; BEIR II, 1977; and BEIR III, 1979. The Rasmussen group used the conclusions of BEIR I in its study.

Among the conclusions of BEIR III are the following: Assume a dose of 1 rem/year delivered as a one-time whole-body dose, which is ten times the average dose, owing to all forms of background radiation. For a population of 1 million men and women the assumed dose would produce, they said, 70 to 353 deaths.[11] In addition, every million live births by parents so exposed would include, they said, 5 to 75 serious genetic disorders caused by the exposure. These numbers exhibit a great deal of uncertainty, as is expected, given the kinds of information available. There is typically a factor of five between the lower and the upper figure, and in some cases the factor is ten or fifteen.

Another remarkable feature of these figures is their comparatively low values.

Using 200 deaths per million exposed people as a convenient average, this is a rate of 25 per 100,000 over a lifetime. By contrast, 50,000 auto deaths a year in a population of 200 million works out to an annual rate of 25 per 100,000. And this is for an assumed dose ten times background and twice the EPA's limit for the general population (although only one-fifth the limit for occupational exposure).

What About Coal?

There have been many studies of the occupational hazards of mining coal, and there have also been studies of the health effects of air pollution. Until the mid-1970s, however, there were no systematic investigations into the health hazards of coal- (or oil-) fired power plants. Moreover, the recent studies of this problem seem to have been undertaken by the nuclear establishment in response to questions about the hazards of nuclear power. The objectivity of such studies, like some studies prepared by the antinuclear lobby, is questionable. Still, some features of the debate are worth noting here.

One healthy feature of the later literature is that it no longer considers just the operation of the power plant itself but the whole fuel cycle: mining, processing, transportation, actual use, and waste disposal. If one is to compare nuclear and coal-fired power plants, whether for economics or health hazards, one must surely take account of the complete fuel cycle. Yet the problems in attempting such studies are enormous.

The hazards of coal mining have been well documented. Similarly reliable data for uranium miners seem not to be available, although a Department of Health, Education and Welfare report on occupational diseases states that as many as one-sixth of all uranium miners will eventually die of radiation exposure (Lowrance 1976, p. 147). There are good data about the numbers of people killed at railroad crossings by trains hauling coal to power plants, but few data on the hazards of transporting nuclear fuel. Conclusions about the health effects of normal emissions from both types of plant are equally uncertain. Furthermore, there is much information about the disposal of ash, but very little about the disposal of nuclear wastes.

In light of these uncertainties, it is surprising to find a 1977 draft report that compares the excess mortality per .8 gigawatt year for coal and nuclear plants. Including both workers and the general public, the numbers given are .47 for nuclear plants, and 15 to 120 for coal (NRC 1977). Most of this difference, roughly 10 to 100, is due to "respiratory failure among the sick and elderly from combustion products from power plants." Lost in the body count is the important value issue of the relative seriousness of the death of an elderly person already suffering from respiratory disease, compared with the cancer death of a healthy young person or the occurrence of a genetic defect in a child.

SOCIAL, ECONOMIC, AND
INSTITUTIONAL KNOWLEDGE

Any comprehensive energy policy will have significant social and economic consequences that are clearly relevant to the decision problem. Knowledge of these consequences is thus required for a good decision. For example, after the first energy crisis of 1973, the demand for electricity began dropping. After a generation in which demand grew at a fairly steady 7 percent a year, thus doubling in ten years, it fell to less than 3 percent in 1979 (Parisi 1980). Not only did the utility companies not foresee the drop, but they also were very slow to realize that the drop was not merely a temporary fluctuation and so continued planning new plants assuming the old growth rate. The result was that by 1980 there was an excess reserve peak capacity of roughly 10 percent and a total reserve peak capacity of roughly 30 percent. It is clear that reliable, comprehensive knowledge of social and economic variables is every bit as important as are the other types of scientific knowledge we have considered.

Finally, there seems to be a serious need for knowledge of the behavior of institutions as well as of individuals. For example, in determining the probable health effects of a nuclear accident, the Rasmussen group assumed that in the event of a really serious accident, people downwind for a radius of twenty-five miles would be evacuated. This assumption contributed to their generally low estimates of death and disease from a containment failure. But one of the lessons of Three Mile Island was that although some evacuation was probably in order, the institutional machinery for making the decision to evacuate and to carry it out simply was not in place. And a later investigation showed that most other populated areas that encompassed nuclear plants did not have adequate evacuation plans. Knowledge of likely institutional failure is as important as is knowledge of likely equipment failure.

One final example: The relatively poor performance of both equipment and personnel at Three Mile Island could at least be partly blamed on the fact that the company rushed to get the plant on line before December 31 in order to gain a significant tax advantage. Knowledge of the operation of various institutional structures thus is relevant to an evaluation of the performance of technological enterprises. Similar comments apply to the whole regulatory process as well.

In sum, in the operation of any technology, the "software" is as important as is the hardware. Thus in evaluating any technology one needs to know as much about the software as about the hardware.

WHAT KIND OF A DECISION PROBLEM WAS IT?

Having surveyed the range of scientific knowledge both available and relevant, what can we conclude about the nuclear power debate? What role might the available scientific knowledge have played in the controversy?

At first sight the situation appears to have been a clear case of decision making with uncertain knowledge. But the problem was not so simple as that. For many possible consequences of the options, the available knowledge provided only a wide range of probabilities, such as those regarding genetic damage due to radiation, whereas for other consequences there were effectively no probabilities at all. Even assuming that one could have attached a single numerical value to each possible outcome, the calculations of expected value would have yielded a wide range of expected values for many quite different options.

Moreover, it is likely that many of these ranges would have overlapped to a considerable extent, leaving no effective means of discriminating among such options. Thus even though there was much knowledge available, it may still have left one operating in a situation of uncertainty with regard to many, importantly different, options.

This perception helps explain why neither side was willing to concede in the face of scientific evidence offered by the other side. Even if we put all of the evidence together and even if we assume no value differences among the two sides, there still remained pronuclear and antinuclear options whose ranges of expected value broadly overlapped. But the true situation was one of effective uncertainty, and in such a situation there are few effective decision rules available.

Dominance

We might construe some antinuclear proponents as having claimed that non-nuclear options dominate nuclear ones. A simple version of this argument has only three types of relevant consequences: availability and cost of electricity, low-level health consequences, and catastrophic consequences. It was, for a long time arguable that both options could provide sufficient power at a comparable cost. Similarly, given all the uncertainties, it was not clear that either option would have worse low-level health effects during normal operations. But nuclear plants are subject to catastrophic failure, whereas coal plants and other sources are not. Thus, no matter how low the probability of a catastrophic failure is, we would be better off with a nonnuclear option that has zero probability for such a negatively valued outcome.

This would have been a fairly good argument were it not for such possibilities as a carbon dioxide catastrophe, that is, the continued burning of fossil fuels, leading to a significant increase in atmospheric carbon dioxide, leading to heating through a greenhouse effect, leading to some melting of the polar ice caps, and so on. One might have replied that such a sequence is improbable, but in so doing one would be retreating from a position of dominance into the morass of uncertain knowledge.

Sufficiency

Another popular antinuclear argument can be understood as an appeal to the principle of sufficiency. This argument concedes that a nuclear option might turn out to be best but insists that it might also turn out to be much worse than other options that are "good enough." It would also be a powerful argument if there were some options that had no possibly disastrous consequences, as a disaster, almost by definition, could not be satisfactory. Perhaps some option combining coal, conservation, and the vigorous development of renewable resources could be genuinely satisfactory.

Minimax

Even if no option has satisfactory outcomes for all possible consequences, still, it has been argued, the worst outcomes with a nonnuclear option are better than the worst outcomes with nuclear options. A cautious, minimax strategy would thus recommend a nonnuclear option. Once again, however, this argument works only as long as there are no potentially disastrous consequences, such as a carbon dioxide disaster, associated with the nonnuclear options. As it is, some nonnuclear consequences could be as bad as, or even worse than, the worst nuclear disaster.

Sequential Strategies

We have yet to consider the possible advantages of *sequential* decision making. This approach requires the formulation of options that can be carried out sequentially in such a way that information gained at earlier stages can be used to improve decisions at later stages. It is thus crucial that sequential options have an information-gathering component. Of course such options are not cost free. Research costs money, and one might end up later with an outcome that could have been achieved earlier—and the delay itself may be costly. But one might avoid a course of action that turns out badly and is very expensive to reverse at a later date.

Any sequential option must include research that can contribute to the formulation of new options—for example, options involving renewable resources, such as solar energy, and decentralized generation. In considering new alternatives, attention needs to be directed at the "end-use problem." For some applications, such as running electric motors, electricity is absolutely essential. But there are other applications for which it is neither essential nor efficient, such as space and water heating. For such applications, solar and other on-site installations may be more efficient. Moreover, as is clear from our earlier discussion, it is not merely the hardware for new options that needs to be developed. The economic, as well as the social and institutional, aspects of new options require equal attention.

Other research must be directed at the consequences of options under consideration. For example, because early estimates of the probability of a serious nuclear accident were apparently mistaken, more should have been known about these possibilities, particularly the operational and institutional factors in such accidents. Similarly, more should be known about air pollution from fossil fuels and about the possibility of a carbon dioxide catastrophe. And more should be known about the effects of low-level radiation, although, as we pointed out earlier, the difficulties in obtaining the relevant dose–response rates are enormous.

In sum, it is possible that a carefully sequenced strategy, including new research and the evaluation of existing information, could sufficiently have reduced the uncertainty so as to have enabled a fairly clear-cut resolution of the issue—unless, of course, the value conflicts remained too severe.

CONCLUSION: KNOWLEDGE AND VALUES

Given a set of options and consequences, the two remaining ingredients required to apply any decision strategy are knowledge and values. In this chapter we have been primarily concerned with the role of knowledge in technological decisions.

Regarding the example of nuclear power, our conclusion is that the variety of knowledge required has been so great that the resulting uncertainty left no clear-cut decision rule. Thus, even in the absence of any value conflicts, it is difficult to defend any particular decision. Arguably, the best strategy would have been to seek sequential options that would have delayed final commitments while further information was being gathered.

Although principally theoretical, a decision theoretic analysis suggests empirical questions. If there was indeed no clear-cut decision theoretic basis for deciding in favor of nuclear power, why was that option chosen by those in positions to make the decision? The analysis suggests several general possibilities. One is that those in policymaking positions regarded the available knowledge as being more definitive than it in fact was. Another is that the relevant policymakers put a very high value on nuclear options, a value high enough to permit nuclear options to dominate nonnuclear options in their thinking. I would speculate that both of these possibilities were operative.

The period following World War II was one of tremendous optimism regarding the power of science to solve technical problems. Policymakers may simply have been overconfident in the ability of science to provide answers when they were needed. They assumed the knowledge was there or that it would be readily forthcoming.

On the value side, there was a general belief that civilization had entered the "atomic age." Nuclear power was part of that ideology. There was also a high value placed on being in the technological forefront of this new age. Thus, managers of public utilities may simply not have wanted to be left behind technologically. This value commitment could override the uncertainties so as to make the nuclear option appear obviously correct.

There is also a more sinister possibility (Ford 1982). Nuclear power was originally subsidized quite heavily by the Atomic Energy Commission (AEC). There were strong connections between the AEC and the Defense Department. The Defense Department had a very strong interest in nuclear power's being accepted by the general population so that there would be correspondingly less fear of nuclear weapons, which were becoming the basis of strategic military planning in the postwar period. Thus values having little to do with electric power as such may have played a large role in decisions promoting the nuclear option. The accuracy of these speculations will have to await, however, more comprehensive histories of nuclear power in the United States.

ACKNOWLEDGMENT

This chapter was originally prepared as part of the Hastings Institute's project on the resolution of scientific and technological controversies. My thanks for providing the opportunity to explore these issues go to the director of that institute, Daniel Calahan; the director of the project, Art Caplan; and the many other participants in the project. The support of the National Science Foundation for research on decision making in science and technology is hereby gratefully acknowledged.

NOTES

1. Mazur (1981) provides a sociological framework for analyzing technical controversies. For more extensive presentations from a decision theoretic perspective, see Elster (1983) and Fischoff et al. (1981).

2. For further background in decision theory, see any standard text such as Luce and Raiffa (1957, chap. 13). For a brief, elementary treatment of the essentials, see Giere (1984, pt. iv). This section owes much to discussions with Lewis Gray and to his dissertation (Gray 1980).

3. A gigawatt year is the amount of power generated by a 1000-megawatt plant operating for one year. This is roughly the capacity of the existing nuclear generators.

4. A similar point is often made regarding economic decision making and the utility of money. A person with a total fortune of $1000 might be quite willing to wager $1 against a fifty–fifty chance of receiving $2 in return but be unwilling to risk the whole thousand in a similar wager. The true value of money for real agents is not measurable simply as numbers of dollars.

5. The notion of "complete dominance" used here is stronger than the notion of dominance usually employed in decision theory. The standard notion assumes that options and consequences are independent in the sense that which option is chosen makes no difference to which consequence occurs. The notion of complete dominance applies even if options and consequences are dependent, as we hope they are in most technological decision problems.

6. The principle of sufficing as a decision-making strategy is due largely to Herbert A. Simon. See, for example, Simon (1957, chaps. 14 and 15; 1983).

7. The assumed model of causal hypotheses is developed more fully in Giere (1979). For a more elementary presentation, see Giere (1984, chap. 9).

8. For an elementary exposition of the differences between randomized experimental studies and prospective designs, see Giere (1984, chap. 12).

9. A study of just this type was begun in 1964 at the government nuclear facility in Hanford, Washington, under the direction of Professor Thomas Mancuso of the University of Pittsburgh. But federal funding for the project was cut off in 1978, ostensibly because of methodological flaws in the design of the study.

10. For a discussion of recent experiments, see Bockris (1977, chap. 15).

11. In BEIR I the comparable figure was 68 to 293, showing that new evidence and reflection do not necessarily decrease uncertainty.

REFERENCES

Arrow, K. J. (1951). *Social Choice and Individual Values*. New York: Wiley.
Atomic Energy Commission (AEC) (1974). *An Assessment of Accident Risks in U.S. Commercial Nuclear Power Plants*. Washington, D.C.: AEC no. WASH–1400.
Bockris, J. O'M. (ed.) (1977). *Environmental Chemistry*. New York: Plenum.
Elster, J. (1983). *Explaining Technical Change*. Cambridge: Cambridge University Press.
Fischoff, B., Lichtenstein, S., Slovic, P., Derby, S. L., and Keeney, R. L. (1981). *Acceptable Risk*. Cambridge: Cambridge University Press.
Ford, D. (1982). *The Cult of the Atom*. New York: Simon and Schuster.
Giere, R. N. (1979). "Causal Systems and Statistical Hypotheses." In *Applications of Inductive Logic,* ed. L. J. Cohen and M. B. Hesse, pp. 251–70. Cambridge: Cambridge University Press.

Giere, R. N. (1984). *Understanding Scientific Reasoning.* 2nd ed. New York: Holt, Rine-
    hart and Winston.
Gray, L. (1980). "A Decision Theoretic Model of Congressional Technology Assess-
    ment." Ph.D. diss., Indiana University.
Lowrance, W. W. (1976). *Of Acceptable Risk.* Los Altos, Calif.: Kaufman.
Luce, R. D., and Raiffa, H. (1957). *Games and Decisions.* New York: Wiley.
Mazur, A. (1981). *The Dynamics of Technical Controversy.* Washington, D.C.: Com-
    munications Press.
Nelkin, D. (ed.) (1979). *Controversy: Politics of Technical Decisions.* Beverly Hills, Calif.:
    Sage.
Nuclear Regulatory Commission. (1977). *Health Effects Attributable to Coal and Nuclear
    Fuel Cycle Alternatives.* Washington, D.C.: NUREG no. 0332.
Parisi, A. J. (1980). "Electricity Use No Longer Soaring." *New York Times,* April 6,
    p. 1.
Simon, H. A. (1957). *Models of Man.* New York: Wiley.
Simon, H. A. (1983). *Models of Bounded Rationality.* 2 vols. Cambridge, Mass.: MIT
    Press.
Wald, M. L. (1988). "Nuclear Plant Drain Put at $100 Billion for U.S." *New York Times,*
    February 1, p. 21.

10

Causing Harm: Epidemiological and Physiological Concepts of Causation

KENNETH F. SCHAFFNER

In this chapter I shall examine the relations between what appear to be two somewhat different concepts of causation that are widely employed in the biomedical sciences. The first type is what I term *epidemiological causation*. It is characteristically statistical and uses expressions like "increased risk" and "risk factor." The second concept is more like the form of causation we find in both the physical sciences and everyday life, as in expressions such as "the increase in temperature caused the mercury in the thermometer to expand" or "the sonic boom caused my window to break."[1] In the physical and the biological sciences, such claims are typically further analyzed and explained in terms of underlying mechanisms. For example, accounts in the medical literature of cardiovascular diseases associated with the ischemic myocardium typically distinguish between the risk factors and the mechanisms for these disorders (Willerson 1982).

Interestingly, both concepts of causation have found their way into the legal arena, the first or epidemiological concept only relatively recently in both case law and federal agency regulatory restrictions. The second, perhaps more typical, notion of causation has turned out to be not so simple on deeper analysis and led Hart and Honoré, among others, to subject the notion to extensive study in their classic book *Causation in the Law*. In another paper (Schaffner 1987), I examined some of these issues, in particular the epidemiological concept of causation as it might apply to recent DES cases such as *Sindell* and *Collins*.[2]

Reflections on the *Sindell* case and on one of its legal precedents, the *Summers v. Tice* case, led Judith Jarvis Thomson to introduce a distinction between two types of evidence that might be adduced to support a claim that an agent caused harm to a person. The two types of evidence parallel the distinction between these two concepts of causation, and I shall introduce them by means of a particularly striking example originally credited to David Kaye (Kaye 1982). Thomson, quoting Kaye, writes:

> There are two taxicab companies in town who have identical cabs except that one has red cabs and the other green cabs. . . . Sixty percent of all cabs in the town are

red. Plaintiff has been knocked down on a deserted street by a cab. He is color blind and cannot distinguish red from green. Shortly after the accident a taxicab driver said over the air, "I just hit someone (at the accident location). I should have seen him, but I was drinking and going too fast." The static was such that no identification of the voice was possible, but the frequency is used only by cabs from that town. Neither company keeps dispatch records so [neither knows] what drivers were in what parts of town. And to make things complete, the garage in which all of the cabs of the two companies [were] housed was burned down the night of the accident. (1984, p. 127)

Thomson adds that "on the facts, Red Cab Company is 0.6 probably the harm causer" and "the familiar 'more probable than not' rule of tort law yields that Red Cab should be held liable for all of the plaintiff's costs" (p. 127). Thomson argues, however, that our intuitions suggest that such a judgment would be unfair. The hypothesis, "(H) A red cab caused the accident," is, according to Thomson, supported in an importantly different way by the external type of statement: "(E) 60% of the cars which might have caused the accident are red cabs," in contrast with the internal type of statement: "(I) A red cab went speeding away from the area moments after the accident."

For Thomson, the key difference between statements (I) and (E) is that the internal statement stands in an "explanatory relation" with the hypothesis (H). That is, the truth of the hypothesis would explain the truth of the internal evidence statement (I), whereas the truth of (H) would not explain the truth of the external statement (E). Further, this is not a matter of temporal asymmetry, as (I) could just as well have been rewritten as "A red cab went speeding into the area moments before the accident." In this case the statement (I) would explain the truth of (H); thus an explanatory relation still holds, but in a more complex manner (it is opposite). Thomson is troubled by this internal–external evidence distinction but thinks it important, primarily because we are disinclined to impose liability unless we have warrant for a causal connection between the tort feasor's act and the harm to the plaintiff, and such causation is not statistical or external but is internal in the preceding sense.

I have spent a good bit of time discussing the Kaye–Thomson example because I think it—and Thomson's distinction between internal and external evidence—is a conceptually significant one, although as I shall argue toward the end of this chapter, it may not make much difference with respect to public policy if an insurance model or a superfund model is adopted to deal with liability and compensation. In order to look somewhat deeper at the difference(s) between a statistical concept of causation and a prima facie nonstatistical concept, I shall turn now to consider how two groups of scientists, the epidemiologists and the physiologists, characterize causation.

I shall begin by citing some of the general epidemiological literature on the epidemiologists' conception of cause. This will lead to a philosophical analysis of that concept of causation, which can then be used to demonstrate the differences between the epidemiologists' approach and the concept of causation we find in physiological (and pathophysiological) mechanisms.

THE EPIDEMIOLOGISTS' NOTION OF CAUSE

Several different groups of epidemiologists such as MacMahon and Pugh (1970), Fletcher, Fletcher, and Wagner (1982) and also Kleinbaum, Kupper, and Morgenstern (1982) have put forth the interesting notion that epidemiologists use a different notion of cause than do, for example, physiologists. MacMahon and Pugh suggested that "a causal association may usefully be defined as an association between categories or events or characteristics in which an alteration in the frequency or quality of one category is followed by a change in the other" (1970, pp. 17–18). Fletcher and colleagues similarly wrote: "When biomedical scientists study causes of disease, they usually search for the underlying pathogenetic mechanism or final common pathway of disease" (1982, p. 190). Although these authors indicate their agreement with the importance of this approach to causation, they add:

> The occurrence of disease is also determined by less specific, more remote causes such as genetic, environmental, or behavioral factors, which occur earlier in the chain of events leading to a disease. These are sometimes referred to as "origins" of disease and are more likely to be investigated by epidemiologists. These less specific and more remote causes of disease are the risk factors. (1982, p. 190)

Finally, Kleinbaum and associates write:

> In epidemiology, we use a probabilistic framework to assess evidence regarding causality—or more properly to make causal inferences. . . . But a probabilistic viewpoint does not automatically negate our belief in a (modified) deterministic world. . . . In other words, we need not regard the occurrence of disease as a random process; we employ probabilistic considerations to express our ignorance of the causal process and how to observe it.
>
> Because of the lack of certainty in our results, epidemiologists generally use the term risk factor instead of cause to indicate a variable that is believed to be related to the probability of an individual's developing the disease prior to the point of irreversibility. (1982, p. 29)

These quotations raise the provocative thesis that the notion of causation may be in some methodological and/or substantive sense different for the epidemiologist. If this point of view can be substantiated, it may help us make sense of some of the epidemiologists' other comments concerning the criteria by which they assess causal claims and assist us in sharpening our analysis of inference from epidemiological studies. In order to answer this question, we must examine a notion of causation developed by several philosophers known as *probabilistic causation*. With the aid of this notion, I believe we can make some progress in clarifying the differences between the way in which epidemiologists, as opposed to other biomedical scientists, approach the notion of cause.

PROBABILISTIC CAUSATION

The notion of epidemiological causation appears to me to be superficially similar, though in an important sense not identical, to what philosophers have termed probabilistic causation. The pioneer in this area is Hans Reichenbach, who characterized the notion of *causal betweenness,* used to establish a linear causal sequence and a time order, in the following way (1956, p. 190):

An event B is causally between the events A and C if these relations hold:

$$1 > P(C|B) > P(C|A) > P(C) > 0 \qquad (10\text{--}1)$$

$$1 > P(A|B) > P(A|C) > P(A) > 0 \qquad (10\text{--}2)$$

$$P(C|A \ \& \ B) = P(C|B) \qquad (10\text{--}3)$$

where the symbol | is used to introduce conditional probability; for example, $P(C \mid B)$ is read as the probability of C given that B is the case. These relations license the causal chain $A \rightarrow B \rightarrow C$ where the arrow $\rightarrow$ denotes probabilistic causation.

Other philosophers of science, such as Good (1961–62) and Suppes (1970), have developed similar ideas, and still others have provided extensive criticism of such concepts, among them Salmon (1980). Suppes's approach to probabilistic causation is perhaps the most widely known and is worthy of a few comments.

Suppes (1970, p. 12) introduces his notion of a probabilistic cause through the idea of a prima facie cause, defined as follows:

The event $B_{t'}$ is a prima facie cause of the event A_t if, and only if,

1. $t' < t$
2. $P(B_{t'}) > 0$
3. $P(A_t|B_{t'}) > P(A_t)$

where t denotes time.

Further elaboration of the notion is required, so that "spurious" causes occasioned by some common cause that produces an artifactual association between two variables are eliminated, and such extensions of the concept can be found in Suppes's 1970 work. Both Reichenbach's and Suppes's proposals still encounter a number of problems, however, including that of the improbable alternative causal path, an issue about which I cannot comment here but about which I have written elsewhere (Schaffner 1983). There also are other conceptual difficulties associated with probabilistic causation, and for an incisive review of these, see Salmon (1980) as well as Cartwright (1979), Skyrms (1980), Otte (1981), Eels and Sober (1983), and Sober (1984, chap. 7).

I shall now turn to a somewhat different approach to the notion of probabilistic causation that I believe is clearer and ultimately more coherent than is Suppes's account. My approach will begin from some useful suggestions made by Giere (1980) concerning the relationship between determinism and probabilistic causation in populations and then examine some of Cartwright's (1979, 1986a, 1986b) views on Simpson's paradox. In particular I shall argue that it is the feature of nonidentical populations that is encountered as part of epidemiological research and inference that introduces the key difference into the account of causation ascribable to such systems.

Some useful distinctions appear in Giere's account that are not explicit in most of the discussions of probabilistic causation with which I am familiar. First, Giere differentiates between deterministic and stochastic systems but permits both types to exhibit probabilistic causal relationships. This is important because it distinguishes two different ways in which determinism may fail to hold for populations. This is useful for my account because it allows a more coherent fit between causation in physiologically characterized biological systems and those that we approach from an epidemiological perspective, though it does not exclude stochastic components in physiological processes. In physiology or pathophysiology we are usually dealing with deterministic systems (at least in ideal cases), whereas in epidemiological contexts, a more probabilistic or stochastic characterization of causation is evident. Giere also invokes a "propensity" interpretation of probability for stochastic systems and, in particular, maintains that this offers a means of directly attributing probabilistic causality to individuals rather than to populations. In my view the propensity interpretation is needed only for irreducible singular nondeterministic causation, which, though it may have much to recommend it in quantum mechanics, is less clearly demanded in biomedical contexts. I favor a frequency interpretation of probability, which I maintain is more natural in epidemiological contexts in which the application of epidemiological conclusions to individuals is concerned, and although I will return to this issue later, it is not a point that I can pursue in any depth here. I shall begin by indicating what elements of Giere's approach I find useful.

In Giere's analysis of deterministic systems, a causal relation between C and E is not necessarily a universal relation. The entities in Giere's account are individual deterministic systems that can differ in their constitution, so that different individuals with the same causal input C may or may not exhibit E. Furthermore, because E may come about from a different cause than C does, some individuals may exhibit E but not have had C as a causal input. An example that Giere often uses is smoking and lung cancer, a useful example for my purposes, as the evidence for such a claim is largely epidemiological, but the same general point could be made using any one of the many cardiovascular studies of risk factors.

For a population of deterministic systems, some number of individuals with input C will manifest $E,$ and some will not. An actual population may yield a relative frequency:

$$\#E/N$$

where N is the number of the individuals in the population and $\#E$ is the number of individuals exhibiting effect E. This fraction has the properties of a probability (though not a propensity), because for any given individual in the population, a universal law $L(C) = E$ is either true or false, depending on their individual constitutions.

Giere prefers to use counterfactual populations for which the outcomes are well defined, and he contends that this idea is what one finds in the typical randomized clinical trial, a point with which I shall disagree later. Thus, by hypothesis, two counterfactual populations that are counterfactual counterparts of an actual population of interest are envisaged, and one is provided with causal factor input C and the other with input $\sim C$. Each counterfactual population exhibits some number of effects $\#E$ that is $<$ its N. Then Giere defines (1980, p. 264) a positive causal factor as

    C is a positive causal factor for E in [population] U if, and only if,

$$P_c(E) > P_{\sim c}(E)$$

A reversed inequality defines a negative causal factor, and equality indicates causal irrelevance.

Note that this notion is almost identical to the definition given earlier for a prima facie cause in Suppes's system. What is different is the interpretation in the context of explicitly deterministic systems, and an explicit counterfactual account.

Giere also introduces a measure of effectiveness of C for E in population U,[3] namely,

$$Ef(C,E) = P_c(E) - P_{\sim c}(E)$$

These notions developed in the context of deterministic systems can be extended to stochastic systems, about which Giere observes, "The interesting thing about stochastic systems is that we can speak of positive and negative causal factors for individual systems and not just for populations" (1980, p. 265). Giere's counterfactual definition (again we say counterfactual because system S may or may not actually have C or $\sim C$ as input) is that

C is a positive causal factor for E in S if, and only if,

$$Pr_c(E) > Pr_{\sim c}(E)$$

where now Pr refers to a probability in the sense of a propensity (an inherent objective tendency) of the stochastic individual system. Although this extension of Giere's to stochastic systems is interesting and may well, in a fuller treatment of epidemiological causation, be significant, it is not essential to my purposes, and thus, I shall not discuss several measures of effect for such systems.

As I have noted, Giere prefers to use counterfactual populations for which the outcomes are well defined, and he maintains that this situation is what one finds in the typical randomized clinical trial. But this is incorrect on Giere's part, as such a trial uses relevantly similar actual populations for which it hopes to control interfering factors (1) by matching and stratification (if the factors are known or suspected to operate) and (2) by randomization (for all other unknown interfering factors). Thus a clinical trial hopes at best only to approximate the major feature of such counterfactual test populations, in that the investigator hopes to have only specified and testable relevant differences between experimental and control groups. The appeal to counterfactuality, however, is essential if Simpson's paradox is to be avoided, and it is precisely this type of appeal that points toward the difference between the epidemiologists' notion of causation and the probabilistic causal notion.

Several essays by Cartwright (1979, 1983) have focused on the counterfactual foundation underlying a probabilistic conception of causality. Cartwright begins her 1979 paper by citing a paradox known as Simpson's paradox (or the Yule-Cohen-Nagel-Simpson paradox to give all who have identified it their due). This paradox shows that any association that holds between two variables in a population that can be used for a relation of probabilistic causation of the type characterized by Reichenbach (1956) and Suppes (1970) "can be reversed in the subpopulations [of that population] by finding a third variable which is correlated with both" (Cartwright 1979, p. 422). Her example is relevant to our discussion.

Suppose that cigarette smoking is a weak cause (in the sense of a risk factor) of my-ocardial infarctions. But suppose, too, that in the population examined, cigarette smoking is associated with physical exercise and, also, that such exercise is strongly preventive of infarctions. Then unless we examine subpopulations in which the exercise factor is held fixed, cigarette smoking will be concluded to be a preventive factor.

Cartwright's solution for Simpson's paradox is to appeal to a counterfactual condition. "C causes E if and only if C increases the probability of E in every situation which is otherwise causally homogeneous with respect to E" (1979, p. 423).[4] This notion of a causally homogeneous situation is elaborated using Carnap's idea of a state description. The elaborated definition of a causal relation between C and E, written symbolically as $C \rightarrow E$, is given by a definition involving four conditions that we cannot describe here.[5] Suffice it to say that her condition (iv), which was initially added so that state descriptions would not hold fixed any factors in the causal chain from C to E, gave Cartwright pause (see Cart-wright 1979, 1986a, 1986b; Eels and Sober 1983) and has subsequently been reevaluated and explored for its important implications for singular causation, about which I shall say more later.

In a formalism that perhaps indicates more explicitly the nature of the counter-factual conditions underlying Cartwright's definition of causality, she writes that "we look not at $P(E \mid C)$, $P(E \mid -C)$ but rather at $P(E \mid C \pm F_1 \ldots \pm F_n)$, $P(E \mid -C \pm F_1 \ldots \pm F_n)$ for each of the possible arrangements of other causal factors $F_1, \ldots, F_n$" (1986a, p. 3). These test factors whose presence and absence in addition to C are presupposed in order to block problems associated with Simp-son's paradox she then cites in a new (but equivalent to her 1979) form of the CC principle:[6]

CC: C causes E iff $P(E \mid C \pm F_1 \ldots \pm F_n) > P(E \mid -C \pm F_1 \ldots \pm F_n)$

I believe that this approach to blocking the paradoxical effects of selecting sub-populations suggests the root of the difference that the epidemiologists perceive between their concept of causation, which we cited, and the pathophysiological notion of causation. I shall argue in the following section that the reason is not that epidemiological research involves probabilistic elements but, rather, that epide-miology as practiced cannot be certain that it has examined homogeneous popu-lations.[7] Pathophysiological causation, on the other hand, deals with idealized models and often experiments with specially bred strains of organisms and can thus be reasonably certain that its populations are homogeneous.

MECHANISTIC AND EPIDEMIOLOGICAL
CAUSATION DISTINGUISHED

In this section I shall use the noncounterfactual analogues of the Giere popu-lations to distinguish between the concept of causation we seek to capture in the basic sciences such as physiology and the kind we find in epidemiology. This distinction is important because it will be used to interpret the way epidemiologists have consciously approached causation in their own recent writings.

I shall begin by considering four different types of exemplars of causation. The first, which we alluded to in the previous section, is a close analogue of the epidemiological example of cigarette smoking's causing lung cancer. For our purposes, assume either that a group of patients who give a history of familial ischemic heart disease (IHD) occurring before age fifty-five and a group of controls who offer no such family history have been identified retrospectively or that two groups are being followed prospectively. Further assume that both groups followed prospectively develop IHD but that the incidence of new cases per year among those with a positive family history is two times that of the controls.

The second exemplar is drawn from muscle physiology. We investigate the cause of the increased force of contraction that skeletal muscle delivers when stretched before the administration of a contracting impulse when compared with the same muscle in an unstretched state. We determine that the increased force is a consequence of additional overlap, generated by the stretch of the muscle, between the two sets of muscle filaments comprising the sarcomere (filament bundle), as this additional overlap increases the number of reactive sites where chemical energy is transformed into the energy of motion. This is known as the sliding filament model of muscle physiology (see Katz 1977, chaps. 5–8).

Our third exemplar is drawn from the germ theory: The TB bacillus causes tuberculosis, in the sense that it a necessary condition for the disease and, in an appropriately susceptible host, will, when inoculated into the host, produce the disease. We assume that such a series of test inoculations in appropriate animal models has been done with a 100 percent success rate.

Our final example comes from neurophysiology.[8] We want an account of how a nerve can stimulate a muscle on which it impinges to contract. It is found that the signal for contraction is carried by chemical neurotransmitter molecules that are probabilistically released from the nerve endings in packets, or "quanta," and that these quanta diffuse across the neuromuscular junction space and cause the opening of microscopic channels on the muscular side, resulting in an electrical signal that causes muscle contraction. Furthermore, "fluctuations in release [of the quanta] from trial to trial can be accounted for by binomial statistics . . . [and] when the release probability p is small . . . the Poisson distribution provides a good description of the fluctuations" (Kuffler, Nicholls, and Martin 1984, p. 257). This tells us that "release of individual quanta from the nerve terminal . . . [is] similar to shaking marbles out of a box" through a small hole (p. 254). This model of a probabilistic process agrees with what is observed in a variety of microexperiments at the neuromuscular junction involving neural signals' causing muscular contractions.

These four examples demonstrate three distinct and, I believe, independent dimensions of interest in the area of causation. First, consider the dichotomy of probabilistic versus nonprobabilistic or, more accurately, stochastic versus nonstochastic. In our examples, both the neuromuscular junction (or the gas cylinder) illustration and the familial history–IHD case appear to involve stochastic elements. Similarly, both the TB and the sliding filament exemplars appear to be deterministic, at least in the manner in which they have been presented for this discussion.

Second, consider what I will call the dimension of *connectivity*. By this notion I mean the extent to which one event follows another in a temporal sequence and in which a general connecting principle can be adduced to explain the connection. Such a connecting principle may be scientifically sophisticated or drawn from "naive physics"—the kind of rough-and-ready generalizations we use in the everyday world to describe simple interactions of bodies. In both the sliding filament model and the neuromuscular junction (or gas cylinder) examples, there is strong connectivity present. In the TB example, there is a gap between the inoculation of the mycobacterium and the production of the disease. In the familial history–IHD example (or in the analogous smoking–lung cancer example), there are many gaps in any causal sequence.

Finally, consider the dimension of the uniformity of the entities in the examples. Even if we can hypothetically state that all cases of ischemic heart disease are the same (or that the lung cancers are of the same type), we know that there is considerable variation in other properties of the subjects. All myofilaments are identical; all neurotransmitter quanta (or gas molecules) are identical;[9] and all mycobacterium are in this case identical (all the highly inbred animals can also be considered relevantly identical).

I want to argue that most people would be willing to interpret as causally warranted accounts each of the preceding exemplars, with the exception of the positive family history–IHD example. The reason that the IHD example seems to be different is only partially because there is a stochastic element present. More important, I would argue, are the features of entity variation and low connectivity. In sampling the Giere type of populations, we run the risk of assembling subgroups that do not represent the populations as a whole; that is, there are likely to be a number of different causally homogeneous subpopulations. If all entities are identical except for the two properties of interest, namely, the cause and the effect, such an error cannot be made.

This suggests quite strongly that the difference between the epidemiologists' concept of causation and the physiologist's or pathophysiologist's notion is not simply that the former involves probability and the latter does not; rather, the difference appears to rest primarily on the homogeneity or lack thereof of the populations studied.

Even if there is variation in the populations and in the sample groups investigated, if the sequence of events between cause and effect shows a high degree of connectivity, this will screen off errors of misattribution of causation.[10] Finally, even if there are variation and low connectivity, if the relation between putative cause and effect is universal, this will screen off any errors of misattribution of causation. A generalization associating a cause with an effect based on populations that have varying characteristics in situations of gappy knowledge and in which the association is not universal (either on the basis of variable initial conditions or purely stochastic grounds) is a claim about population causality in which causal connections apply to populations in which averaging has occurred and in which homogeneity is not guaranteed or even strongly warranted. This is a different kind of claim than causal claims that are grounded on entity identity, high connectivity, or universality. In those situations causal generalizations apply to each individual at the level of the individual, whereas a population-level causal claim is valid only for groups over which variations have been averaged. For individuals for which the relation is probabilistic, however, some propensity interpretation is probably reasonable.

This view does not preclude us—as we may adopt an analogue of the Reichenbach and Salmon notion of "weight"—from fictitiously or "hypothetically" ascribing a kind of putative individual causation based on population causation through the probabilistic concept of odds or risk. This tack allows some degree of control over the environment and enables predictions, though these are actuarial types of predictions and are, strictly speaking, empirically applicable to averaged populations. Because of the type of prediction and control that such actuarial generalizations permit and because the generalizations may well associate properties temporally; that is, E follows temporally after C, such generalizations exhibit important analogies to causal generalizations. It is important to keep in mind, however, that the forms of putative individual causation, population causation, and true causation (i.e., the type meeting Cartwright's CC condition) are only analogous and that they have their domain of application at two different levels, hold in different populations, and are based on different test conditions.

I believe these differences are the source of Thomson's discomfort (noted at the beginning of this chapter) with the use of statistical evidence to attribute responsibility. It is exactly the fact that the population under consideration, namely, all cabs, is *un*homogeneous with respect to relevant properties that relate strongly to the liability-generating action—hitting the pedestrian—that leads one to want to withhold attributing the responsibility to the Red Cab Company. Properties such as being in the area, speeding, having a driver who has been drinking, constitute the F-types of properties cited by Cartwright in her CC principle. It is not the fact that uncertainty or probability is involved but, rather, that the conditions of entity homogeneity are not met and that neither high connectivity nor universality are available to override the probabilistic causal construal and to warrant the attribution of liability to the Red Cab Company. In fact, the "internal" type of evidence cited by Thomson is tantamount to providing "connectivity," in the light of our generally accepted background generalizations about the behavior of cab drivers, and thus supports individual causation. Finally, had there been only one cab company in town, "universality" would have screened off any issue of uncertainty.

I should add and emphasize that the actuarial probabilistic notions of causation can fairly be used in connection with an cost spreading or "insurance" approach to liability. This will depend on a prior agreement to spread costs and liabilities, so that for cases in which external evidence alone is available, the burdens will fall in the same proportion as will the probabilistic evidence. (Thus in the Kaye–Thomson example, the Red Cab Company will pay the plaintiff 60 percent and the Green Cab Company 40 percent of his costs. This is, in fact, the solution imposed *post hoc* in *Sindell,* and Thomson agrees with its fairness in such special cases.) A superfund approach is similar.

SINGULAR CAUSATION

It now remains to address one more issue raised in particular by Cartwright (1979) and especially (1986a, 1986b), and this is the problem of "singular caus-ation." Cartwright maintains that an adequate characterization of the **CC** prin-ciple requires modification to a **CC** form, in which an appeal is made to singular causes, as opposed to the generic causes required in a Humean regularity view. (The problem arises in those situations when cause C produces E only through an intermediate factor $K,$ in which K can sometimes occur as a result of an alternative antecedent C' and produce $E.$) Specifically, Cartwright adds the following condition: "Test situations should be subpopulations homogeneous with respect to $K,$ except for those individuals where K has been produced by $C.$ These individuals belong in the population where they would otherwise have been were it not for the action of C" (1986a, p. 4). Thus the individual causal history of the production of K by a spatiotemporal singular descriptor becomes a component of the defining condition for a cause.[11]

I am not persuaded by this argument, which I believe raises several problems. Although Cartwright has distinguished philosophical company in her defense of singular causation (see, e.g., Mackie 1974 and also Anscomb 1975), such a thesis evokes the following dilemma: Either we fail to distinguish causes from co-incidences in individual situations, or we must appeal to an intuition of some perceived "causal power" that operates in causal but not in coincidence situa-tions. It also seems to me that the standard response to just those situations that Cartwright cites as raising the need for her condition—namely, intermediate events such as K that lie on the final common pathway to event E—is resolved by controlling for those antecedents that produce K when C is not operating. Finally, I do not see the "empiricist" remedy that Cartwright proposes—the "one-shot" physics experiments—as warranting singular causation. I interpret such experiments as very narrow event classes, such that if the same initial conditions were to be reconstituted, the same result would occur. But this is a (potential) generalization, and in fact Cartwright's example (1986b) indicates that the Einstein–de Haas experiment was retried a number of times with varying conditions.[12]

SUMMARY

There are two different concepts of causation in the ideas of epidemiological causation (involving, for example, risk factors) and more typical scientific caus-ation (involving, for example, physical, physiological, or pathophysiological mechanisms). I related this distinction to the parallel problem of two ways in which statistical and causal evidence might be used to assign liability, by citing the Kaye–Thomson example. I quoted from some of the general epidemiological literature on the epidemiologists' conception of cause. This led to a philosophical analysis of that concept of causation, which I used to demonstrate the differences between the epidemiologists' approach and the concept of causation we find in pathophysiological mechanisms. I found the essential difference between the epidemiologists' notion of cause and the pathophysiologists' conception in the uncontrolled individual variation in actual populations. I argued that the probabilistic feature of epidemiology was not the source of the difference in causal notions and attempted to show that by indicating how a probabilistic example would be perceived as causal in the strong nonepidemiological sense. I also indicated the extent to which precise (theoretical) characterizations of probabilistic cause that were immune from the effects of Simpson's paradox required a counterfactual component and distinguished this theoretical notion, exemplified in Giere's and Cartwright's definitions of cause, from the epidemiologists' notion, which is not counterfactual.

NOTES

1. Note that I am dealing with two concepts of causation. Whatever the metaphysics of causation may be, the existence of two concepts does not require the existence of two different ontological types of causation: One will suffice. I thank Ron Giere (in his comments on this chapter at the March 1987 APA Pacific Division Meetings) for urging that I make this distinction more explicit.

2. The two legal cases are *Sindell* v. *Abbott Laboratories* (26 Cal. 3d 588, 163 Cal. Rptr. 132, 607 P. 2d 924 [1980]), and *Collins* v. *Eli Lilly* (342 N.W. 2d 37 [Wis.] p. 45 ([1984]).

3. The measure of effectiveness introduced here is essentially identical to what the epidemiologists have termed a *measure of effect* for *attributable risk*. Giere does not comment on this notion, but see Fletcher, Fletcher, and Wagner (1982, p. 101).

4. I think that Cartwright's definition of cause here is essentially identical to Giere's introduced earlier, with the somewhat superficial difference that Cartwright formally includes homogeneity in her definition, whereas Giere introduces the idea implicitly into his counterfactual interpretation.

5. The conditions formally define a principle of causal connection **CC** and are as follows:

CC: $C \rightarrow E$ iff $P(E \mid C.K_j) > P(E \mid K_j)$ for all state descriptions K_j over the set $\{C_i\}$, where $\{C_i\}$ satisfies

1. $C_i \in \{C_i\} \Rightarrow C_i \rightarrow \pm E$
2. $C \notin \{C_i\}$
3. $\forall D \ (D \rightarrow \pm E \Rightarrow C \text{ or } D \in \{C_i\})$
4. $C_i \in \{C_i\} \Rightarrow \neg(C \rightarrow C_i)$

Here the expression $C_i \rightarrow \pm E$ represents a complete set of causal factors C_i for E such that either $C_i \rightarrow +E$ (i.e., E occurs) or $C_i \rightarrow -E$. K_j stands for the 2^n state descriptions (after Cartwright 1979).

6. Exactly what factors can be allowed into such a formulation is somewhat complex. Sober points out that unless some restrictions are imposed, probabilities may become ill

defined. In addition, the factors may need to be temporally indexed (see Sober, 1984, chap. 7).

7. Randomization is a device for eliminating or reducing bias, but it does not offer any guarantees of homogeneity; rather, it hopes to have matched the control and experimental populations for unknown interfering factors.

8. A simpler example from the physical sciences employing similar probabilistic elements can be found in an elementary form of the kinetic theory of gases. We want to know, from that theory's perspective, what causes the pressure pushing against a cylinder compressing gas in a tube. The account given is that gas molecules are randomly bombarding the cylinder and causing innumerable microscopic collisions. The net sum of all the tiny collisions is an averaged-out constant pressure.

9. In the experimental preparations used, we have identity holding. If one goes to another muscle type (e.g., cardiac muscles, as opposed to skeletal muscle) or to a different neurotransmitter, there may be changes.

10. This is the Reichenbach–Salmon notion of "screening off."

11. Let me emphasize that singular causation should be distinguished from individual causation in that the former involves what I, at least, see as an epistemologically stronger claim. Individual causation can be permitted by Humean or Millean regularities.

12. Much of what Cartwright presents in her 1986a and 1986b manuscripts has been published in more developed form in her 1989 book. The modified form of her *CC* principle appears on pp. 55–56, and additional features of the principle are discussed throughout Chapter 2 of her 1989 book. Chapter 3 is a detailed account of singular causation, and also discusses "one shot" physics experiments, such as that of Einstein and de Hass. This book reached me while this essay was in press, so I have been unable to discuss Cartwright's more developed views here.

REFERENCES

Cartwright, N. (1979). "Causal Laws and Effective Strategies." *Nous* 13:419–37.
Cartwright, N. (1983). *How the Laws of Physics Lie*. Oxford: Oxford University Press.
Cartwright, N. (1986a). "Regular Associations and Singular Causes." Unpublished manuscript.
Cartwright, N. (1986b). "An Empiricist Defense of Singular Causes." Unpublished manuscript.
Cartwright, N. (1989). *Nature's Capacities and Their Measurement*. Oxford: Clarendon Press.
Eels, E., and Sober, E. (1983). "Probabilistic Causality and the Question of Transivity." *Philosophy of Science* 50:35–57.
Fischer, J. M., Ennis, R. H., and Kagan, S. (1986). Comments (and J. J. Thomson, Response). Thomson's "Remarks on Causation and Liability." *Philosophy and Public Affairs* 15 (Winter).
Fletcher, R. H., Fletcher, S. W., and Wagner, E. H. (1982). *Clinical Epidemiology—The Essentials*. Baltimore: Williams & Wilkins.
Giere, R. (1980). "Causal Systems and Statistical Hypotheses." In *Applications of Inductive Logic,* ed. L. J. Cohen and M. B. Hesse, pp. 251–70. Oxford: Oxford University Press.
Giere, R. (1984). "Causal Models with Frequency Dependence." *Journal of Philosophy* 81:384–91.
Good, I. J. (1961–62). "A Causal Calculus I–II." *British Journal for the Philosophy of Science* 44:305–18, 45:43–51, 49:88.

Katz, A. M. (1977). *Physiology of the Heart.* New York: Raven Press.

Kaye, D. (1982). "The Limits of the Preponderance of the Evidence Standard: Justifiably Naked Statistical Evidence and Multiple Causation." *American Bar Foundation Research Journal* 2 (Spring):487–88.

Kleinbaum, D. G., Kupper, L. L., and Morgenstern, H. (1982). *Epidemiologic Research.* London: Wadsworth.

Kuffler, S. W., Nicholls, J., and Martin, A. (1984). *From Neuron to Brain.* 2nd ed. Sunderland, Mass.: Sinauer.

MacMahon, B., and Pugh, T. F. (1970). *Epidemiology: Principles and Methods.* Boston: Little, Brown.

Mayo, D. (1986). "Understanding Frequency-dependent Causation." *Philosophical Studies* 49:109–24.

Otte, R. (1981). "A Critique of Suppes' Theory of Probabilistic Causality." *Synthese* 48:167–81.

Reichenbach, H. (1956). *The Direction of Time.* Berkeley: University of California Press.

Salmon, W. (1980). "Probabilistic Causality." *Pacific Philosophical Quarterly* 61:50–74.

Schaffner, K. F. (1983). "Explanation and Causation in the Biomedical Sciences." In *Mind and Medicine*, ed. L. Laudan, pp. 79–124. Berkeley: University of California Press.

Schaffner, K. F. (1987). "Causation in Medicine and the Law." In *Medical Innovation and Bad Outcomes,* ed. M. Siegler et al., pp. 71–99. Ann Arbor, Mich.: Health Administration Press.

Skyrms, B. (1980). *Causal Necessity.* New Haven, Conn.: Yale University Press.

Sober, E. (1982). "Frequency-dependent Causation." *Journal of Philosophy* 79:384–91.

Sober, E. (1984). *The Nature of Selection.* Cambridge, Mass.: MIT Press.

Suppes, P. (1970). *A Probablistic Theory of Causality.* Amsterdam: North-Holland.

"Symposium on Causation in the Law of Torts." (1987). *Chicago–Kent Law Review* 63:397–680.

Thomson, J. J. (1984). "Remarks on Causation and Liability." *Philosophy and Public Affairs* (Spring):101:33.

Thomson, J. J. (1986). "Liability and Individualized Evidence." In *Rights, Restitution, and Risk,* ed. W. Parent, pp. 60–66. Cambridge, Mass.: Harvard University Press.

Willerson, J. T. (1982). "Acute Myocardial Infarction." In *Cecil Textbook of Medicine,* 16th ed., ed. J. B. Wyngaarden and L. H. Smith. Vol. 1, pp. 247–56. Philadelphia: Saunders.

11

Reductionist Approaches to Risk

KRISTIN SHRADER-FRECHETTE

Many Americans, sensitized by the media to the dangers of cigarette smoking, have been appalled to discover on their visits to the Far East that most adult Chinese smoke. The Chinese, on the other hand, consume little alcohol and have expressed bewilderment about the hazardous and excessive drinking in the West. Differences in risk acceptance, however, are not merely cross cultural. Within a given country, some persons are scuba divers, hang gliders, or motor-cyclists, and some are not; there are obvious discrepancies in attitudes toward individual risk. At the level of societal risk—for example, from nuclear power, toxic dumps, and liquefied natural gas facilities—different persons also exhibit analogous disparities in their hazard evaluations.

In this chapter I shall argue that two of the major accounts of societal risk acceptance are highly questionable. Both err because of fundamental flaws in their conception of knowledge. This means that to understand the contemporary controversy over societal risk, we need to accomplish a philosophical task, that is, to uncover the epistemologies assumed by various participants in the conflict. Proponents of both positions err, in part, because they are reductionistic and because they view as irrational the judgments of citizens who are risk averse. After showing why both views are built on highly doubtful philosophical pre-suppositions, I shall argue in favor of a middle position that I call *scientific proceduralism*. An outgrowth of Karl Popper's views, this account is based on the notion that objectivity in hazard assessment requires that risk judgments be able to withstand criticism by scientists and lay people affected by the risks. Hence the position is sympathetic to many populist attitudes toward involuntary and public hazards. Although scientific proceduralism is not the only reasonable view that one might take regarding risk, I argue that it is at least a rational one. And if so, then rational behavior should not be defined purely in terms of the assessments of either the cultural relativists or the naive positivists. Most importantly, risk experts should not "write off" the common person.

Because hazard assessment is dominated by these two questionable positions, it is reasonable to ask whether criticizing them threatens the value of quantified risk analysis (QRA). Indeed, many of those allegedly speaking "for the people," as I claim to be doing, are opposed to scientific and analytic methods of assessing environmental dangers. I am not. I have argued elsewhere that although QRA is in practice flawed, it is in principle necessary to the rational, democratic assessment of risks (Shrader-Frechette 1985b).

THREE PHILOSOPHIES CONCERNING RISK

Contemporary controversy over the rational assessment of risks is, at the core, a conflict over values. Cultural relativists Douglas and Wildavsky, authors of the best-selling *Risk and Culture* and two of the world's most prominent social scientists, have enraged environmentalists and given industrial risk imposers a justification for continuing "business as usual." They overemphasize value judgments in hazard assessment and claim that "risk is a collective construct." I shall argue that in so doing, they reduce the third stage of hazard analysis (risk evaluation) to sociology (Douglas and Wildavsky 1982, p. 186).[1] Naive positivists, like Starr and Whipple, authors of numerous hazard assessments, ignore value judgments in risk analysis, claiming that risks can be objectively measured (Hafele 1976; Okrent and Whipple 1977). I shall show that by virtue of their position, they reduce the first two stages of hazard analysis (risk identification and risk estimation) to natural science (although it can be argued that even natural science is not objective in their sense) (Laudan 1984; Scriven 1982).[2]

Risk evaluation is not merely a social construct, as this chapter will show. Risks may be defined primarily in terms of probabilities, and probabilities are always (to some degree) theoretical. But this does not make their evaluation a mere construct. Constructs do not kill people, but faulty reactors and improperly stored toxics do. At least some hazards are real, and many of these are measurable, albeit within limits. Hence the cultural relativists are wrong to overemphasize value judgments in risk assessment.

The naive positivists also are wrong, I will argue, because risk estimates are not purely objective and capable of being determined by experts. That they are not is evidenced by the fact that the current technological landscape is littered with the bodies of victims of various hazards. These were risks that experts allegedly objectively measured, catastrophes that were not supposed to happen. They include radwaste dumps whose contents were "objectively" and definitively said not to be able to migrate (Shrader-Frechette 1988) and toxic chemicals that were "objectively" claimed not to cause harm to humans (Kraybill 1975).[3] From Chernobyl to Bhopal, they provide evidence that public skepticism regarding the position of the naive positivists is, in some cases, warranted.

Cultural relativists and naive positivists attack many public attitudes toward risk. Although their views are extreme, uncovering their flaws is useful because many assessors, at least to some degree, fall victim to them. The naive positivists attempt to discredit citizens by alleging that they are ignorant of the science and mathematics essential to hazard evaluation and by assuming that technical expertise, alone, is sufficient for rational risk assessment. The relativists attempt to discredit citizens by alleging that their faulty attitudes toward risk arise because they are victims of a biased "sectarian" social framework.

But if risk is neither merely a social construct nor purely objective, then what is it? According to at least one account, which I call *scientific proceduralism,* it is true to say that although hazard assessments can never be wholly value free (as many naive positivists claim), nevertheless it is false to assert (as many cultural relativists do) that any perspective on risk can be justified (Douglas and Wildavsky 1982). According to the account I shall defend, some risk judgments are more warranted than others are, although none is value free, and many hazard evaluations of the public are both rational and ethically defensible, contrary to what the relativists and naive positivists claim. Later I shall specify some Popperian criteria for democratically reaching more objective risk judgments.

CULTURAL RELATIVISM ACCORDING TO DOUGLAS AND WILDAVSKY

Cultural relativists like Douglas and Wildavsky begin from an astute (although not original) insight. They recognize that risk is not wholly objective, as many positivists claim, and they criticize assessors for their repeated error of assuming that lay evaluations of risk are mere "perceptions," whereas expert analyses are "objective." Clearly, the cultural relativists are to be commended for discussing this error, as they are correct at least in this: Both experts and the public have risk perceptions. No one has privileged access to the truth about risk acceptability, merely because she is an engineer rather than a housewife. Even though the preponderance of technical errors made in assessing hazards likely lies with lay people, not experts, both engineers and housewives make mistakes, especially in judging risk acceptability. Hence the cultural relativists come close to being correct: All judgments about hazards or risks are value-laden.

Douglas, Wildavsky, and others go wrong, however, in assuming that because everybody can be wrong, all risk evaluation is merely a social construct. They err, in part, because they reserve their harshest criticisms for the public. They claim that "risks are social constructs" (Douglas and Wildavsky 1982, p. 186), but they single out environmentalist laypersons (as opposed to technical experts) as having particularly irrational constructs. They even admit that they have a "bias" against the risk attitudes of the public (whose views they identify as "sectarian," rather than market or hierarchy- oriented) (Douglas and Wildavsky 1982). They allege that laypersons are averse to technological risk because their social frameworks dictate attitudes that cause them to be obsessed with "purity," opposed to industry, and negative toward the powerful persons and institutions that often are responsible for pollution and environmental hazards (Douglas and Wildavsky 1982).

Of course, because Douglas and Wildavsky are cultural relativists who believe that "no one is to say that one [risk judgment] is better or worse" (1982, p. 188), they are inconsistent in arguing that environmentalists have risk positions that are more questionable than those of the "centrists" (persons whose value systems, in their scheme, derive from the market or from hierarchical organizations). Let us examine why they espouse a position of cultural relativism and how they go wrong.

Douglas, a sociologist, and Wildavsky, a political scientist, have attempted to describe current attitudes toward technological and environmental dangers.

They divide all persons into two groups, "the danger establishment" or "border" and the "industrial establishment" or "center." They claim that those in the first group are egalitarian, antigovernment, antiindustry persons who are typically removed from centers of power and influence and are averse to environmental and technological risks. Douglas and Wildavsky maintain that those in the second group are more authoritarian, less democratic persons who are close to cultural centers of power and influence and are less averse to environmental and technological risks. Their attitudes are in part a result of their beliefs that environmental damage has been overemphasized and that the earth has repair mechanisms that will enable us to avoid the most dangerous consequences (e.g., desertification, groundwater pollution) of human actions.

Using these two groups, border and center, Douglas and Wildavsky divide all persons on the basis of membership in three risk "portfolios": sect, market, and hierarchy. They claim that persons in the sect category uphold border values, whereas those in the market and hierarchy groups support center values. Members of sects, they maintain, fear risks to humans and the rest of nature. Persons in the market category fear economic rather than environmental or technological risks. Finally, they allege that those in the hierarchy group fear political risks that could threaten their administrative power.

Douglas and Wildavsky claim that "on other issues [not related to risk evaluation] they [people] use their private discretion." In the areas of risk, however, they say that citizens do not make personal decisions. Rather, they assert that "having chosen for the sect," market, or hierarchy, people "slide their [risk] decisions onto . . . institutions" that support, respectively, the sect, market, or hierarchy (p. 174). In other words, "the selection of risk is a matter of social organization" (p. 198).

By subscribing to cultural relativism regarding risk evaluation (the third stage of hazard assessment, after risk identification and risk estimation), Douglas and Wildavsky attempt to substantiate their position by means of at least five arguments:

1. Increased knowledge and additional reasoning about risks do not make people more rational about hazards.
2. Risk assessments are like judgments in aesthetics.
3. Any form of life, including risk behavior and attitudes, can be justified, as everyone is biased in her perceptions of danger, including experts who disagree about hazard analysis.
4. Modern persons are no different from primitives in that social structures dictate their views of, and responses to, alleged hazards.
5. More specifically, environmentalists' views of risk are a result of their "sectarian problems."

The First Argument of Douglas and Wildavsky

In their first argument, the statement that additional knowledge and reasoning about risk do not make people more rational about hazards, Douglas and Wildavsky make it clear that they do not believe that risk acceptance and aversion have anything to do with rational criteria. They state explicitly that "objective knowledge about technology in general is not going to take us very far," both because "each method [of risk assessment] is biased" and because "better measurement opens more possibilities, more research brings more ignorance to the light of day." Hence, they conclude, "thinking about how to choose between risks, subjective values must take priority" (1982, pp. 63–64, 71, 73, 194).

The main presupposition of this first argument is that because there are always things that we do not know, research, measurement, and increased information about risk are not valuable. This is a version of the "all-or-nothing argument": Because I cannot know everything about something, any knowledge about it is not very valuable, and therefore I must rely merely on "subjective values." To see why all-or-nothing arguments fail, one need merely consider several examples of them: It is impossible for most people to become perfect, or even concert-level, pianists; therefore playing the piano is not important or will not take them very far. Or, it is impossible for anyone to be a perfect parent, therefore parenting is not important.

Such all-or-nothing arguments presuppose that only perfection in a certain area (whether in making risk judgments, playing the piano, or parenting) is valuable and that if perfection is not attainable, imperfect accomplishments have only subjective value. Clearly, however, these presuppositions are wrong, and for several reasons. First, the value of many activities lies in the process of accomplishing them, rather than in their outcome, such as mastering some skill. Second, it is obviously better, for example, to be an imperfect but loving parent than to be an imperfect parent who is a child abuser. In other words, we can make correct and noncontroversial judgments about better and worse instances of a thing, even in the absence of a perfect example.

The argument also fails to take account of the many ways in which knowledge has radically changed our risk judgments. Back in the 1950s, most people were ignorant, for example, of the hazards of ionizing radiation. As a result, many shoe stores determined correct fit by taking x-rays of their customers' feet inside new shoes. Several decades later, we discovered that ionizing radiation is harmful, and this knowledge has eliminated its earlier, shoe-store uses. Likewise, scientific knowledge about the hazards of lead has caused us to avoid the use of lead pipes, leaded gasoline, and paint containing lead. Similar examples of how knowledge about hazards has changed our assessment of risks have occurred throughout this century in cases such as those concerning asbestos, vinyl chloride, DDT, chlordane, contraceptives, and food additives. To claim otherwise, as do Douglas and Wildavsky, is counterfactual and anti-intellectual.

The Second Argument of Douglas and Wildavsky

Douglas and Wildavsky's second relativistic argument also discounts the role of knowledge and reasoning in evaluating hazards. They contend that risk assessments are like judgments in aesthetics:

> Public perception of risk and its acceptable levels are collective constructs . . . like language and . . . aesthetic judgment . . . perceptions of right and truth depend on

cultural categories created along with the social relations they are used for defending
. . . there is no reasoning with tastes and preferences. . . . In the end, we either favor
a centrist view of the human predicament, or we favor the sectarian view, or we
prefer not to choose. (1982, pp. 187, 191)

Like the first argument for the relativity of risk judgments, this second one
also has the all-or-nothing form. Its major presupposition is that because our
assessments of hazards and our evaluations of "right and truth" are laden with
cultural values; therefore "there is no reasoning" about them. Obviously, however, as we earlier indicated, one can admit that risk judgments are not perfect
and yet claim that it is possible to reason about them. For example, a number
of courts have reasoned about value-laden risk judgments in the area of liability.
Likewise, one can admit that, for example, the risk judgment "innocent people
ought not have their rights jeopardized without their informed consent and
without due process of law" is value laden. Despite this value ladenness, one
can reason about the thesis in question. For instance, one can reason that if
denials of due process are wrong, then it is more serious to deny due process
in situations in which victims have received serious bodily injuries than in situations in which they have sustained minor property damage. Moreover, even
if one believes that due-process rights are merely value-laden legal conventions,
one can reason that we are nevertheless ethically bound to honor such conventions as long as we have accepted the contract of citizenship and the duties that
it imposes.

It could be argued further that once one has contracted or promised to do
something (e.g., obey all laws, including those safeguarding due process), then
doing it is not merely a matter of taste, preference, or social construction; it is
also a matter of moral obligation. Moreover, one can appraise even conventions
according to how well they lead to certain goals, for example. But if so, then
Douglas and Wildavsky's second argument misses at least two key points: (1)
One can reason about value-laden risk judgments, just as I have illustrated. (2)
If the Fifth and Fourteenth Amendments to the U.S. Constitution give citizens
legal rights to due process and equal protection, then citizens' claims to due
process and equal protection from technological risks are not based on mere
preferences but on legal obligation. Even at worst, legal obligation is not purely
subjective but is something about which we have reasoned and continue to
reason.

Moreover, although we obviously speak of civil rights and legal obligations
in terms of our own categorizations of the world, such categories do not render
our judgments relative in any but a trivial sense. That people conceptualize or
categorize differently does not mean that there are no rational or ethical grounds
for their judgments. Likewise, just because people observe differently does not
mean that they observe relativistically.

Another problem with the second argument (that risk evaluations are "like
aesthetic judgments" about which "there is no reasoning," as they concern
"tastes or preferences") is that there are strong disanalogies between aesthetic
and risk-related judgments. One of the strongest disanalogies concerns the fact
that aesthetic judgments rarely have life-and-death consequences in the way that
technology-related judgments do. The latter can cause, for example, a Bhopal

toxic leak. Moreover, one can often check or substantiate one's calculated risk probabilities. But there is no similar way to check one's aesthetic judgments, as facts are less relevant. Hence, to talk about risk claims as if they were mere aesthetic judgments is not only to be insensitive to the many deaths that technological risks often impose but also to ignore the relevance of quantitative parameters, like probability, to the determination of risk. Similar quantitative parameters are not relevant to reasoning in aesthetics.

Douglas and Wildavsky's Third Argument

Having seen Douglas and Wildavsky's first two relativistic arguments, that reasoning is irrelevant to risk claims and that judgments about hazards are like aesthetic preferences, their third argument is not surprising: "Any form of life [including risk behavior and attitudes] can be justified." In their concluding chapter, they describe the point to which their analysis has brought them. They have argued that everyone is biased in hazard evaluation and that risk assessors disagree among themselves. Because of this disagreement and bias, they conclude that there are no objective risk judgments: "The center takes one view of risk, the border takes another. Is there any judgment possible between their views? Any form of life can be justified; no one is to say that one is better or worse" (1982, p. 188).

Not surprisingly, this argument fails for many of the same reasons as does the previous one. It ignores the fact that risk judgments can be more or less in keeping with morally binding laws or civil rights, or more or less in keeping with confirmed accident frequencies. This means that not all risk attitudes can be justified. For example, suppose that one is morally bound by the law of the land to recognize rights to due process. If so, then risk judgments that fail to recognize that right, all things being equal, are not as justifiable as are those that affirm it. Likewise, risk judgments predicated on probabilities that are wildly inconsistent with observed accident frequencies are also less justifiable, all things being equal, than are those that are more consistent with them.

If Douglas and Wildavsky's third argument were correct, that any position on risk can be justified and that "there is . . . no correct description of the right behavior [regarding risk]" (1982, p. 187), then several problematic consequences would follow. One is that ignorant, fearful people who attempt to avoid all risks of any kind are no less rational in their risk decisions than are well-informed persons who attempt only to avoid all unnecessary or unreasonable risks. But this conclusion is obviously false. Some risk behavior is more rational than other such behavior is because some activities are more dangerous than others are. Even classical economists recognize this fact and have taken great pains to argue that there ought to be a compensating wage differential (CWD), that is, that the riskier the job is, the higher the pay should be, all things being equal (Shrader-Frechette 1985a). The CWD suggests that market economists, among others, realize that any view of risk cannot be justified because some activities are more hazardous than others are. Moreover, if Douglas and Wildavsky are correct that any position is justifiable and that there is "no correct description of the right behavior [regarding risk]," then insurance companies and actuaries would go out of business; that they do not indicates that there are better and worse accounts of risk. But if so, then Douglas and Wildavsky are wrong to say that any risk can be justified.

Most importantly, if Douglas and Wildavsky were right, then unscrupulous persons could, with impunity, involuntarily impose whatever risks they wished on others. They could, for example, avoid the cost of pollution controls by claiming that there was no correct description of the relevant risk. Then they could simply pass on environmental threats to innocent members of the public. Obviously, however, a minimal sense of fairness indicates that one should not impose on others whatever hazards he or she chooses, under the guise that any risk behavior can be justifiable. It would be easy to argue that rights to life and to security would prohibit such an imposition (Shrader-Frechette, forthcoming). But if so, then some risk attitudes are more justifiable than others are.

Furthermore, if Douglas and Wildavsky are right, then it would be impossible to regulate technology and industry in any rational way. One could merely charge that because all regulations were justifiable, none were. Similarly, one could not morally justify the imposition of sanctions on those who violated laws regarding certain hazards, even violations that resulted in significant losses of life. The reason is that the offenders could quite reasonably claim that because there was no correct description of risk behavior, they were not negligible in preferring one position (regarding some hazard) to another, even though its acceptance resulted in many deaths.

A final problem is with Douglas and Wildavsky's inconsistently affirming that there is no correct description of risk behavior yet admitting that they have a "bias toward the center" (1982, p. 198). If there is no correct view of risk behavior and any position can be justified, then being biased toward only one of those accounts is not reasonable. In a situation in which one alleges that there is no correct view, one cannot argue that one's own position is correct, without being obviously inconsistent.

Apart from the fact that this third claim is inconsistent with Douglas and Wildavsky's admitted bias toward the center, their argument runs into other difficulties with consistency. Within two pages of affirming their thesis of relativism—"no one is to say that one [position on risk] is better or worse" (1982, p. 187)—they assert that "there is nothing relativistic in this exercise [their theory as articulated in *Risk and Culture*]." Their basis for denying the relativism in their account is that they have merely described "with impartial care the deducible consequences of preferring one form of social organization over others" (p. 187). However, when one claims that any risk position can be justified and that none is better than another, as Douglas and Wildavsky do, then one is not making an "impartial," first-order, descriptive claim. Instead, one is making a second-order, normative assertion about evaluating risk positions. Douglas and Wildavsky err here because they seem to confuse both first- and second-order claims as well as descriptive and normative assertions. Their own words commit them to an ethical and epistemological relativism, to the thesis that no one "view of risk . . . is better or worse" than another. That is, their own words commit them to a second-order claim about how to evaluate risk descriptions—relativistically. Their response to the charge of second-, or meta-, level relativism is to claim that their first-order accounts of social organization have been given with impartial care. However, the charge of relativism is not directed against their first-order social descriptions (of hierarchy, market, and sect) but against their second-order judgments about criteria for comparing and evaluating all social descriptions. Hence they do not succeed in meeting the charge of relativism directed against them. Moreover, if "everyone . . . is biased" in her risk judgments, as Douglas and Wildavsky assert (pp. 80–81), then their own claims are biased as well, and there is no such thing, in their own terms, as describing consequences "with impartial care."

The Fourth Argument

Douglas and Wildavsky's assertion that they are merely describing various positions (on risk) "with impartial care" is especially problematic in the light of their claim that there is little difference between the risk responses of modern persons and primitives: "Everyone, expert and layman alike, is biased. No one has a social theory above the battle. . . . Thus the difference diminishes between modern mankind and its predecessors" (1982, pp. 80–81).

This fourth argument is formulated after Douglas and Wildavsky present a string of anthropological examples concerning the pollution fears of other cultures. They claim that the contemporary aversion to environmental contaminants, with its blame of industry, is like the fear of "pollution" in primitive cultures, in which powerless individuals blame powerful persons and institutions for some impurity. They maintain that the inhabitants of the South African Transvaal, for example, fear drought and believe that their queen's anger causes it (1982, p. 39).

After providing a number of similar examples of how primitives irrationally "blame" those in power for environmental contamination and hazards, Douglas and Wildavsky conclude that

> an internal social problem about guilt and innocence is stated in the pollution belief. Defilement is also a well-pointed metaphor of moral stain . . . many pollution fears are associated with sex. . . . Cultural analysis shows us that ideas about pollution are not sufficiently explained by the physical dangers . . . laying blame at someone's door makes sure it will be invoked whether it is believed or not. (1982, pp. 37–38)

In short, they are claiming that contemporary environmentalists are averse to risks because they, like primitives, want to "blame" those in power for "impurity."

The main problem with this fourth argument is that despite Douglas and Wildavsky's assertion that they are merely describing consequences of various risk positions "with impartial care," their allegations about the similarities in pollution beliefs, among primitives and contemporary people, are hardly mere descriptions. Instead, they are attributing intentions (e.g., the need to blame someone) to persons, even though it is impossible to substantiate these charges with any precision. They are also postulating causes for environmentalists' risk aversion, causes that go beyond the desire to avoid the physical dangers associated with hazards such as toxic chemicals and nuclear power plants. This means that Douglas and Wildavsky are in a precarious position when they claim to be "describing" situations, even though persons whose behavior is so "described" would not agree with how they are depicted. For this reason, Douglas and Wildavsky must bear the burden of proof for the intentions and motives that they attribute to environmentalists. They give no arguments for their attribution of motives but simply assert that pollution fears serve the same "function" ("blame") among all cultures. Their argument, therefore, begs the question.

There are other difficulties with their fourth argument. One is that there are strong reasons for doubting that contemporary risk aversion can be explained so simply. And if not, there is reason to doubt that the fears of primitive and advanced people are relevantly similar. The most obvious basis for doubt is that although primitives have virtually no scientific understanding of risks and their causes, contemporary laypersons typically have quite sophisticated technical knowledge about the aetiology of various hazards. Most persons in developed countries usually do not believe that demons cause illnesses or that the gods' anger causes flooding. Because their understanding of various harms diverges dramatically from that of members of primitive cultures, there is little reason to suppose that their motives for attributing pollution should converge. Why should a scientifically sophisticated American have the same grounds for aversion to technological hazards as would a scientifically ignorant primitive? They are not even explaining the same thing, as their conceptions of its causes differ so dramatically. What basis is there for postulating that pollution beliefs have the same source, given that the respective pollution causes are understood in such radically different ways? In the absence of an explanation, it appears that Douglas and Wildavsky have merely proposed a primitivist metaphor for, not a description of, contemporary risk judgments.

Apart from whether primitive and more technologically sophisticated people are alike in their hazard aversion and pollution beliefs, Douglas and Wildavsky's fourth argument is problematic in claiming that risk attitudes are determined solely by sociology or "social organization." If Fischhoff, Slovic, and other psychometric researchers are correct, many risk judgments are a function of both personal psychology and ethical beliefs, for example, in the equity of risk distribution (Fischhoff, Slovic, and Lichtenstein 1980; Fischhoff et al. 1978; Kasper 1980). They are not merely a function of social organization, as Douglas and Wildavsky claim. Fischhoff and his fellow researchers have also established that laypersons' risk responses can be predicted almost completely, purely on the basis of one attribute of a risk's severity of consequences (Fischhoff et al. 1978; Green 1977). If this is so, then the hazards themselves, not the characteristics of the social group responding to probable harm, are the key determinants of attitudes toward risk aversion. Douglas and Wildavsky reduce all causes of risk behavior to social causes and thus employ a "sociological reductionism." But surely this reduction is erroneous, at least in part because it is so simplistic, because it ignores other plausible causes of risk attitudes, and because it fails to predict important (if not most) instances of hazard evaluation and behavior.

If this sociological reductionism were true, then Douglas and Wildavsky would be unable to explain a number of facts. One such anomaly is how members of the same social group (market, hierarchy, or sect) have divergent views about risk, or how those who share the same views about hazards can be members of different social groups. For example, Douglas and Wildavsky admit that the Sierra Club has a hierarchical organization (1982, p. 132), even though they allege that environmentalists belong to sectarian, not hierarchical, groups. Likewise, they claim that environmentalists tend to be removed from the center of power and influence (and therefore presumably are Democrats rather than Republicans, for example). Yet the Sierra Club membership shows the same percentages of persons, by party affiliation, as does the general population. The membership is also primarily white-collar, middle-class professionals, not blue-collar, lower-class hourly-wage workers (p. 130).

Another anomaly inexplicable in terms of Douglas and Wildavsky's sociological reductionism is their claim that environmentalists are group oriented, democratic, and averse to personal (as opposed to group) power. Numerous environmentalists in the Teddy Roosevelt tradition are rugged individualists, entrepreneurial, and oriented toward survival activities in the wilderness (Nash 1973). In other words, many environmentalists belong to no voluntary associations, and the group's social structure (that Douglas and Wildavsky allege) determines their risk attitudes is not even present in their lives. They often are loners who like to commune with nature and carry only a backpack. At best, they are apolitical and uninfluenced by any specific sectarian social structure, particularly by a voluntary environmental group. Douglas and Wildavsky cannot explain the environmentalism of such loners as a product of any social structure or voluntary association to which they belong. Hence it is reasonable to suspect, once again, that in ignoring the influence of ethics, personal psychology, and the characteristics of the danger itself on risk judgments, Douglas and Wildavsky provide too simplistic a scheme. All reasons for attitudes toward hazards cannot be reduced to social determinants.

Apart from the epistemological problems with this fourth argument, reducing the determinants of risk judgments to social and political causes, Douglas and Wildavsky face a number of undesirable practical consequences of their position. Their view is oversimplified and reductionistic in ascribing divergent hazard judgments to social structures and to politics, rather than also to different risk philosophies or alternative methodological assumptions. For example, people might disagree in their risk evaluations, not only because they are members of different social groups, but also because they make different assumptions about risk methods. Some persons might assume that purely economic methods are correct, whereas others might reject this presupposition. If one assumes that conceptual and methodological issues never play a role in risk controversy, then one is ignoring a subtle and therefore dangerous source of conflict over hazards. Moreover, to assume that social structures and political clout alone cause risk disagreements prejudges the issues that often divide reasonable persons. It is also anti-intellectual, because methodological and conceptual disagreements (e.g., as to the accuracy of risk probability projections or the reliability of distributed methods of risk–benefit analysis) often are at issue. This is so regardless of whether political disagreements (e.g., as to whether it is in industry's interest to compute the social costs of hazards) are present. For all these reasons, it seems premature, at best, for Douglas and Wildavsky to assume that only social structures determine risk evaluations.

The Fifth Argument

If the fourth argument is, at best, premature and, at worst, false, the fifth argument is suspect as well, as it is one variant of the earlier claim that risk judgments are determined by social structure. Here Douglas and Wildavsky maintain that environmentalists' attitudes toward hazards are a result of their "problems" as members of sectarian, rather than hierarchical or market-oriented, groups. Douglas and Wildavsky characterize sectarians as pessimistic, anti-institutional, prone to believing conspiracy theories, and eager to find enemies and impurities to condemn (Wildavsky and Douglas 1982, pp. 102–74). They allege that these group-induced sectarian characteristics explain why environmentalists both criticize commerce and industry as purveyors of pollution and are quick to complain about the supposed impurities that manufacturers impose on the population.

Like the previous charge, this fifth argument invokes hidden causes of risk aversion and attributes questionable intentions and motives to environmentalists, both without substantiation. This lack of support is all the more significant because up to 90 percent of cancers are environmentally induced and so theoretically preventable (Lashof et al. 1981). Thus there are often real reasons for risk aversion. One need not invoke aberrant motives and bizarre intentions to explain why people are afraid of many pesticides, food additives, and energy sources. Their aversion can be explained very simply: They want neither to be a cancer statistic nor to be endangered without their knowledge and consent. Moreover, even if the allegations about group characteristics were accurate, it would be highly questionable to assume that only social structures, rather than personal, psychological, and ethical reasons, contribute to judgments about risk.

In the absence of evidence supporting the claims that environmentalists are sectarians and that sectarians may be characterized as pessimistic, paranoid, and the like, Douglas and Wildavsky appear to have begged the question of the social determinants of risk aversion. They have invoked an *ad hominem* argument against environmentalists who supposedly have "problems" because of their socially determined sectarian views. They need to establish that environmentalists are more risk averse than the facts about hazards dictate and to show that the causes of this aversion are their alleged group-induced pessimism, paranoia, and anti-institutional sentiments. It will not do for them to say merely that "life is growing longer not shorter; health is better not worse" and therefore that citizens have no right to complain or to be averse to technological risks. For an adequate argument, Douglas and Wildavsky need also to substantiate highly sophisticated ethical and scientific claims that apart from whether things are getting safer, environmentalists have no grounds for affirming that this progress ought to be faster or ought to be achieved with more consent or more equity. This they have not done.

In reducing the reasonableness of risk aversion to the question of whether or not life is becoming safer, Douglas and Wildavsky have again ignored the ethical, psychological, and scientific issues that make risk acceptability more than a matter of mere quantity of safety. They appeal to alleged facts about hazards, that is, that life is becoming safer, when they wish to discredit the environmentalists' aversion to risk. Yet throughout the remainder of their book, they deny the existence of any such "facts" about hazards and claim that "risk is a collective construct" (1982, p. 186), that it is immeasurable (p. 184), that it is determined by social structures, and that no hazard judgment is better or worse than another (p. 188). If their sociological reductionism and relativism are correct, however, they are inconsistent in making the risk judgment that life is getting safer. Their attacks on environmentalists' risk aversion (as contrary to the assertion that life is getting safer) also are inconsistent, as a consequence of their own position is that there are no "facts," only conventions, about risk. Likewise, if their complaints about citizens' ignoring the facts are correct, they are incorrect in their claims about sociological relativism and reductionism.

NAIVE POSITIVISTS: STARR AND WHIPPLE

Interestingly, in reducing risk determinants to social structures, members of the cultural relativists' camp, like Douglas and Wildavsky, share a common error—reductionism—with the naive positivists. Just as the cultural relativists attempt to reduce risk evaluations to sociological constructs, so the naive positivists attempt to reduce hazard estimates to allegedly objective calculations determined by scientists. The cultural relativists overemphasize values in risk assessment, and the naive positivists underemphasize them. Both groups err, however, in believing that categorical value judgments are purely relative and matters of taste. The naive positivists ignore such values precisely because they believe that they are subjective. The cultural relativists embrace alleged subjective values because they believe that relativism is unavoidable. Let us see where and why they go wrong.

The naive positivists believe that risk judgments can and should be both wholly neutral and objective and that they can be reduced to the terms of natural science. (Admittedly, however, one can argue that natural science is neither wholly neutral nor wholly objective.) Let us see how this account of objectivity and neutrality goes wrong. First, the naive positivists base their position on a number of doubtful presuppositions, such as the pure-science ideal and the fact–value dichotomy. Second, acceptance of their account of risk would lead to serious public-policy consequences. Third, their view amounts to proscribing the role of the normative scholar in risk assessment.

The Principle of Complete Neutrality

The naive positivist belief that risk judgments should exclude normative (ethical and methodological) components is based on several assumptions that comprise (what I shall call) a "principle of complete neutrality." One assumption is that hazard assessment can and should be wholly objective and value free. Another assumption is that any foray into applied ethics or methodological criticism represents a lapse into advocacy and subjectivity. Representative statements of some such principle of neutrality, central to the naive positivist account of risk assessment, have been given by the National Academy of Engineering, the U.S. Office of Technology Assessment (OTA), and a variety of scholars and risk assessors, such as Starr and Whipple (Burnham 1979; Hafele 1976; Kasper 1980; Lawrance 1976; Okrent and Whipple 1977; Rowe 1977; Starr and Whipple 1980). The OTA, for example, says that assessments must be "free from advocacy or ideological bias" and must be "objective" (OTA 1976b, 1977). Dorothy Nelkin claims that no scholar has any grounds for talking about "the rights and wrongs of policy choice" but instead should "understand . . . interpret . . . draw a coherent picture of what is going on" (Nelkin 1981, pp. 16–17).[4] So great is the naive positivist faith in the principle of complete neutrality that some scholars have even stated that different persons, coming from alternative vantage points, should "come to the same conclusion" in their risk judgments (Carpenter 1972, p. 42). Such assertions overlook the unavoidable presence of categorical and instrumental value judgments in science.

Although the principle of complete neutrality represents a noble aim and an important effort to keep both science and policy evenhanded and empirically relevant, rather than prejudiced, biased, and superstitious, the principle is nevertheless open to question. It is clearly wrong if it is construed to mean that science and risk judgments can and should be wholly value free. (To their credit, Douglas and Wildavsky recognized that such value freedom is unattainable. Hence, despite the failure of their relativistic program, they were correct in showing some of the sociological reasons that risk judgments can never be completely value neutral.)

Part of the reason that the naive positivists have assumed the legitimacy of the principle of complete neutrality is that they have failed to distinguish among different types of values. Some of these values occur in science and risk judgments, whereas others can be avoided. According to Longino's classification, they can be divided into three types: bias values, contextual values, and constitutive values—a classification (she admits) that is neither mutually exclusive nor exhaustive (Longino 1983).

Bias values occur in risk judgments whenever one deliberately misinterprets or omits data, so as to serve one's own purposes. These types of values obviously can be avoided in risk assessment and in science generally, as eliminating them is merely a matter of avoiding a deliberate, erroneous manipulation of risk data. The naive positivist view is correct insofar as it stipulates that bias values can and should be kept out of science and risk assessment.

Contextual values, however, are more difficult to avoid. Risk assessors subscribe to particular contextual values whenever they include personal, social, cultural, or philosophical emphases in their judgments. For example, one uses contextual (philosophical) values when deciding to follow an atomistic, rather than a field-theoretic, paradigm in high-energy physics research, or when deciding to use certain existing probabilistic data for a given risk, on the grounds that one does not have the research money to do one's own (new) studies on a particular accident frequency.

Although in principle it might be possible to avoid contextual values—especially pragmatic ones associated with "cultural baggage" and financial con-straints—in practice it would be almost impossible to do so, in either science or risk assessment. This difficulty is illustrated by a typical case: contextual values in the study of human interferon. As Longino points out, industrial microbiology as practiced by small firms of biochemists has been heavily influenced by cultural and financial values, such as the profit motive. These values have determined both the way that the interferon research results are tested and the way that they are announced (Longino 1983). Likewise, Korenbrot showed that the se-lection of health risks to be considered when testing oral contraceptives was heavily influenced by the contextual value of checking population growth; be-cause of this value, researchers overemphasized the benefits of the contraceptives and underestimated the risks (Longino 1983).

One reason that contextual values have played such a large role in many areas of scientific activity and risk assessment is that they often fill the gap left by limited knowledge. Because any research is hampered by some type of in-complete information, there is always an opportunity to determine scientific procedures and risk policy by means of contextual values. Moreover, these contextual values often have little to do with the factual adequacy of the risk procedures and policies.

Constitutive values are even more difficult to avoid in both science and risk assessment than are contextual and bias values. Indeed, it is impossible to avoid them, as one makes a constitutive value judgment whenever one follows one methodological rule rather than another. Following a methodological rule, even in pure science, represents adherence to constitutive values because choosing it means rejecting other methodological principles. Indeed, even collecting data requires using constitutive value judgments because one must make evaluative assumptions about what is studied, so as to know what data to collect and what to ignore, how to interpret the data, and how to avoid erroneous interpretations.

Constitutive value judgments are required even in pure science because per-ception does not give us pure facts; the knowledge, beliefs, values, and theories we already hold play a key part in determining what we perceive. The high-energy physicist, for example, does not count all the marks on his cloud-chamber photographs as observations of pions, but only those streaks that his theories indicate are pions. Even in the allegedly clear case of observational "facts" contradicting a particular theoretical account, a researcher need not reject the theory (laden with values); she could instead reject the facts and hold on to the theory. Earlier in this century, for example, when beta decay was observed, it was not accepted as a counterinstance to the theory of conservation of energy and momentum. Rather, the value-laden conservation theory was accepted, and beta decay was treated, not as a hard-and-fast fact but as a problem that scientists needed to solve. All this suggests that values structure and organize experiments, determine the meaning of observations, and tell us which ones are relevant, how to attack problems, and what counts as solutions to them (Brown 1977; Shrader-Frechette 1980).

Not all constitutive or methodological values are created equal, however, and although the naive positivists reject some of them, they accept others. This means that to see precisely where and why the naive positivist account of risk judgments goes wrong, we need to clarify further the nature of constitutive value judgments. Scriven and McMullin do a good job, noting that values can be emotive (what we have called bias), pragmatic (a subset of what we have called contextual), or cognitive (what we have called constitutive). They claim that only emotive or bias values have no place in science.

Evaluating and Valuing

Within the cognitive or constitutive value judgments, McMullin distinguished evaluating from valuing, instrumental value judgments from categorical value judgments. We can make a largely factual judgment and evaluate the extent to which a particular thing possesses a characteristic value; for example, a theory about risk assessment possesses explanatory or predictive power. Or we can make a largely subjective judgment and value an alleged property, for example, assess the extent to which a characteristic, such as simplicity, is really a value for a scientific theory or a theory about risk aversion (McMullin 1983; Scriven 1980).

Like Hempel's "instrumental value judgments," what McMullin calls "evaluating" judgments assert that if a specified value or goal is to be obtained, a certain action is good. For example, if the value of predictive power is to be obtained, then one should develop and test all relevant implications of the theory. On the other hand, like Hempel's "categorical judgments," what McMullin calls "valuing" judgments state that a certain goal is prima-facie good, independent of particular circumstances (Hempel 1965, 1979, 1980, 1983; McMullin 1983; Scriven 1980).

McMullin, Scriven, and other postpositivists accept both instrumental and categorical value judgments in science (Hempel 1965; Nagel 1961; Scriven 1980). Hempel and the naive positivists, along with a number of risk assessors like Starr and Whipple, believe that the latter have no place in science and science-related activities like risk assessment. Their complaint against categorical value judgments is that because they cannot be confirmed empirically, they are subjective (Hempel 1979, pp. 45–66; 1980, p. 263; 1983, pp. 73–100).

In making empirical confirmability a criterion for judgments in science and risk assessment, and thereby excluding categorical judgments of value, naive positivists like Carnap, Reichenbach, Starr, and Whipple appear to err for at least two reasons. First, Aristotle claimed that wise persons realize the reliability characteristic of different kinds of judgments and that they demand only that assurance appropriate to the particular type of investigation. The naive positivists demand an inappropriate level of assurance because their requirement of empirical confirmability would allow neither pure scientists (if there is such a thing) nor risk assessors to decide on criteria for choosing a theory, gathering and interpreting data, or rejecting hypotheses. Such judgments could not be empirically confirmed because each of them relies on at least one categorical judgment of value, for example, an assumption about the prima facie importance of some criterion for choosing a theory or hypothesis. Yet, as examples such as beta decay and pion observation indicate, scientists do make judgments about weighting criteria to evaluate theories, observations, and hypotheses, and they do so all the time. This means that scientific practice does not confirm the naive positivists' notion of the role of confirmability in science and risk assessment (Shrader-Frechette 1985b).

Second, if all judgments in science had to be empirically confirmed, then science as we know it would come to a halt. Scientists could never make judgments about theory choice, as such choices rely in part on unconfirmable categorical judgments of value. The same must hold true for risk analysis. As a policy-related activity, it is even more value laden than is science. If the naive positivists are correct in requiring empirical confirmability of all judgments in hazard assessment, then they have brought it to a halt. Moreover, if risk estimation were as objective as Starr and Whipple have claimed, then it would never progress, and impotent hypotheses used to estimate risks would never be discarded. But risk estimation does progress, and impotent theories often are discarded. Scientific revolutions occur, even in hazard analysis. Hence not all judgments in science and risk assessment are empirically confirmed. They cannot be if science and risk assessment are to continue as they have.

The Fact–Value Dichotomy

Perhaps one of the most basic reasons that many risk assessors and scientists have erroneously believed that it was possible to make only confirmed judgments, about either risks or science, is that they subscribe to the fact–value dichotomy, a famous tenet of naive positivism (Hanson 1958; Kuhn 1962; Polanyi 1958; Toulmin 1961). This is the belief that facts and values are completely separable and that some facts are not value laden. Applied to hazard assessment, this claim is that risk analysis should consist of factual and neutral risk estimates, although the policy decisions made as a consequence of them may be evaluative. Many hazard assessors take exactly this position (Foell 1978; Lovins 1977; OTA 1976b).

Despite the fact that scholars such as John Stuart Mill, Lionel Robbins, Milton Friedman, and others have affirmed the fact–value dichotomy (Hare 1979; Tool 1979), there are a number of reasons for doubting it and the pure-science ideal associated with it. First, belief in the dichotomy is incompatible with the formulation of any scientific theory or analysis to explain causal connections among phenomena. This is because the formulation of any theory requires one to make epistemic value judgments, including categorical value judgments. Second, presenting alternative accounts of one's options can never be a purely descriptive or factual enterprise. This is because one has to use evaluative criteria to select which options to present, and these normative criteria are outside the scope of allowable inquiry (Tool 1979). Third, to subscribe to the fact–value dichotomy is to believe that there can be pure facts and presuppositionless research. As we indicated earlier in regard to bias, contextual, and constitutive values, (including evaluating and valuing), there is no presuppositionless research, although everyone should attempt to avoid bias. All facts are laden, in principle, with constitutive values and, in practice, with contextual values. Hence even pure science, if there is such a thing, cannot be achieved merely by collecting data, as there are no pure or brute facts.

Value-free Observations Are Not the Only Guarantees of Objectivity

Admittedly the traditional empiricist motivation behind the naive positivist belief in the fact–value dichotomy is a noble and important one. It is noble because value-free observations, if they existed, would guarantee the objectivity of one's research. Values threaten objectivity, however, only if they alone determine the facts. A great many philosophers of science (myself included) maintain that both our values and the action of the external world on our senses are responsible for our perceptions, observations, and facts. Perception is not passive observation, and its objects are not created out of nothing. Rather, perception is both structured by the world outside us yet susceptible to the imprint of our own presuppositions, values, and theories. This means that although observations or facts may be seen in different ways, they may not be seen in any way. Just because facts are value laden, it does not mean that there is no sufficient reason for accepting one theory over another. One theory may have more explanatory or predictive power, or unify more facts, or be more coherent.

Quark theory provides a dramatic example of this point. Isolated quarks are in principle unobservable, yet there are numerous good reasons for accepting quark theory. The value-ladenness of facts (whether in science or in risk assessment) implies the cultural relativism of Douglas and Wildavsky only if one accepts the positivists' presupposition that the observation of value-free data alone provides reasons for accepting one theory rather than another. Because there are other good reasons for theory acceptance (e.g., external consistency, predictive power), one need not accept this positivist presupposition. Interestingly, however, despite their divergent views of risk assessment, both the naive positivists and the cultural relativists accept this erroneous presupposition, that the observation of value-free data alone provides reasons for accepting one theory rather than another. This means that both groups fail to understand the epistemological grounds for theory acceptance, whether in science or risk assessment.

It Is Undesirable for Risk Judgments to Be Wholly Neutral

In response to the arguments that both relativists and naive positivists misunderstand the grounds for theory acceptance and that presuppositionless research is impossible in both science and risk assessment, many naive positivists make an important point: They maintain that although complete objectivity and neutrality cannot be achieved, it still should be a goal or ideal of science and risk assessment. In other words, they state that even if we cannot make presuppositionless judgments, we should try to do so (Skolimowski 1976).[5]

On the contrary, there are good grounds for arguing that it is not even desirable for risk judgments to be wholly neutral, in the sense of avoiding advocacy or criticism of particular positions on risk. For one thing, to avoid normative criticism in risk assessments is to assume that one should never analyze or criticize the constitutive, contextual, or cognitive values embedded in the data or methodology underlying a particular risk judgment. But if such values cannot be avoided, then to refrain from criticizing them is merely to endorse a risk judgment that sanctions those values, regardless of the errors to which they might lead. This means that if neutrality or objectivity is identified with avoiding normative analysis, objectivity is therefore identified with sanctioning received or status-quo values. But if so, those who argue for pursuing the ideal of objectivity are inconsistent. They are implicitly arguing for sanctioning received or status-quo values, even as they explicitly claim to avoid normative work.

A second reason that a risk assessor should not pursue even the ideal of complete neutrality is that neutrality (with its presupposition that one should not engage in advocacy or normative criticism) presupposes that any judgment about hazards is as good as another. Yet, as we argued earlier in connection with Douglas and Wildavsky, one risk judgment is clearly not as good as another.

Third, far from contributing to objectivity, pursuing the ideal of neutrality (by describing widely diverse risk options in an equally plausible way) would, in many instances, fail to serve authentic objectivity. Consider, for example, the case in which largely political reasons for accepting a particular risk were widely publicized by a special-interest group. For example, suppose nuclear utilities urged the acceptance of the breeder technology because they desired a return on their investment, having spent huge sums in nuclear research. To support their arguments for the breeder technology, suppose the utility used implausible assumptions in order to argue that future energy demand would be so high that it could not possibly be met by means of any energy source except breeder reactors. Because some assumptions about electricity usage are more reasonable (e.g., cost-induced elasticity of energy demand) than others are (e.g., increasing energy demand, regardless of price), authentic neutrality would not be served by presenting all assumptions about energy demand in an equally plausible or neutral way. This is because various assumptions are based on different constitutive values, not all of which are equally plausible. But if not, then failing to criticize questionable constitutive values in a given risk assessment is unwise, and for at least two reasons. First, someone else might fail to do so. Second, accepting such values (in the name of neutrality) could lead to disastrous public-policy and economic consequences. The most objective thing to do, in the presence of questionable methodological assumptions or constitutive values, is to be critical of them and not to remain neutral.

Admittedly, a problem arises when one erroneously criticizes a risk judgment in the name of objectivity. Linking objectivity with advocacy and with criticism, as has been done here, opens the door for extremism and ideology, both practiced in the name of risk assessment and science. To this danger, however, there is one important reply. All activity of any kind would cease if a necessary precondition for it were that one never erred by engaging in it. The antidote to moralistic excesses and absolutistic pronouncements is not to prohibit criticism but to provide a risk-assessment framework in which alternative risk judgments can be developed, compared, and criticized. (I have tried elsewhere—1985b— to indicate the main features of such a framework.)

A fourth reason that one should not pursue the ideal of complete neutrality and should not reject normative analyses in making risk judgments is that ethical and methodological criticisms often contribute to better public policy. Criticizing the assumptions and values related to risk judgments enables policymakers and the public to grasp the corollaries and consequences of their decisions about societal hazards. This, in turn, enables them to determine whether they can live with their choices (Weiss and Bucuvalas 1980). Fifth, to accept the ideal of complete neutrality is to ignore the reality that risk judgments affect public policy and that public-policy judgments are made in a political environment (Quade 1975). Policy is not made by a hypothetical decision maker pursuing the public interest but by interaction among a plurality of partisans (Lindblom and Cohen 1979). This being so, pursuing the ideal of complete neutrality would limit both the quality and the quantity of these interactions necessary for creating public policy.

Sixth, those who pursue the ideal of complete neutrality are inconsistent. By denouncing all forms of ethical and methodological advocacy and criticism in risk assessment, they are adopting an unsubstantiated advocacy for the naive positivist presupposition. This presupposition is that one can and should be purely objective and neutral in one's risk judgments.

Seventh, the real issue concerning value judgments in hazard assessment is not whether one should speak normatively or critically in one's risk judgments, as values, theories, and norms unavoidably structure both science and risk analysis. The real issue is whether a given judgment is normative in a way that is misleading, incoherent, incomplete, question begging, or implausible. In other words, the epistemological presuppositions of the naive positivists lead them to ask the wrong question. They ask whether a particular judgment is normative, rather than whether it is normative in a way that is not substantiated or not reasonable. Because they ask the wrong question, their level of epistemological analysis in hazard assessment is too crude to shed much light on the reasons that risk judgments are appropriate or inappropriate.

Finally, dangerous consequences would follow if one accepted the naive positivist presupposition of the ideal of complete neutrality: (1) sanctioning ethical relativism, (2) accepting the status quo, and (3) failing to see the real sources of controversy over risk acceptability. The first consequence follows because if one accepts the naive positivists' claims that only facts are neutral and objective, there are no rational grounds for deciding questions of value. For them, all values are relative. Ethical relativism is a dangerous position because it can lead to catastrophic and unconscionable consequences, as was suggested earlier in our analysis of Douglas and Wildavsky. For example, if one claimed that avoiding nuclear war were only a relative or subjective value, inestimable harm could occur. The judgment (of relative value) would probably render war more likely. Similarly, the belief that all value judgments were relative could lead one to remain silent while great risks or great abuses were showered on innocent people.

Sanctioning the naive positivists' view of neutrality, as was already mentioned, could also lead one to accept the status quo. If one made a wholly neutral judgment about a risk policy that promoted racism or sexism or violated civil liberties, for example, then one's neutral judgment would be taken as supporting the questionable risk policy. In fact, this is exactly what happened in 1933 when

Albert Einstein condemned Adolf Hitler's violations of civil liberties. The Prussian Academy of Sciences denounced Einstein for not remaining neutral. Yet by their denunciation, the academy was supporting the status quo, that is, the violations of civil liberties (Einstein 1949, 1954; Frank 1974).

Analogously, if one never makes normative judgments about which risks are acceptable, under the guise of being neutral, this silence about alleged errors, evils, or misrepresentations will simply help legitimate whatever risk policy is dominant or currently being followed. However, when dealing with a questionable policy with the potential to cause great harm, it can even be argued that on grounds of beneficence (Ross 1930) one has a moral obligation not to be neutral in judging it. Albert Camus made the same point: "We are guilty of treason in the eyes of history if we do not denounce what deserves to be denounced. The conspiracy of silence is our condemnation in the eyes of those who come after us" (Camus 1965, p. 146). Abraham Lincoln said something similar: "silence makes men cowards" (see Primack and von Hippel 1974).

A third consequence of subscribing to the ideal of complete neutrality is that it might blind one to real sources of controversy over risk acceptability. If one subscribes to this ideal and therefore emphasizes the factual side of risk judgments, this can only mean that the evaluative aspects of hazard assessment are underemphasized. And if the evaluative aspects are underemphasized, there is little chance that the real (evaluative) sources of controversy over risk acceptance will be recognized and therefore little chance that the debates they fuel will be settled. As a consequence, public policy regarding environmental hazards is likely to be far less rational and successful than it might be. In fact, this is exactly what has happened. Risk assessors such as Starr, Whipple, Cohen, Maxey, and Lee all subscribe to the naive positivist faith in complete neutrality and objectivity. As a consequence, they miss the numerous evaluative presuppositions in their own work (which they view as objective). Moreover they are insensitive to the merits of alternative presuppositions in the work of those who disagree with them. They assume that those who hold different positions about the desirability of nuclear power, for example, are simply mathematically and scientifically "illiterate" laypersons who do not know "the facts" about nuclear-risk probabilities (Shrader-Frechette 1982).

Their assumption is highly questionable. As we already mentioned, psychometric researchers have confirmed that nuclear opponents and proponents make roughly the same judgments about nuclear-risk probabilities; that is, the technical aspects of their claims are not statistically different. Where they disagree, however, is on the acceptability of nuclear-risk distribution, possible catastrophic accidents, and the alleged benefits to be exchanged for the risk (Fischhoff, Slovic, and Lichtenstein 1980; Fischhoff et al. 1978; Kasper 1980). In other words, psychometric research confirms that the nuclear-risk debate is primarily over values, not risk probabilities. The naive positivists, however, miss the real sources of the disagreement. Their misperception, in turn, is likely to discourage resolution of the controversy and to encourage the formulation of a risk policy that is not capable of addressing the real issues.

A MIDDLE PATH BETWEEN NAIVE POSITIVISM
AND CULTURAL RELATIVISM: TOWARD A NEW
NOTION OF OBJECTIVITY IN RISK JUDGMENTS

If public policy concerning risk is to address the real sources of lay unrest and expert disagreement, that policy must be based on an accurate account of the reasons for the controversy. At least part of this unrest and disagreement, I have argued, can be attributed to misunderstanding the role of values in hazard assessment. Cultural relativists overemphasize values and reduce all risk judgments to mere sociological constructs. Naive positivists underemphasize values and reduce all risk estimates to pure science, to alleged objective determinations of quantities such as risk probabilities (even though pure science is not objective in the sense they presuppose). The most plausible view, as is often true, lies between these two extremes. To substantiate this middle position, which I call scientific proceduralism, we must develop a new notion of objectivity, one not based on the naive positivists' criterion of empirical confirmability.

According to the view known as scientific proceduralism, the rationality appropriate to risk analysis is best defined in terms of three propositions: (1) that, following Hempel, there is at least one general, universal criterion for choosing a theory or paradigm in both science and risk assessment, that is, explanatory power as tested by prediction; (2) that most of the remaining criteria for theory choice, although evaluated and interpreted in terms of how well they function as a means to an end or goal of explanatory power as tested by prediction, are both situation specific or determined largely by practice; and (3) that if some of these criteria allow intelligible debate and criticism by the scientific community, they will guarantee what I shall call *scientific objectivity*.

Because I cannot defend here all three of these propositions, I will rely in part on the arguments of McMullin and Hempel to assert that there is at least one general, universal criterion for choosing a theory in science and risk assessment, explanatory power as tested by prediction (Cornman 1980; Hempel 1979, 1983; Sellars 1967). To see how this general criterion might operate, one need only recall some of our earlier arguments. For example, I claimed that the theory of risk behavior adopted by Douglas and Wildavsky fails in its explanatory power because it does not provide an account of hazard aversion arising from personal and psychological variables. In ignoring these factors, Douglas and Wildavsky cannot make accurate predictions of behavior regarding societal hazards. For example, in their view, members of the Sierra Club should be more left-wing, liberal, and democratic than other members of society are. Yet the Sierra Club is perfectly representative of the entire population with respect to its Democratic–Republican, liberal–conservative, left-wing–right-wing membership.

Explanatory power as tested by prediction also functions as a universal criterion of theory choice in the area of risk estimation. To the degree that existing risk assessments alleged that the Chernobyl accident was "impossible," they failed both to explain and to predict the actual occurrence of the accident. To say that Soviet hazard estimates fail with respect to explanatory and predictive power, however, is not the same thing as saying that the assessments have not been empirically confirmed. If a risk is small, often its magnitude cannot be confirmed. That is why explanatory power, tested by prediction, sometimes functions as a goal of science and risk assessment, not as a criterion for it. To understand how explanatory power as tested by prediction can function as a universal criterion or goal for theory choice in science and in hazard analysis,[6] we must understand why objectivity does not always require empirical confirmability.

Moving Beyond Empirical Confirmability

It is not reasonable to require empirical confirmability of all risk judgments, because it is not the only test of objectivity, either in science or anywhere else. For example, we often call a judgment objective, even in science, if it is not obviously biased or subjective. Objectivity, in this sense, is not tied to empirical confirmability so much as it is linked to evenhanded representation of the situation. What I call scientific objectivity (and I use the word *scientific* not because the objectivity is unique to science but simply because it is characteristic of science and science-related activities, like making risk judgments) is closely related to this sense of objectivity, as evenhandedness.

Presumably one could be blamed for failing to be objective, in the sense of not being evenhanded, if one were biased in a particular risk judgment. For example, if one drew a conclusion about risk acceptability after deliberately misrepresenting some of the data on the probability of the hazard, then one could be said to have violated scientific objectivity (and to have fallen victim to bias values, as earlier explained). Because we often do blame people for not being objective, in a sense close to that about which I am speaking, it is clear either that objectivity in this sense must be attainable or that one can be more or less objective, contrary to what the cultural relativists believe.

But how might one guarantee scientific objectivity in the sense of evenhanded representation of the situation? It will not do to say that a risk judgment is objective if it fits the situation or the facts, because (1) we do not want to beg the realism question regarding the status of any alleged facts, (2) we might not have all the facts, and (3) every risk situation will be different. Thus it is virtually impossible to specify ahead of time what an evenhanded representation of a particular situation might be. But if so, we may not be able to come up with any characteristic of objective scientific activity applicable to risk judgments.

Instead, we may be able only to do two things. First, we may be able to provide a general criterion for scientific rationality and objectivity in its application to risk assessment. This general criterion is likely to be explanatory power as tested by prediction. For example, we might say that a particular theory of risk response is rational if it can explain the behavior in question (e.g., aversion to commercial nuclear fission) and predict future instances of such behavior. Or we might say that a particular theory of risk estimation is rational if it can explain the magnitude and distribution of a particular risk and predict both the magnitude and distribution of such risks in the future. Second, in addition to providing this general criterion, we may be able to contribute to rationality and objectivity by specifying how to avoid the charge of violating scientific objectivity in our risk judgments.

One way to avoid such a charge might be to subject our hazard assessments to review by affected laypersons and the scientific community. These expert and lay judgments might be said to possess scientific objectivity, in some minimal sense, if they can be subjected to criticism, debate, and amendment. According to this account, although scientific rationality might be guaranteed by a risk assessor working individually, by pursuing a goal of explanatory power tested in part by prediction (Hempel 1979, 1983; McMullin 1983; Sellars 1967), scientific objectivity could be guaranteed only by scientists and affected laypersons working together to criticize the proposed accounts of risk acceptability. According to this view, scientific rationality is a necessary condition for scientific objectivity, and scientific objectivity characterizes the middle position (between cultural relativism and naive positivism), which I call *scientific proceduralism*. The position is procedural because it relies on democratic procedures, that is, criticism by lay people likely to be affected by the risk judgments, as well as by scientists.

Also, according to this account, if epistemic or cognitive value judgments in risk assessment (e.g., probability A has more predictive power than does probability B) were subjected to the criticism and evaluation of the lay and scientific communities, even these value judgments could be said to possess scientific objectivity. This is not as implausible a use of the term *objectivity* as it sounds, for several reasons.

First, when I make an epistemic value judgment about two accident probabilities or, for example, about the acceptability of a particular risk, I am not talking merely autobiographically and subjectively. I am talking about characteristics of possible, public, external events that are capable of being known and understood by other people. Second, the skills associated with making these judgments are a function of training, education, experience, and intelligence. Objectivity does not require having an algorithm guaranteeing the correctness of risk judgments. Third, empirical factors, such as observed accident consequences or actual accident frequencies, could change the likelihood that our risk judgments are correct; hence such judgments cannot be purely subjective. Fourth, to make empirical confirmability (instead of the ability to be subjected to the criticism of the scientific community and affected laypersons) a necessary condition for the objectivity of all risk judgments would be to ignore the way that reasonable people behave. Reasonable people accumulate observations and inferences about judgments until the probability of those judgments is so great that they do not doubt them. They make assumptions when their inferences and evidence support them, but they do not demand empirical confirmation for everything. Only if one were searching for objectivity that came close to avoiding all error could one complain about well-supported risk judgments that met the criteria for scientific objectivity just outlined (Nagel 1986; Scriven 1980).[7] Because even science has never claimed objectivity that came close to avoiding all error, it is not reasonable to demand more than these criteria as standards for risk judgments (Scriven 1980).

This new sense of scientific objectivity also seems plausible because it relies on the social and critical character of science and science-related activities (like risk assessment), as Popper realized (Popper 1950, 1962, 1965) and as James, Wittgenstein, and Wisdom suggested. If they are correct, then the naive positivists make too strict a demand on scientific objectivity by requiring confirmability.

Moving Beyond Infallibility

The cultural relativists also make too strict demands on scientific objectivity, in their case, by presupposing that objective risk judgments must be infallible and universal. Douglas and Wildavsky (1982) believe that because every risk judgment is value laden and because both experts and lay people disagree about hazards, no risk judgments can be said to be objective. Because of the lack of consensus on risk evaluation, they claim that none is certain, that any of them can be said to be justified, and that one is as good as another.

Admittedly, cultural relativists are correct in recognizing the inapplicability of the positivists' model of rationality and in rejecting their presupposition that all judgments in science must be empirically confirmed. Douglas, Wildavsky, and others are wrong, however, to leap prematurely to the conclusion that because the positivists' criterion of empirical confirmability fails (e.g., in the case of categorical judgments of value), risk judgments are arational and purely relative. They assume that because all judgments are not infallibly certain and universal (therefore precluding disagreement about them), they must be relative.

In searching for a certainty that appears to transcend the possibility of error and in presupposing that objectivity requires infallibility and universality, Douglas and Wildavsky share the presuppositions of other cultural relativists like Feyerabend (1977). They assume that because there is no perfect judgment, all risk attitudes are equally imperfect. Yet neither from the fact that all judgments have been falsified historically, nor from the fact that it is impossible for any judgment to escape falsification, would it follow that all risk assessments are equally unreliable (Kulka 1977). Why not? Any alleged historical falsifications of judgments would provide only necessary, not sufficient, conditions for the correctness of claims that there are no objective risk assessments and that only infallible judgments guarantee objectivity.

Furthermore, there is no obvious reason that judgments or rules about risk assessment must be exceptionless in order to be useful or rational. Rules and judgments used to justify risk policy ought not be exceptionless; any type of justification is always complex and context dependent (Hellman 1979; Hempel 1965; Quine and Ullian 1978), and both risk judgments and rules of scientific method need be merely acceptable to rational persons in the situation at hand, not universal (Scriven 1980). Both scientific and legal inference establish that something is prima facie true, that it is reasonably probable, or that there is a presumption in its favor, not that it is infallibly true. Hence there is no reason to assume that risk judgments need be more than prima facie true if they are said to be objective.

Great differences in scientific and policy-related behavior are compatible with objective risk judgments. Disagreements over how to analyze or evaluate a given hazard do not mean either that there are no rules of risk assessment or that any rule is as good as another. Why not? Those (e.g., the cultural relativists) who deny the existence of universal rules or values in science and risk assessment appear to do so because they fail to distinguish three different questions: (1) Are there general principles (e.g., that postulate risk probabilities that are consistent with observed accident frequencies, that postulate risk acceptability as proportional to the degree of public consent to the societal danger) that account for the rationality of hazard assessment? (2) Are there particular procedures, or instantiations of the general principles (e.g., accident frequencies should be observed for a period of at least five years before concluding that they are consistent with postulated risk probabilities), that account for the rationality of risk judgments? (3) Does risk assessment in fact always illustrate either the general principles or the specific procedures (e.g., did the authors of WASH–1400 observe accident frequencies for a period of at least five reactor years before postulating the per-reactor-year probability of a core melt)?

Cultural relativists like Douglas, Wildavsky, and Feyerabend appear to assume that if one answers the second and third questions in the negative, then the answer to the first question is also negative. This is false. Debate about the second question does not jeopardize the rationality and nonrelativism of risk judgments, in the sense of suggesting there is no good answer to the first question. In fact, debate over the second question must presuppose rationality in the sense of the first question, or the discussion would be futile (Rudner 1966; Siegel 1985).

Insights from Moral Philosophy

Another way to argue for this principle-versus-procedure or hierarchical conception of rationality, applicable to both science and risk assessment, is to incorporate some insights from moral philosophy. In moral philosophy, as both natural-law philosophers and contemporary analysts such as R. M. Hare recognize, there is a hierarchy of methodological rules and value judgments, with different degrees of certainty appropriate to the different levels of generality in the hierarchy. In science, risk assessment, and ethics, the most general rules are the most certain and the most universal—for example—postulate risk probabilities consistent with observed accident frequencies, or do good and avoid evil. The least general rules or value judgments are the least certain and the least universal (Hare 1981).

In moral philosophy, as well as in science and risk assessment, one must make a number of value judgments, especially at the lower levels of universality and generality, in order to interpret and to apply the rules from the most universal, most general level. In other words, in a specific situation, one must make very specific value judgments about what is "doing good," and about what is "greater risk acceptability." Just because there is no algorithm applicable to all situations for deciding what is doing good or what has greater risk acceptability, this does not mean that certain practice-based ethical, scientific, or risk-assessment judgments are no better than any others are. Some are better than others, as is evidenced by how well they fare when they are evaluated as means to the end or goal of explanatory power and risk acceptability. Moreover some risk-assessment rules are better than others are at reducing uncertainty, even though they do not guarantee infallibility. Or, for example, some rules are better than others at achieving equity. For instance, it is clearly better, given the problem of compensating persons for societal risks imposed on them, to follow the methodological rule of compensating involuntarily imposed risks at a higher rate than that for voluntarily chosen ones (all things being equal), than to compensate all risks at the same rate, whenever their expected utility is the same.

Douglas, Wildavsky, and Feyerabend miss the fact that some judgments reduce uncertainty because they focus on the infallibility of the scientific or risk assessments. They ignore scientific objectivity. This particular sense of objectivity relies on a number of insights by Popper, Wisdom, and Wittgenstein (Newell 1986). It anchors objectivity with actions and practices, as well as explanatory and predictive power; it does not define objectivity in terms of an impossible, perhaps question-begging, notion of justification. In securing objectivity—in part by means of the criticisms made by the scientific and affected lay community rather than by specific rules—this account presupposes that in their final stages, rationality and objectivity require an appeal to particular cases as similar to other cases believed to be correct. This partially naturalistic appeal to cases and general methodological values rather than to specific rules is required (1) in order to avoid an infinite regress of justification of risk judgments, (2) because decisions about rules cannot ultimately rest on rules, and (3) because specific criteria would be too dogmatic, in all cases, to take account of counterinstances and the peculiarities of particular risk cases.

CONCLUSIONS

More generally, if relativistic arguments requiring that all objective risk judgments be based on universal and stable rules were correct and if Carnap's, Starr's, and Whipple's arguments requiring that all judgments in science and risk assessment be empirically confirmed were correct and if these arguments were extended to and used in other areas of epistemology and policy, then they would invalidate most of our knowledge claims (Hellman 1979).

The cultural relativists and the naive positivists appear to have conceived of risk judgments in a highly unrealistic way, either (respectively) as infallible and universal or as empirically confirmable. Moreover, they have confused general principles of risk-assessment methods with specific procedures. I have argued that a more realistic way to conceive of them is in terms of the rationality and objectivity appropriate to scientific proceduralism.

ACKNOWLEDGMENT

I am especially grateful to Deborah Mayo for constructive criticisms of an earlier draft of this chapter. Whatever errors that remain are my responsibility.

NOTES

1. Hazard assessors typically divide their discipline into risk identification, estimation, and evaluation. They usually assert that only the last—evaluation—stage is incapable of being value free. For further discussion of the three stages of risk assessment and the degree of objectivity appropriate to each, see Shrader-Frechette (1985a, chap. 2).

2. By naive positivism I mean the belief that there is a fact-value dichotomy and that the only objects of knowledge are facts. Because naive positivists claim that facts are the only objects of knowledge, it follows that they believe that all knowledge claims must be empirically confirmed and that one can have no knowledge of values. Both of these tenets correspond to early or naive positivism, as defined by Abbagnano; later positivists, such as Hempel, rejected the belief that facts are the only objects of knowledge, and hence they are not naive positivists. According to Abbagnano, "the characteristic theses of positivism are that science is the only valid knowledge and facts the only possible objects of knowledge; that philosophy does not possess a method different from science. . . . Positivism, consequently . . . opposes any . . . procedure of investigation that is not reducible to scientific method" (1967, p. 414). As Laudan points out, because they have disallowed talk about developments in "metaphysics, logic [and] ethics . . . 'positivist' [sociologists,] philosophers, and historians of science who see the progress of science entirely in empirical terms have completely missed the huge significance of these developments for science as well as for philosophy" (1977, pp. 61–62).

3. Kraybill, for example, defends DDT. In 1942 Paul Mueller received patents for DDT, and experts awarded him the Nobel Prize for his discovery, because it saved people from malaria and typhus. In the 1940s and 1950s, at least, DDT was not viewed as seriously harmful to humans.

4. Inasmuch as Nelkin (1981) condemns philosophical discussion of the "rights and wrongs" of science policy, she disallows talk about the ethics of science policy. Her disallowing this talk about ethics is tantamount to subscribing to a positivistic model of the role of philosophy and of studies about science and technology.

5. Skolimowski points to a similar condemnation of applied ethics and normative policy analysis. He cites an article by Genevieve J. Knezo, "Technology Assessment: A Bibliographic Review," published in the first issue of *Technology Assessment*. He says that Knezo condemns all normative literature as "emotional, neoluddite, and polemic" in nature and "designed to arouse and mold mass public opinion." At the same time, says Skolimowski, Knezo praises all alledgedly "neutral" work and claims it "usefully serves to inform the public, through a responsible press, of the pros and cons of a public issue of national importance."

6. Although I believe that there is at least one, general, universal criterion for theory choice, my account of scientific rationality and objectivity does not fall victim to the naive positivist account—as does that of Carnap, for example—for at least four reasons: (1) I am not committed to a priori rules of scientific method, although I am committed to a normative account of it. (2) I am not committed to specific rules of method that are universal, but only to general goals of method that are universal. (3) I am committed to the belief that just as observation and experiment often underdetermine theory, so also does this universal methodological goal (explanatory power as tested by prediction) underdetermine specific methodological rules, and therefore the rules must be dictated largely by practice, largely by the particular situation. (4) I make a naturalistic appeal to the criticism of the scientific community and ultimately an appeal to cases, rather than to specific rules, in order to help define scientific objectivity. For a fuller account of scientific proceduralism, see Shrader-Frechette (1991, chaps. 3 and 12).

7. For a related account of rationality and objectivity in the risk-assessment context, see Rip (1985, pp. 94–110). Rip defends a philosophy that he calls "pragmatic rationalism," seeking robust solutions to environmental and risk problems, solutions that will withstand the social, political, and scientific pressures to which they will inevitably be exposed.

REFERENCES

Abbagnano, N. (1967). "Positivism." In *The Encyclopedia of Philosophy,* vol. 6, ed. P. Edwards, pp. 414–19. New York: Macmillan.

Abbagnano, N. (1977). *Progress and Its Problems.* Berkeley: University of California Press.

Brown, H. I. (1977). *Perception, Theory and Commitment.* Chicago: University of Chicago Press.

Burnham, J. (1979). "Panel: Use of Risk Assessment." In *Symposium/Workshop . . . Risk Assessment and Governmental Decision Making,* ed. Mitre Corporation. Mclean, Va: Mitre Corporation.

Camus, A. (1965). *Notebooks.* Trans. J. O'Brien. New York: Knopf.

Carpenter, R. A. (1972). "Technology Assessment and the Congress." In *Technology Assessment: Understanding the Social Consequences of Technological Applications,* ed. R. G. Kasper, pp. 29–47. New York: Praeger.

Cornman J. W. (1980). *Skepticism, Justification, and Explanation.* Boston: Reidel.

Douglas, M., and Wildavsky, A. (1982). *Risk and Culture.* Berkeley: University of California Press.

Einstein, A. (1949). *The World as I See It.* Trans. A. Harris. New York: Philosophical Library.

Einstein, A. (1954). *Ideas and Opinions.* Trans. S. Bergmann. New York: Crown.

Feyerabend, P. (1977). "Changing Patterns of Reconstruction." *British Journal for the Philosophy of Science* 28:351–69.

Fischhoff, B., Slovic, P., and Lichtenstein, S. (1980). "Facts and Fears." In *Societal Risk Assessment,* ed. R. Schwing and W. Albers, pp. 181–216. New York: Plenum.

Fischhoff, B., Slovic, P., Lichtenstein, S., Read, S., and Combs, B. (1978). "How Safe Is Safe Enough? A Psychometric Study of Attitudes Towards Technological Risks and Benefits." *Policy Sciences* 9:127–52.

Foell, W. K. (1978). "Assessment of Energy/Environment Systems." In *Environmental Assessment of Socioeconomic Systems,* ed. D. F. Burkhardt and W. H. Ittelson, pp. 183–202. New York: Plenum Publishing Co.

Frank, P. (1974). *Einstein: His Life and Times.* Trans. G. Rosen. New York: Knopf.

Green, H. (1977). "Cost–Benefit Assessment and the Law." *George Washington Law Review* 45:901–10.

Hafele, W. (1976). "Benefit–Risk Tradeoffs in Nuclear Power Generation." In *Energy and the Environment,* ed. H. Ashley, R. Rudman, and C. Whipple, pp. 141–84. New York: Pergamon Press.

Hanson, N. R. (1958). *Patterns of Discovery.* Cambridge: Cambridge University Press.

Hare, R. M. (1979). "Contrasting Methods of Environmental Planning." In *Ethics and the Problems of the 21st Century,* ed. K. M. Goodpaster and K. M. Sayre, pp. 63–78. Notre Dame, Ind: University of Notre Dame Press.

Hare, R. M. (1981). *Moral Thinking: Its Levels, Methods and Point.* Oxford: Oxford University Press.

Hellman, G. (1979). "Against Bad Method." *Metaphilosophy* 10:190–207.

Hempel, C. (1965). *Aspects of Scientific Explanation*. New York: Free Press.

Hempel, C. (1979). "Scientific Rationality." In *Rationality Today,* ed. T. Geraets, pp. 45–58. Ottawa: University of Ottawa Press.

Hempel, C. (1980). "Science and Values." In *Introductory Readings in the Philosophy of Science,* ed. E. Klemke, R. Hollinger, and A. Kline, pp. 254–68. Buffalo, N.Y.: Prometheus Books.

Hempel, C. (1983). "Valuation and Objectivity in Science." In *Physics, Philosophy, and Psychoanalysis,* ed. R. Cohen and L. Laudan, pp. 73–100. Dordrecht: Reidel.

Kasper, R. (1980). "Perceptions of Risk." In *Societal Risk Assessment,* ed. R. Schwing and W. Albers, pp. 71–80. New York: Plenum Press.

Kidd, C. V. (1972). "Technology Assessment in the Executive Office of the President." In *Technology Assessment: Understanding the Social Consequences of Technological Applications,* ed. R. G. Kasper, pp. 123–48. New York: Praeger.

Kraybill, H. F. (1975). "Pesticide Toxicity and the Potential for Cancer." *Pest Control* 43:9–16.

Kuhn, T. S. (1962) *The Structure of Scientific Revolutions*. Chicago: University of Chicago Press.

Kulka, T. (1977). "How Far Does Anything Go? Comments on Feyerabend's Epistemological Anarchism." *Philosophy of the Social Sciences* 7:277–87.

Lashof, J. C., et al., Health and Life Sciences Division of the U.S. Office of Technology Assessment (OTA) (1981). *Assessment of Technologies for Determining Cancer Risks from the Environment*. Washington, D.C.: OTA.

Laudan, L. (1977). *Progress and Its Problems*. Berkeley: University of California Press.

Laudan, L. (1984). *Science and Values*. Berkeley: University of California Press.

Lindblom, C. E., and Cohen, D. K. (1979). *Usable Knowledge: Social Science and Social Problem Solving*. New Haven, Conn.: Yale University Press.

Longino, H. (1983). "Beyond 'Bad Science': Skeptical Reflections on the Value-Freedom of Scientific Inquiry." *Science, Technology, & Human Values* 8:7–17.

Lovins, A. B. (1977). "Cost–Risk–Benefit Assessments in Energy Policy." *George Washington Law Review* 45:911–43.

Lowrance, W. (1976). *Of Acceptable Risk*. Los Altos, Calif.: Kaufman.

McMullin, E. (1983). "Values in Science." In *Philosophy of Science Association 1982,* vol. 2, ed. P. D. Asquith, pp. 3–28. East Lansing, Mich.: Philosophy of Science Association.

Nagel, E. (1961). *The Structure of Science*. New York: Harcourt, Brace.

Nagel, T. (1986). *The View from Nowhere*. New York: Oxford University Press.

Nash, R. (1973). *Wilderness and the American Mind*. Rev. ed. New Haven, Conn.: Yale University Press.

Nelkin, D. (1981). "Wisdom, Expertise, and the Application of Ethics." *Science, Technology & Human Values* 6:16–17.

Newell, R. (1986). *Objectivity, Empiricism, and Truth*. New York: Routledge & Kegan Paul.

Okrent, D., and Whipple, C. (1977). *Approach to Societal Risk Acceptance Criteria and Risk Management*. Report no. PB–271264. Washington D.C.: U.S. Department of Commerce.

Polanyi, M. (1958). *Personal Knowledge*. New York: Harper & Row.

Popper, K. (1950). *The Open Society and Its Enemies*. Princeton, N.J.: Princeton University Press.

Popper, K. (1962). *Conjectures and Refutations*. New York: Basic Books.

Popper, K. (1965). *The Logic of Scientific Discovery*. New York: Harper.

Primack, J., and von Hippel, F. (1974). *Advice and Dissent: Scientists in the Political Arena.* New York: Basic Books.

Quade, E. S. (1975). *Analysis for Public Decisions.* New York: American Elsevier.

Quine, W. V. O., and Ullian, J. (1978). *Web of Belief.* 2d ed. New York: Random House.

Rip, A. (1985). "Experts in Public Arenas." In *Regulating Industrial Risks,* ed. H. Otway and M. Peltu, pp. 94–110. London: Butterworth.

Ross, W. D. (1930). *The Right and the Good.* Oxford: Clarendon Press.

Rowe, W. (1977). *An Anatomy of Risk.* New York: Wiley.

Rudner, R. (1966). *Philosophy of Social Science.* Englewood Cliffs, N.J.: Prentice-Hall.

Scriven, M. (1980). "The Exact Role of Value Judgments in Science." In *Introductory Readings in the Philosophy of Science,* ed. E. Klemke, R. Hollinger, and A. Kline, pp. 269–91. Buffalo, N.Y.: Prometheus Books.

Sellars, W. (1967). *Philosophical Perspectives.* Springfield, Ill.: Thomas.

Shrader-Frechette, K. S. (1980). "Recent Changes in the Concept of Matter: How Does 'Elementary Particle' Mean?" In *Philosophy of Science Association 1980,* vol. 1, ed. P. D. Asquith and R. N. Giere, pp. 302–16. East Lansing, Mich.: Philosophy of Science Association.

Shrader-Frechette, K. S. (1982). "Economics, Risk–Cost–Benefit Analysis, and the Linearity Assumption." In *Philosophy of Science Association 1982,* vol. 1. ed. P. D. Asquith and T. Nickles, pp. 217–32. Ann Arbor, Mich.: Edwards.

Shrader-Frechette, K. S. (1985a). *Risk Analysis and Scientific Method.* Boston: Reidel.

Shrader-Frechette, K. S. (1985b). *Science Policy, Ethics, and Economic Methodology.* Boston: Reidel.

Shrader-Frechette, K. S. (1988). "Values and Hydrogeological Method: How Not to Site the World's Largest Nuclear Dump." In *Planning for Changing Energy Conditions, Energy Policy Studies,* vol. 4, ed. J. Byrne and D. Rich, pp. 101–37. Oxford: Transaction Books.

Shrader-Frechette, K. S. (1991). *Risk and Rationality.* Berkeley: University of California Press.

Siegel, H. (1985). "What Is the Question Concerning the Rationality of Science?" *Philosophy of Science* 52:517–37.

Skolimowski, H. (1976). "Technology Assessment as a Critique of Civilization." In *Philosophy of Science Association 1974,* ed. R. S. Cohen et al., pp. 459–65. Boston: Reidel.

Starr, C., and Whipple, C. (1980). "Risks of Risk Decisions." *Science* 208:1114–19.

Tool, M. C. (1979). *The Discretionary Economy: A Normative Theory of Political Economy.* Santa Monica, Calif.: Goodyear.

Toulmin, S. (1961). *Foresight and Understanding.* New York: Harper & Row.

Office of Technology Assessment (OTA) (1976a). *Annual Report to the Congress for 1976.* Washington D.C.: U.S. Government Printing Office.

Office of Technology Assessment (OTA) (1976b). *Technology Assessment Activities in the Industrial, Academic, and Governmental Communities.* Hearings Before the Technology Assessment Board of the Office of Technology Assessment. 94th Cong. 2d sess., June 8–10 and 14. Washington D.C.: U.S. Government Printing Office.

Office of Technology Assessment (OTA) (1977). *Technology Assessment in Business and Government.* Washington D.C.: U.S. Government Printing Office.

Weiss, C. H., and Bucuvalas, M. J. (1980). *Social Science Research and Decision Making.* New York: Columbia University Press.

12

Sociological Versus Metascientific Views of Risk Assessment

DEBORAH G. MAYO

In this chapter I shall discuss what seems to me to be a systematic ambiguity running through the large and complex risk-assessment literature. The ambiguity concerns the question of separability: can (and ought) risk assessment be separated from the policy values of risk management? Roughly, risk assessment is the process of estimating the risks associated with a practice or substance, and risk management is the process of deciding what to do about such risks. The separability question asks whether the empirical, scientific, and technical questions in estimating the risks either can or should be separated (conceptually or institutionally) from the social, political, and ethical questions of how the risks should be managed. For example, is it possible (advisable) for risk-estimation methods to be separated from social or policy values? Can (should) risk analysts work independently of policymakers (or at least of policy pressures)? The preponderant answer to the variants of the separability question in recent risk-research literature is no. Such denials of either the possibility or desirability of separation may be termed *nonseparatist* positions. What needs to be recognized, however, is that advocating a nonseparatist position masks radically different views about the nature of risk-assessment controversies and of how best to improve risk assessment.

These nonseparatist views, I suggest, may be divided into two broad camps (although individuals in each camp differ in degree), which I label the *sociological* view and the *metascientific* view. The difference between the two may be found in what each finds to be problematic about any attempt to separate assessment and management. Whereas the former (sociological) view argues against separatist attempts on the grounds that they give too small a role to societal (and other nonscientific) values, the latter (metascientific) view does so on the grounds that they give too small a role to scientific and methodological understanding. Examples of those I place under the sociological view are the cultural reductionists discussed in the preceding chapter by Shrader-Frechette. Examples of those I place under the metascientific view are the contributors to this volume themselves. A major theme running through this volume is that risk assessment cannot and should not be separated from societal and policy values (e.g., Silbergeld's uneasy divorce). However, these calls for nonseparatism have been accompanied by calls for greater scientific and methodological understanding, albeit an understanding that allows for a critical (or metascientific) scrutiny of the uncertainties involved. The problem I am raising results because nonseparatists of the metascientific stripe are lumped along with nonseparatists of the sociological stripe.

Although they are not often put forward as such, I believe that each of the two camps of nonseparatists corresponds to a different underlying philosophy of scientific knowledge and scientific rationality. I grant that the distinction I am seeking is neither clear-cut nor uniform, which is doubtless part of the reason it has largely gone unrecognized. However, failing to see this key difference between those who argue against separatism has, in my opinion, so obscured the debates over how to improve risk assessment that even an oversimplified partitioning of nonseparatists seems necessary.

Processes of Risk Assessment

The term *risk assessment* may be used more broadly[1] than I intend here, to include risk perceptions (or even risk management). To avoid a further ambiguity that often infects discussions of risk-assessment controversies, I shall delineate at the outset what I understand risk assessment to cover. It then will be possible to ask, without begging key questions, whether the components of risk assessment do or do not include these broader aspects. Following the characterization spelled out in a 1983 report by the National Academy of Science (p.18), I understand risk assessment to include the following four steps:

1. *Hazard identification:* Characterizing the nature and strength of the causal evidence as to whether exposure to an agent can increase the incidence of a health condition (cancer, birth defect, etc.) in humans, lab animals, or other test systems.
2. *Dose Response Assessment:* Estimating the incidence of an effect as a function of exposure in various populations of interest, extrapolating from high to low dose and from animals to humans. It should describe and justify the methods of extrapolation used and should characterize the statistical and biological uncertainties.
3. *Exposure Assessment:* Measuring or estimating the extent of human exposure to an agent that exists or would exist under specified circumstances in various subgroups.
4. *Risk Characterization:* Estimating the incidence of health effect under the various conditions in the specified subgroups. This combines the previous steps and includes a description of the uncertainties at each step.

The NAS–NRC Report and Separability

Far from being of merely conceptual interest, the issue of separability is fundamental to debates directly affecting actual practices of government regulation of hazards. An extremely rich source of material in which to trace recent arguments concerning separability is the period from the early 1980s to the present.

I shall focus on the practices and the philosophy of the Environmental Protection Agency (EPA) during 1981 and 1982, referring to it as the "Gorsuch EPA," as Ann Gorsuch (later Ann Burford) was the EPA administrator at the time.

The problem that arose in case after case[2] (often resulting in the Gorsuch EPA's being brought to trial) was that many widespread procedures for risk assessment were being repudiated in favor of "scientific" assessments (and in some cases reassessments) that fairly blatantly reflected the antiregulatory policy favored by the affected industry and by the Reagan administration generally. As an outgrowth of such concerns, Congress requested a study of these problems and some proposals for reforms. This resulted in a report by the National Academy of Science–National Research Council (NAS–NRC) in 1983. The concern, as the report states, is this:

> With a scientific base that is still evolving, with large uncertainties to be addressed in each decision, and with the presence of great external pressures, some see a danger that the scientific interpretations in risk assessments will be distorted by policy considerations, and they seek new institutional safeguards against such distortion. (p. 14)

To avoid such distortion, the proposed reforms suggest that risk assessment be separated from risk management.

Among the institutional reforms suggested, the NAS–NRC report focuses on two: reorganization, to ensure that risk assessments are protected from inappropriate policy influences, and the development and use of uniform guidelines for carrying out risk assessment. Although the report did not recommend institutional separation, it did believe it important to strive to enable risk analysts to work independently of policy pressures and to distinguish risk assessment from risk management conceptually:

> We recommend that regulatory agencies take steps to establish and maintain a clear conceptual distinction between assessment of risks and consideration of risk management alternatives; that is, the scientific findings and policy judgments embodied in risk assessments should be explicitly distinguished from the political, economic, and technical considerations that influence the design and choice of regulatory strategies. (p. 7)

To implement its recommendations the NAS–NRC proposed setting out uniform inference guidelines for interpretating scientific and technical information relevant to risk assessment. The use of such guidelines, it believes, "will aid in maintaining the distinction between risk assessment and risk management" (p. 7). It also recommends an independent Board on Risk Assessment Methods whose main function would be to assess critically the evolving scientific basis of risk assessment and to make explicit the underlying assumptions and policy ramifications of the inference options (p. 8).

The NAS–NRC report, then, was the basis for implementing separation at regulatory agencies such as the EPA (Environmental Protection Agency) and OSHA (Occupational Safety and Health Administration). William Ruckelshaus, who replaced Gorsuch after this controversy erupted, made it clear in his 1983 statement that he was relying on the NAS–NRC report in stressing the importance of separating risk assessment and risk management. Even in the face of conflicting political pressures, Ruckelshaus declared that "risk assessment at

EPA must be based only on scientific evidence and scientific consensus. Nothing will erode public confidence faster than the suspicion that policy considerations have been allowed to influence the assessment of risk" (Ruckelshaus 1983, pp. 1027–28).

The attempt to improve risk assessment through such separation has been challenged both in principle and in regard to the form it has actually taken. It is often argued in work in the social studies of science that any attempt to enhance the neutrality of risk assessment by separating risk assessment and risk management is wrongheaded because it presupposes a view of risk assessment as a matter of objective, impartial, empirical fact, while viewing risk-management policy as invested with social values, subjective, emotional or aesthetic feelings not adjudicable in a "rational" manner. What is being contested is the view underlying Ruckelshaus's remark that "although there is an objective way to assess risk, there is, of course, no purely objective way to manage it" (Ruckelshaus 1983, p. 1028). Not surprisingly, challenges to the objectivity and rationality of scientific risk assessment, and the corresponding arguments against separatism, are closely connected with the more general challenge from philosophers, historians, and sociologists to a certain image of scientific rationality and objectivity. The image being questioned is the naive positivist or what I shall call the *old image* of scientific rationality, in which science is thought to be value free.

Overview

I shall begin by considering how the challenge to the plausibility of separating assessment from policy relates to a more general philosophical challenge to what may be called the old image of scientific rationality, and I shall outline the argument that leads the sociological view to reject separability. Then I shall compare the implications of the metascientific view, which also rejects separability, with those of the sociological view and two other positions. My conclusion is that what matters most is not whether or not a view espouses separatism but whether it adheres (implicitly or explicitly) to the old image of science or, alternatively, sets the stage for a new or postpositive image, in which scientific scrutiny—though neither algorithmic nor value free—can nevertheless appraise objectively the adequacy of risk assessments. By "appraising objectively" I mean determining (at least approximately) what the data do and do not say about the actual extent of a given risk. Last I shall illustrate these issues and arguments by considering a controversial ruling during the Gorsuch EPA pertaining to the significance of the risks associated with formaldehyde.

THE OLD IMAGE OF SCIENTIFIC RATIONALITY AND THE SOCIOLOGICAL CHALLENGE TO SEPARABILITY

The Old (or Naive Positivist) Image of Scientific Rationality

According to the old image of science, stemming from the positivist tradition that prevailed from 1930 to 1960, the rationality of science rested on objective rules for appraising hypotheses and adjudicating between competing hypotheses. Demonstrating the rationality of science required articulating objective, value-free rules to assess hypotheses.

Philosophers, however, failed to articulate such rules satisfactorily, and the entire view that science follows impartial algorithms came under challenge by Kuhn (1962) and others. Actual scientific debates often last several decades and are not adjudicated simply by empirical rules; "extrascientific" values enter as well. This has been taken to show the untenability of the old image of scientific rationality. But some philosophers, presumably holding the old image ideal to be the only type of rationality worth having, have taken this as grounds for abandoning altogether the view that science is characterized by rational methods. Instead they hold that such extrascientific values—metaphysical beliefs, goals, subjective interests and the like—play a greater role than does empirical evidence in evaluating hypotheses and adjudicating scientific disputes. According to extreme versions of this view (e.g., Feyerabend) beliefs about the world are not constrained at all by what we learn from empirical evidence in science. Sociology of knowledge has further strengthened this "antipositivist" sentiment by revealing how social contexts have influenced scientific theory appraisal in specific instances.[3]

Based on the old image, if scientists disagree in the face of the same empirical data—one group concluding that a substance is carcinogenic, say, and another group concluding it is not—there is a violation of scientific rationality. Such disagreement seems to illustrate, as Hamlin (1986) notes, that "experts on one side, or even both sides, are falling so much under the sway of 'interests' that they violate central norms of impartiality, emotional neutrality, universality . . . and the like . . . their behavior has been seen as the prostituting of science, as the selling of credibility to the highest bidder" (p. 486). Were it possible to avoid or somehow to neutralize these interests, adjudication of the scientific disagreement would be forthcoming. According to this old image perspective, it makes sense to seek reforms by means of new alignments among experts and policymakers, for example, the type of independent board recommended in the NAS–NRC report.

But this perspective has been questioned in both philosophy of science generally and the social studies of science. As Hamlin continues:

> A growing body of literature in the social studies of science takes issue with this formulation. It suggests that it is meaningless to think about a disinterested science; especially in such policy-relevant areas as ecology and public health, it will be impossible to impose any unarbitrary separation of scientific from social issues, and unreasonable to expect that "unbiased" assessors will have no interest of their own to represent. In this view there can be no disinterested parties, but simply parties with different, more-or-less conflicting or compatible interests. (1986, p. 487)

The Sociological View

Such questioning of the old image is taken as the basis for variants of what I shall call the sociological view. An example of such a view is put forward by Brian Wynne, who claims:

The debates over the safety of many environmental pollutants are structurally conditioned by the fact that underlying the overt technical discourse is a symbolic discourse in which those who have previously committed themselves to a particular scientific point of view, with particular policy implications, are attempting to defend their long-term credibility. (1982, p. 133)

The meaning of such "symbolic discourse" (which he compares with poetry, art, and religion) is "socially malleable," in contrast with the "cold steel of hard empirical fact" assumed in the old or positivistic view to be characteristic of science. Wynne still employs the term "scientific rationality," but now it is seen as "intrinsically conditioned by social commitments" (p. 139).

Another type of sociological view stems from a variant of this criticism of the old image. The criticism begins with the recognition that facts are not only necessarily theory laden but also fail unequivocally to pick out a best hypothesis, that is, facts underdetermine hypotheses. Thus, if decisions about hypotheses are reached despite these knowledge gaps, they must be the result of "subjective judgments" influenced, if not wholly determined by, social and ethical values. This is one of the tacks employed by Douglas and Wildavsky (1982) (see Chapter 11, this volume) in their sociocultural theory of risk. They maintain that "the risk assessors offer an objective analysis. We know that it is not objective so long as they are dealing with uncertainties and operating on big guesses. They slide their personal bias into the calculations unobserved" (p. 80). The knowledge gaps, they continue, must be filled with educated guesses, and the remainder of their book is devoted to showing "that the kinds of guesses about natural existence depend very largely on the kinds of moral education of the people doing the guessing" (p. 80).

Douglas and Wildavsky go even further: In their view, not only do such social values enter to fill knowledge gaps in reaching risk assessments, but the very methods, models, and interpretations are themselves social constructions. They are led to the extreme position that reduces risk assessment to cultural or moral judgment—a "social reductionism." As S. O. Funtowicz and J. R. Ravetz put it, social reductionism may be seen as "a methodological position which assumes that every debate over technological risks is really a conflict among contradictory 'ways of life' and that the awareness of this would be enough to settle the question" (1985, p. 223).[4]

For our discussion, we can place under the heading of the sociological view of risk assessment two views (both of which come in degrees):

1. Social factors necessarily enter into risk assessment because where there are knowledge gaps, there is nothing to fill them with except social and moral judgments (social relativism).
2. Risk assessment is entirely the outcome of socially determined methods and judgments, which are social constructs (social reductionism).

What matters for my argument is that both of these views lead to similar consequences for risk assessment: Risk-assessment judgments (at least those containing uncertainty) are policy judgments, and risk-assessment disagreements largely reflect disagreements about policy (including moral, social, economic, or other nonscientific[5]) values. Thus, risk-assessment judgments and the adjudication of disagreements necessitate going beyond empirical evidence, scientific criteria, and analytic methods to extra scientific (social and policy) considerations. In the sociological view, the primary issue is not estimating objective physical risks, but instead social and politically conditioned attitudes toward physical risk. Wynne adds: "Determining objective physical risks will still be valid of course, but the lingering tendency to start from this scientific vantage point and add social perceptions as qualifications to the 'objective' physical picture must be completely reversed" (1982, p. 138).

The result is that the methods of science are given little, if any, role in an unbiased adjudication of disagreements over risk assessments. How could they, after all, if each view is as biased as the next? According to Douglas and Wildavsky, "everyone, expert and layman alike, is biased. No one has a social theory above the battle" (1982, p. 80). If a judgment is biased as long as it is the output of a human or of a method that humans articulate, then, of course, all judgments are biased. But then for a risk assessment to be biased becomes trivially true: By definition there is no way to criticize an assessment as biased. There would be no sense in criticizing a risk assessment as a misinterpretation of data, in the sense of incorrectly asserting what the data indicate about the actual extent of a given risk. Because in the sociological view, interpreting scientific results is necessarily colored by social and political contexts, such criticism would be simply a criticism of the social and moral views of the assessor. The claim is not just that science cannot tell us which risks to find acceptable— that much is not controversial. The claim is that science cannot tell us the extent of a given harm, because given the uncertainties such an assessment is always a matter of one's ethical and policy values. No wonder, then, that Douglas and Wildavsky conclude: "Science and risk assessment cannot tell us what we need to know about threats of danger since they explicitly try to exclude moral ideas about the good life" (p. 81). Can this conclusion be avoided?

The Sociological Argument Recapitulated: Premise (P)

Let us recapitulate the link between rejecting the old image of rationality and the view that risk-assessment views reflect (to varying degrees) prior policy or social commitments. The first part of the argument is that a strict demarcation between risk assessment and risk management (or of "facts" and "values") is plausible only if the old image of science can be maintained—that is, only if there are methods for a strictly factual appraisal of hypotheses—in this case hypotheses about the actual risks caused by a given substance. The old image is unattainable; therefore strict separability is unattainable. Let us accept the argument thus far as sound. But how does this lead to the further claims (of the sociological view) that the risk estimate is largely or solely determined by prior social policy positions, that disagreements about estimates reflect different extrascientific values, and that there can be no unbiased scientific court of appeals? The additional premise required is something like this:

> *P:* If strict separability is unattainable, then empirical, technical, and scientific methods cannot provide unbiased risk assessments or adjudicate objectively between conflicting assessments.

Again, by an objective adjudication I mean one that reflects what is actually the case about a risk, regardless of one's policy preferences.

However, to assume premise (P) is tantamount to assuming that the old image of scientific rationality is the only one possible or worth having. Although it is not typically recognized, workers in social studies of science are led to this sociological view not as a consequence of repudiating the old image of scientific rationality but by taking it all too seriously. Laudan (1989) makes an analogous point concerning the postpositivists Kuhn, Feyerabend, and Quine, as does Shapere (1986) with regard to work in social studies of science. For what supporters of the sociological view must be arguing is that unless risk assessment can be accomplished by (value-free) logical rules or algorithms, it is outside the domain of science proper (falling instead into the domain of extrascientific policy), and we are led to dethrone science as an adjudicator in assessment disputes. That is, their grounds for denying that risk assessment consists of applying scientific methods separable from policy stem from implicitly holding to the old image philosophy in which scientific methods must be dictated by neutral, logical rules.

So prevalent is the view underlying premise (P) that it seems to be assumed with little argument. Indeed, in the first issue of *Risk Analysis,* as Funtowicz and Ravetz (1985) note and criticize, Kaplan and Garrick (1981) assume from the outset that risk is radically relative and that there is no difference between risk and perceived risk. This view is a corollary of premise (P).

If premise (P) were true, then anyone rejecting the separability of risk assessment and policy (i.e., any nonseparatist) must embrace (P)'s consequent and repudiate the adjudicating power of science. And because (P) is so widely assumed, every contribution to the literature that takes a nonseparatist line tends to be regarded as giving yet further support for this repudiation. But this is a mistake. There is another view that, like the sociological view, challenges the tenability of separating risk assessment and policy, yet, unlike the sociological view, denies premise (P). This second view also argues against a rigid separation of risk assessment as cold hard facts divorced from risk management; it too objects to the old image. Where this second view differs from the sociological view is in upholding the ability of scientific methods and criteria to criticize objectively those risk-assessment judgments involving knowledge gaps. For reasons to be explained later, I shall refer to this second, critical view as the *metascientific* view.

THE METASCIENTIFIC CHALLENGE TO SEPARATISM AND THE DENIAL OF PREMISE (P)

My understanding of the term *metascience* should be distinguished from the term *transscience.* The latter is defined by Weinberg (1972) as being outside science proper, but metascience is within science, at least if science is understood in a genuinely postpositivist (new image) manner, which I shall be discussing. The term "metascience" seems appropriate because the critical scrutiny of the uncertainties in risk assessment involves a critical reflection that might be seen as one level removed from the process of risk assessment itself—much as we term metalogic the scrutiny of the philosophical foundations of logic.

The metascientific view need not actually be put forward in terms of an argument in favor of any general philosophy of science. In practice, in fact, it is typically presented within a repudiation of the actual form that attempts at separation have taken, as in the period since the NAS–NRC report. A major criticism is that such an attempt hampers communication between risk assessors and risk managers, thereby hindering rather than helping ensure adequate assessments (e.g., Jasanoff and Silbergeld, Chapters 2 and 5, this volume). The result is that managers are left to use assessments without knowing the underlying methodological assumptions and their associated uncertainties. The suggestion is that the problem can be ameliorated or at least improved by a better understanding of the uncertainties underlying choices of risk-assessment estimates. In striking contrast with the sociological view, these calls for nonseparatism are calls for greater scientific and methodological understanding, albeit an understanding that is based on a critical (or metascientific) scrutiny of the uncertainties involved.

Thus, although both the sociological and the metascientific views are nonseparatist—they both hold that risk assessment involves both science and policy—each sees this nonseparability as having very different consequences for the role of science in risk assessment. The proponent of the sociological view argues against separatist attempts because he considers it impossible to perform adequately the science of assessment without being involved in the policy and ethical values of risk management. The metascientist argues against separatist attempts because he considers it impossible to perform risk management adequately without understanding the science underlying the assessments. To clarify the distinction between the two positions, we need to begin by asking where the policy input into risk assessment lies according to the metascientific view. Although in the sociological view the answer is essentially "everywhere," what I am calling the metascientific view is concerned with the entry of policy considerations only in the specific places that the NAS–NRC report labels *risk-assessment policy (RAP)* judgments.

Risk-Assessment Policy (RAP)

Risk-assessment policy refers to the various judgments and decisions, sometimes called *inference options,* that are required to carry out risk-assessment estimates. Because these judgments include choices with no unequivocal scientific answers and because these choices have policy implications, they are intertwined with policy. The NAS–NRC report (1983, pp. 29–33) offers a useful delineation of more than fifty junctures at which, owing to uncertainty, an inferential or analytical choice must be made in the course of making risk assessments. Some of those relevant to the metascientific standpoint are the following:

1. Some RAP questions under different components of hazard identification:
   a. Epidemiologic data

- What weight should be given to studies with different results? Should a study be weighted in accordance with its statistical power?
- What weight should be given to different types of studies (prospective versus case control?)
- What statistical significance level should be required for results to be considered positive?

b. Animal-bioassay data
  - What degree of confirmation of positive results should be necessary?
  - Should negative results be disregarded or given less weight?
  - Should a study be weighed according to its statistical power?
  - How should the occurrence of rare tumors be treated? (Should it be considered evidence of carcinogenicity even if the finding is not statistically significant?)
  - What models should be used to extrapolate to humans?

2. Some RAP questions concerning dose response assessments
  a. Epidemiological data
    - What dose response models should be used to extrapolate from observed doses to relevant doses?
  b. Animal bioassay data
    - What mathematical models should be used to extrapolate from experimental doses to human exposures?
    - Should dose response relations be extrapolated according to best estimates or according to upper confidence limits? If the latter, what confidence limits should be used?

Although science and policy are intermingled in selecting among inference options at each stage of risk assessment (hazard identification, dose response assessment, exposure assessment, etc.), the policy entry here is intended to be distinguished from the broader social, ethical, and economic policy decisions in risk management. According to the NAS–NRC report: "At least some of the controversy surrounding regulatory actions has resulted from a blurring of the distinction between risk assessment policy and risk management policy" (1983, p. 3). To see how this blurring occurs and to consider whether it may be avoided, we need to be clear on how policy enters in RAP.[6]

How Policy Enters into Risk-Assessment Policy (RAP) Judgments

Policy enters into RAP judgments as follows: Insofar as there is more than one scientifically acceptable answer to these RAP questions, there will be latitude for choice among possible plausible responses. Each choice influences the risk assessment and so has a policy implication. In particular, the choice will influence the likelihood that a substance will be judged to pose a significant hazard to human health. The more an inference choice increases the likelihood that a substance will be judged a significant risk, the more protective or conservative it is. For example, deciding to use positive results from animal data as indicative of human risk is more conservative than, say, requiring positive human data. Although animal data provide imperfect predictions of human risk, deciding to use them makes it more likely that a human risk will be uncovered.

Because RAP options differ in their degree of protectiveness, it is possible to advance one's policy values—one's view of how protective an estimate should be required to be—by making suitable choices of these inference options. The criticism lodged against the Gorsuch EPA (and other agencies during the early Reagan administration) was that antiregulatory (i.e., less protective) RAP choices were systematically allowed to influence risk assessments made by agency scientists. The result was to politicize the agency. Science in the Gorsuch EPA was science in support of the policy of freedom from regulation (dubbed "regulatory relief"). Under the guise of demanding stringent scientific evidence, these less protective RAP choices made it extremely unlikely that a substance would be deemed a significant human risk. Indeed, as Silbergeld (Chapter 5, this volume) notes, scientists who did not toe the antiregulatory line were excluded from serving on the agency's scientific staff or as advisers. This resulted in the explicit effort by Ruckelshaus in 1983 to separate risk assessment from risk management (the new separatism). But these separatist reforms failed to have their intended effect, which is the basis of the objections from the metascience corner.

We can now make plain what the metascience view finds objectionable about attempts to separate risk assessment from management. The problem in a nutshell is that in all separatist models, the purely scientific components of risk assessment are to be performed by scientists, and the components in which policy enters are to be performed by policymakers. Once it is recognized that risk assessment involves policy in the form of RAP choices, it follows that such choices should be the policymaker's task. But this allows two undesirable consequences: It permits policymakers (1) to fall into all manner of misinterpretations of the assessment evidence and (2) to introduce, either consciously or unconsciously, the very same biases in assessments that led to the separatist "reforms" in the first place.

In order to avoid such hidden biases in assessments, it is necessary to recognize the specific policy influences and implications of specific RAP choices—for example, the implications for protectiveness of choosing a maximum likelihood estimate, in contrast with an upper 95 percent confidence interval. But understanding the protectiveness of the RAP choices requires understanding the evidential meaning of the different types of risk estimates. And this requires understanding the scientific uncertainties involved. If RAP judgments are made by nonscientist policymakers, they are likely to be made by persons divorced from the original issues and uncertainties underlying the different risk estimates. At the same time, the scientist is limited to presenting possible RAP choices but is involved neither in making them nor in bringing out the implications for protectiveness.[7] For example, if the scientific work ends after reporting two possible estimates that may be used, say, a maximum likelihood estimate and an upper 95 percent confidence-bound estimate, then the scientist will not be around to explain how far off each is likely to be from the actual risk and why. Under the Gorsuch EPA, the risk assessor could choose either, supporting his or her choice as good science. Under the new separatist reforms, the risk assessor could also choose either, supporting his or her choice as selecting from among two scientifically plausible options. Neither case requires the assessor to artic-ulate and defend the standard of protectiveness effectively being held by choosing a given option.

Recapitulation of Four Positions:
Some Shared Consequences

As with any attempt to analyze positions that differ in degree and often exist without explicit articulation, my description of four positions will be somewhat oversimplified. However, I believe it will bring to light their key consequences and philosophical underpinnings. The four positions are those of (1) the Gorsuch EPA, (2) the separatist reform that it engendered (the new separatism), (3) the sociological view, and (4) the metascientific view.

Although risk assessment in the Gorsuch EPA was thought to be in need of the separatist reforms, it could be said to have been espousing a separatist framework (though it did not actually succeed in following one). In the Gorsuch EPA, as in the new separatism, the "scientific" aspects of risk assessment were seen as separable from the value-laden management decisions. But the line between these two was drawn in different places. The Gorsuch EPA—as its defenders' arguments indicate—reflected the view that the decisions that we would call RAP decisions were within the province of science. What was effec-tively a call for less protective standards was typically couched as a call for better and more rigorous science.

The challenges to the Gorsuch EPA (analogous to the challenges to the positivist image) essentially claim that the cutoff between science and policy needs to be moved: The judgments needed to generate and interpret data in order to arrive at risk assessments also permit the entry of policy values (hence the term risk-assessment policy or RAP judgments). Thus, in the new separatist reforms, the judgments that go beyond the "hard facts" of science enter the domain of policy judgments: RAP judgments are moved from the domain of science to the domain of policy. So science *qua* science cannot help in deciding among RAP options. But allowing RAP to be performed by policymakers with-out a critical scientific oversight, we said, permits the original problem to recur. The consequence is that RAP judgments are made without bringing out the different implications for the protectiveness of different choices.

Before illustrating how this occurs, it should be noted that this consequence has implications beyond the separatist views. The same consequence follows from seriously embracing the sociological view. The reason is this: If going beyond the hard facts introduces subjective policy judgments, then any judg-ments beyond pure science belong to the realm of policymaking and ought to be made by policymakers, not scientists. This would follow for both the new separatists and for adherents to the sociological view. If according to the soci-ological view, policy judgments about risk estimates are more like "symbolic discourse"—more like moral or aesthetic judgments—then the judgment se-lected is a matter of subjective preference. I am not saying that the new separatist reforms explicitly endorse this sociological view of RAP judgments, for they do not. Rather, I am saying that the operationalization of these separatist reforms in fact yields the same results willingly embraced by the sociological view.[8]

That the separatist view embodied in the reforms and the sociological view should have this shared consequence is not surprising if one remembers that the latter implicitly holds to the old image of science, as does the former only more explicitly. What follows is that the real difference in views turns not on whether they espouse separatism or nonseparatism but on whether or not they are based on the old image of science.

The philosophical underpinnings of a metascientific approach, I claim, genuinely depart from this old image of science to a postpositivist or "new image" of science. In contrast with the old image of science, the metascientific view acknowledges the lack of value-free, universal, algorithmic methods for reaching and evaluating claims about the world (in our case, risk-assessment claims). But far from understanding this to preclude objectivity, an explicit recognition of how value judgments can influence the statistical risk assessments (in RAP) can—according to the metascientist—be used to interpret assessments more objectively. One way to recognize the policy influences and implications of RAP judgments is to evaluate their corresponding protectiveness for the case at hand.[9] This requires critical metascientific tools.

I shall now turn to the question of why a metascientific approach is more adequate than the alternatives just considered are. As is appropriate for appraising any postpositivist view, I shall judge its adequacy not by a priori arguments but by considering how well it would fare (in contrast with the other views outlined) in an actual risk-assessment controversy.

THE FORMALDEHYDE CONTROVERSY

The formaldehyde controversy at the EPA illustrates the problems that arose from the agency's politicization of science, and the attempted cure—the new separatism.

The RAP Controversy in Assessing Formaldehyde

In order to arrive at a risk assessment in the case of formaldehyde—that is, to determine whether and the extent to which formaldehyde increases the risk of cancer in humans—many of the key RAP judgments discussed earlier had to be made. This is an unusually instructive example of disagreements at each of these decision points and of policy controversies that erupted from such disagreements. My focus will be on the RAP option around which the bulk of the risk-assessment controversy centered: the importance of positive animal studies in the face of negative or inconclusive epidemiological studies. As is often the case with carcinogenic risk assessment, information from prospective randomized treatment-control experiments was available only on animals, that is, in the case of formaldehyde, rats. Epidemiological studies on humans, in contrast, allowed only a retrospective analysis of cancer rates in various occupations. On the basis of the statistically significant increases in (nasal) cancer among treated rats, the Chemical Industry Institute of Toxicology (CIIT) reached the assessment that formaldehyde is carcinogenic in laboratory rats and reported this to the Environmental Protection Agency in November 1980. A panel of eminent scientists convened by the National Toxicology Program confirmed this hazard assessment and concluded that "formaldehyde should be presumed to pose a risk of cancer to humans."[10] The lengthy document detailing the formaldehyde risk assessments was entitled the "Priority Review Level 1" (PRL-1) and dated February 1981.

On the basis of hazard assessments, the EPA staff reached the policy decision to designate formaldehyde as a priority chemical under the EPA provision known as 4(f). On May 20, 1982, the U.S. House of Representatives, Subcommittee on Investigations and Oversight of the Committee on Science and Technology, held a hearing on this matter. Its review, entitled *Formaldehyde: Review of the Scientific Basis of EPA's Carcinogenic Risk Assessment* (hereafter referred to as *Hearing*) concluded:

> EPA believes that formaldehyde has met the criteria for 4(f) for the following reasons. First, the results of a recently reported bioassay study demonstrate that formaldehyde is carcinogenic in rats. A National Toxicology Program (NTP) panel, which evaluated the 18-month data, concluded that formaldehyde should be presumed to pose a risk of cancer to humans. . . . Second, review of the available information on the use of formaldehyde and resulting human exposure suggests that large numbers of people are potentially exposed to harmful concentrations of formaldehyde. Accordingly, the Agency finds that there may be a reasonable basis to conclude that formaldehyde presents a significant risk of widespread harm to humans from cancer. (pp. 5–6)

This last sentence is important because statute 4(f) requires only that there may be a reasonable basis and not that there is a reasonable basis to conclude that a significant risk exists.[11] In itself it does not call for any regulation but is simply a call for closer scrutiny based on an indication that there may be a significant cancer risk.

Then there was a change in administration; the Reagan administration entered, and along with it a new EPA administrator, Anne Gorsuch, and some new staff. In fact, formaldehyde was the first 4(f) recommendation brought before the new administrator for signing. But instead of signing it she had members of the new EPA staff carry out a reassessment of the hazard data in the PRL-1. The new and revised version of the data became the Todhunter memorandum, named for John Todhunter, a new EPA assistant administrator. Some of the changes included blatant erasures of the highest-risk estimates that had been given in the PRL-1. The original document read, "For most of identified subpopulations, the estimated risks are equal to or greater than 1 in 10,000, however in some instances the risks are in the range of 1 in 10 . . . 1 in 100 and 1 in 1,000." The Todhunter memorandum cut out everything after "1 in 10,000" and reported, "For most of identified subpopulations, the estimated risks are equal to or greater than 1 in 10,000."[12] (At least one witness at the hearing (referred to earlier) that resulted testified that he quit rather than make these changes.) Among the other most important changes was deciding to de-emphasize the positive rat studies and to emphasize the negative epidemiological studies on humans. Todhunter concludes, for example, "There does not appear to be any relationship, based on the existing data base on humans, between exposure [to formaldehyde] and cancer. Real human risk could be considered to be low on such a basis" (*Hearing,* p. 260). This hazard reassessment was given as the basis of a changed policy. On September 11, 1981, the original EPA staff recommendation to designate formaldehyde as a 4(f) priority chemical was reversed, and the opposite policy choice was made. Whether or not this shift in hazard assessment was justified was the subject of an enormous controversy leading to the aforementioned congressional hearing on formaldehyde. As the title of the report of the hearing makes plain, it was intended as a "review of the scientific basis of the EPA's carcinogenic risk assessment." (It makes for fascinating reading.) A key question for the subcommittee, as its chairman, Senator Albert Gore, Jr., states, is, "To what extent has EPA and Dr. Todhunter departed from the long standing principles for carcinogenic risk assessment and given the wide acceptance of these principles is there an accepted scientific basis for such departure?" (*Hearing*, p. 3).

Science Politicized in the Gorsuch EPA

RAP judgments, in the Gorsuch EPA, overtly reflected the agency's predetermined policy—to hold a very stringent standard before deeming a risk to be significant. As Ashford, Ryan, and Caldart (1983) note in their insightful article on this case:

> EPA's formaldehyde deliberations powerfully illustrate the ease with which matters of policy may be confused with matters of science. The agency's technical analysis hides significant procedural deficiencies. Whether intentional or not, the result is an invidious one: the analysis purports to justify, in the name of science, a risk assessment policy far less protective of human health than the agency's prior policy. (p. 342)

How was the political agenda (antiregulation) masked as science? We can answer this by referring to our points concerning the RAP decisions. Each choice has implications for the protectiveness of the risk assessment (the chance that it will be considered a significant hazard to humans). With an EPA purged of scientists (save those leaning toward avoiding regulation), there was plenty of leeway for them to choose, consistently, the inference option most likely to have a less protective outcome. (In addition to giving less weight to the well-documented positive animal results and interpreting negative epidemiological studies as indicating low or no increased human risk, they endorsed other less protective choices in the formaldehyde assessment, such as holding to the existence of a threshold for carcinogenicity of formaldehyde, discounting benign tumors, and preferring maximum likelihood estimates over upper confidence level estimates.)

After all, if RAP judgments are viewed as part of the science of risk assessment—as they were in the Gorsuch EPA—then it is appropriate for agency scientists to choose among RAP options. And when the choice among options is not determined by "hard scientific fact," then each option may be presented as scientifically plausible. In his attempt to demonstrate that each of the RAP options in the formaldehyde dispute were scientifically plausible, Todhunter strove to document "expert scientific support" of the positions he favored. It was precisely this tendency, in this case and others, that led to the suspicion that "expert scientific support" was being enlisted to justify any position that the EPA favored. This suspicion of the EPA—and the need to avoid it—led to the formaldehyde hearings. The chairman, Senator Albert Gore, Jr., remarked in his opening statement: "We have witnessed . . . a belief, becoming widespread, that industry has special access to EPA and that EPA is becoming a captive to the industries it was established to regulate" (*Hearing,* p. 1).

One of the main reasons for this suspicion was that the new administration did not base its decision against a 4(f) designation on any new data beyond the PRL-1 document that had been the basis for the original, and opposite, recommendation to designate formaldehyde under 4(f)—though, as we mentioned, it did conceal some evidence supporting a 4(f) designation. Rather, the new administration proceeded to hold a series of secret meetings restricted only to certain scientists and lawyers from the Formaldehyde Institute, the Formaldehyde Trade Association, and the EPA staff. These meetings were claimed to consist solely of scientific discussions. In these "scientific" meetings the participants reinterpreted the data and came to a conclusion opposite to the one that had been reached and approved by numerous scientists and agencies. As one attorney with the Natural Resources Defense council (Jacquelin Warren) testified:

> There are no new data to support the reversal, only a reinterpretation which has been advocated by and is quite favorable to the interests of the formaldehyde industry. Those new assumptions, as we have heard, depart radically from accepted principles of cancer risk assessment. They lack a sound scientific basis and leave the public subject to cancer risks that . . . regulatory statutes were designed to protect against. In our view, this has been an effort to get the Government off the back of the formaldehyde industry. (*Hearing,* p. 188)

The usual peer review was absent in the reassessment they carried out, and Todhunter denied that any such review was called for.[13]

Todhunter's defense reveals most clearly how policy may be masked as science and why it might be expedient to hold a view of risk assessment that permits such masking. Todhunter and his supporters urged the view that each of the RAP options in the formaldehyde dispute was scientifically plausible and that the choices made—in support of not triggering 4(f)—reflected not an implicit bias in favor of an antiregulatory outcome (favorable to the Formaldehyde Institute) but, rather, legitimate scientific disagreement. Consider the following portion of an exchange between Robert Walker—who is defending Todhunter—and Senator Gore:

> *Mr. Walker:* . . . we cannot even come up with good enough science to decide whether or not we ought to ban this substance, so I think that there is plenty of room for some discussion about the science involved in all of this and that there can be a variety of interpretations about the science that is available.

> We have one scientist here who is disagreeing with other
> scientists. That is entirely acceptable within the science com-
> munity, or I did not realize that science was walking in lock
> step on all of these things.

Mr. Gore: ... Dr. Todhunter has stated ... that the epidemiological evi-
> dence indicates that there is no increased risk to human beings.
> There is not really much of an argument about that. There are
> two people who say that, Dr. Todhunter and the Formaldehyde
> Institute. The rest of science takes a contrary view.

Reluctant to accept this, Walker continues:

Mr. Walker: ... a study that the Du Pont Co. now has that indicates that
> they would agree with Dr. Todhunter's assessment, so there
> seems to be considerable disagreement within the scientific
> community. ... It seems to me that what we are into here is
> the politics of science and whose decision base you are going
> to use. (*Hearing*, pp. 137–38)

Walker, like the others defending Todhunter, is led in this last sentence to
claim that the choice of data and their interpretation are inevitably political.
(Only Todhunter, it seems, sticks to the position that his interpretation is free
of policy bias.) In the belief that such intermingling of science and policy is a
perversion of good science and of how science is supposed to operate at agencies
such as the EPA, the hearing and subsequent proposed reforms called for separation. As one scientist (Roy Albert) testifying at the formaldehyde hearings
put it: "It is absolutely essential to the soundness of regulatory decisions that
their underlying scientific assessments of health risks be shielded from the political forces that operate at the level of the regulatory offices" (*Hearing*, p. 36).

The sociological view, we saw, rejects such an ideal of neutrality, which
assumes that science can be separated from policy. But in taking this rejection
as grounds for the various sociological positions outlined earlier, I shall argue
that the sociological view effectively strengthens arguments that could be and
were offered in defense of the Gorsuch EPA.

The Sociological View and the Defense of the
Gorsuch EPA

In the sociological view, one cannot hope to appeal to science to arbitrate
between the judgment of scientists claiming that the evidence warrants a 4(f)
ruling and those such as Todhunter (and representatives of industry) who claim
that it does not. For in this view, the conflict is inevitably an *ad hoc* attempt to
justify a previously held social policy: All views are biased. If it is true that the
acceptability of an interpretation of data leading to risk assessments largely
reflects policy values, then the main ground for criticizing an interpretation is
that it leads to policies deemed unfavorable, such as triggering 4(f). For example,
to offer a criticism of the original assessment that formaldehyde may pose a
significant harm on the grounds that a 4(f) designation would be economically

undesirable (as lodged by the lobby for the Formadehyde Institute), becomes as legitimate a criticism of the assessment as is an argument against its evidential warrant. This view just endorses as inevitable the position that conflicts over hazard assessments are largely conflicts over competing policy values. But if disagreements over hazard assessments are primarily disagreements over policy values, there would seem to be little justification for the allegations of a number of scientists at the hearings that Todhunter's assessment was irresponsible or incompetent on objective scientific grounds—for example, on grounds that it misconstrues what the data say about the actual extent of cancer risk. Indeed, the existence of objective grounds for criticism is what the sociological view wishes to deny.

Unsurprisingly, but of interest to us, this was precisely the argument given by those defending Todhunter's assessment against such allegations of incompetence and invalidity. The committee's minority member, Walker, expressed vehement opposition to the hearings altogether because they were framed as dealing with a scientific issue, whereas in his opinion "what we are dealing with here is a regulatory issue" (*Hearing,* p. 73). If this automatically entails that disagreements simply reflect differences in subjective policy values, one can perhaps understand Walker's portraying the hearings as "an obvious result of the efforts of a few disloyal and disgruntled employees of the EPA who" simply because they disagree with the decision not to place formaldehyde under 4(f) "feel justified in waging guerrilla warfare against the Agency and those in positions of authority" (*Hearing,* p. 4).

One of the few scientists who defended Todhunter's assessment against charges of poor science, Sorell Schwartz, did so by denying that there was anything scientific in the Todhunter memo. If it is not science, it cannot be poor science. The exchange between him and the chairman of the subcommittee, Senator Gore, is interesting:

Mr. Gore:	Do you accept the scientific judgments in the Todhunter February 10 memo? Do you think it is good science?
Dr. Schwartz:	I have not read anything in the Todhunter memo—
Mr. Gore:	Oh—
Dr. Schwartz:	No, no. I have read the Todhunter memo. However, I have never read anything in the Todhunter memo which to me involves science. I think what Dr. Todhunter did was act as a nonscientist, and that is to make a determination whether the risk as presented by the data was significant enough to be unacceptable under 4(f).
Mr. Gore:	That is a little odd. He is the chief scientist responsible for the EPA toxics program and the agency published this document and released it to the public as the scientific justification for the decision. It discusses scientific judgments throughout. How can you say that it is written as a nonscientist, as a nonscientific document? (*Hearing,* p. 239)

When pressed, Schwartz admitted to disagreeing with the memo's statements on level of risk. He declared, for example, "I do not favor the idea that negative epidemiologic data signifies negative human effect." But he defended the Todhunter memo on the grounds that it dealt with "not the level of risk but the acceptability of that risk" (p. 239). However, Schwartz's defense misconstrues 4(f). The 4(f) priority unambiguously states that it is not about risk acceptability—which explicitly takes benefits into consideration (*Hearing*, pp. 491, 773). Rather, it is to indicate only whether there may be a reasonable basis for concluding that a substance presents a significant risk of human harm. The few other defenses of the Todhunter memo were also based on equating evidential issues with those of policy preferences. This seems to illustrate the rather disturbing point of Roberts, Thomas, and Dowling (1984): "Too many of the participants have good reasons *not* to distinguish scientific evidence from policy preferences, *not* to analyze carefully the various sources of technical disagreement and *not* to accept responsibility for some decisions or judgments" (p. 120).

The sociological view, we have seen, offers no ammunition with which to fight the type of politicization of science characteristic of the Gorsuch EPA. Unwittingly or not, it offers a basis for defending the status quo of that period. Although it is unclear whether holders of the sociological view would take this as a weakness of their view, it is clear that those proposing the new separatist reforms were striving to alter the status quo and to help ensure that risk assessment was free of political bias. Thus, if it turns out that separatist reforms also permit politicization to go unchecked, the reforms will have failed to reach their goal.

The new separatism recognizes that deciding which way to interpret data involves policy, that is, RAP judgments. Thus it places such decisions under the policy rubric. For example, after giving the scientific report of the positive rat studies and the negative epidemiological studies in the formaldehyde case, the RAP decision as to how to weigh them falls under policy. In this way, the sociological view, based on a general challenge to scientific rationality, undermines both the importance of and the ability to discern the actual physical risks. The new separatist reforms do the same, not by means of any general philosophical arguments, but by allowing the RAP choices to be made by nonscientists who can not or do not articulate the implications for protectiveness of the given inference option. This is the basis for the metascientific criticism of the new separatism.

The Metascientific (or Metastatistical) Approach

How would the metascientific approach deal with this case? The metascientific view, we said, denies premise (P): It denies that the inability to divorce science from policy entails the inability to adjudicate objectively among risk assessments. The metascientist grants that the judgments required to reach risk assessments may reflect policy values, conventions, and the like and that political interests may be adduced to explain particular choices of RAP judgments (as the formaldehyde case clearly shows). Nevertheless, the metascientist holds that the question whether a given risk assessment is warranted by the evidence is not a matter of social and political values; it is a matter of what the risk actually is.

One may hold, in other words, that what counts as "good science" in a given context may in fact be negotiated, while still maintaining that whether evidence and inferences based on that evidence are acceptable are not negotiable. Instead, the acceptability of evidence is constrained by the extent to which it actually warrants given risk inferences based on it. The question of whether a risk inference is warranted is a question of how well it reflects what is really the case about the causal effect of the risky substance or the practice in question. The latitude in choosing RAP options does not preclude the objective scrutiny needed to answer this question.

An analogy with a weighing instrument is useful. My interest in whether I have gained as little as one-tenth pound may be a matter of my subjective values. But whether a scale with a digital readout in whole pounds, say, is a good tool for finding this out is not a matter of my subjective values. Given the scale chosen, whether or not a gain is detected depends on how much I have actually gained! And understanding the scale enables one to determine (at least approximately) what a given reading indicates about how much this is. Analogously, a critical understanding of tools used for estimating risks enables an understanding of the actual extent of risk that is or is not indicated by a given piece of evidence. This in turn enables one to determine the protectiveness of a given RAP judgment. It allows one to answer the question: According to the standard being required, what extent of risk must be fairly clearly indicated before it is taken to point to a significant human risk? Answering this question requires critical metascientific tools. What sort of tools can accomplish this in the case of formaldehyde?

My focus is on the RAP judgment in interpreting negative epidemiological results. One of the main questions raised at the hearings was: Does a failure to find a positive result (i.e., a statistically significant increase in risk) indicate that there is little or no risk? Todhunter and the Formaldehyde Institute say yes—that the negative epidemiological results are evidence that there is no increased risk to human beings. Yet Todhunter's own epidemiologist on the staff responsible for this work, wrote: "Before leaving [the EPA], I would again like to emphasize that the available epidemiologic data from studies on formaldehyde exposure are inconclusive and not supportive of no association, as purported by the Formaldehyde Institute" (*Hearing,* p. 137). What is the nature of this dispute? Only by understanding the principles of the statistical reasoning involved were the critics able to tell. What was pivotal was understanding how to interpret negative statistical results, as will be generally true in judicial (or other) reviews of decisions not to regulate or prioritize. But lacking this understanding, the courts may be unable to decide correctly between conflicting "expert" assessments. In order to rectify this situation the metascientist urges that we clarify what the disagreement really amounts to, that we go back and unearth what negative results do and do not warrant. Even without seeking an algorithm for risk assessment in general, understanding the implications of RAP judgments does admit systematization. In our example, what is called for is a way to systematize the reasoning for criticizing certain uses of statistical instruments (which unfortunately lend themselves to *un*critical use).

Neyman–Pearson (NP) Tests

The type of statistical test standardly used in reaching risk assessments is the Orthodox or Neyman–Pearson test (NP test). In the formaldehyde case, the hypotheses tested are assertions about a parameter that I will call Δ, the increased cancer risk in the population of humans. The NP test splits the possible parameter values into two: one representing the test (or null) hypothesis H and the other the set of alternative hypotheses J. The test hypothesis H in the formaldehyde case asserts that formaldehyde does not cause an increase in a person's risk of dying from cancer of a given type. That is, it asserts that there is a zero increase in the risk rate: $\Delta = 0$. The alternative hypotheses assert that formaldehyde causes a positive increase: $\Delta > 0$. Because we are looking for positive discrepancies from zero, this is a one-sided test, which I call test T+:

> TEST T+: Test (null) hypothesis H asserts $\Delta = 0$ (no increased risk),[14] and alternative hypothesis J asserts $\Delta > 0$ (a positive increased risk).

The test considers a test statistic that describes an aspect of the outcome of interest. One statistic in testing formaldehyde is D, the difference in cancer rates between the subjects exposed and those not exposed to formaldehyde:

> TEST STATISTIC D: the difference in cancer rates between the subjects exposed and those not exposed to formaldehyde.

Corresponding to each observed difference, D_{obs}, is its level of statistical significance, defined as follows:

> The *statistical significance level* of an observed difference, D_{obs}, is the probability that so large a difference arises, assuming that the null hypothesis H is true; $\mathrm{Prob}(D \geq D_{obs}$, given that H is true).

A good way to see significance levels is as standard measures of distance from H, except with this inversion: The larger (and more significant) the distance is, the smaller its significance level will be.

An NP test consists of a rule that specifies, before the observation is made, how statistically significant (i.e., how improbably far) an observed difference must be before it should be taken to reject H. The maximum significance level chosen beyond which D_{obs} is taken to reject H is called the *size* of the test and is denoted by α. Thus, test T+ with size α consists of the following rule:

> TEST T+ (with size α): Reject H if, and only if, the observed difference D_{obs} is statistically significant at level α.

Observed differences that are not large enough to reach this preset size are taken to accept H. These are *negative results*. In this way the test maps the possible

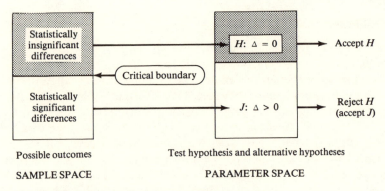

Figure 12.1 Neyman–Pearson Test T+ as a mapping rule.

outcomes—the *sample space*—into either reject *H* (and accept *J*) or accept *H*. The partitioning that results from the test is shown in Figure 12.1.

As long as there is variability in the effect (e.g., not all who are exposed get cancer, and not all who get cancer are exposed) and as long as only a sample from the population is observed, there is a chance that the test will make an error. Two types of errors are considered: First, the test rejects *H* (accepts *J*), even though *H* is true (a Type I error); second, the test accepts *H*, even though *H* is false (a Type II error). A test with size α rejects *H* when in fact *H* is true—that is, it commits a Type I error—with a probability of no more than α. The smaller the test's size α is, the less frequent the Type I error will be. But by making α smaller, the test suffers an increase in the frequency with which it accepts *H*, even when in fact *H* is false (and so should be rejected)—that is, an increase in the frequency of a Type II error. The probability of a Type II error is denoted by β, and α and β are the test's *error probabilities*.

Because these two error probabilities cannot be simultaneously minimized, the NP model instructs one to first fix α, the size of the test, at some small number, such as .05 or .01. (In other words, the test is specified so as to ensure that it is very improbable for the test to reject *H* when the hypothesis *H* is true.) One then seeks out the test that at the same time has a small β; thus, $1 - β$ is the corresponding *power* of the test. Because in our case, alternative hypothesis *J* contains more than a single value of the parameter; that is, it is *composite,* the value of β varies according to which alternative in *J* is assumed to be true. I shall refer to a specific simple alternative hypothesis as Δ′.

When used as automatic procedures for rejecting or accepting substantive hypotheses, NP tests result in common misinterpretations, such as automatically equating rejections of *H* (statistically significant differences) with finding substantively important discrepancies from *H*, and automatically equating failures to reject *H* with finding zero or unimportant discrepancies. To avoid misinterpretations we need a more critical understanding of the statistical results. My focus will be on interpreting failures to reject *H*, that is, negative results, as that was the main issue in the formaldehyde case. (For a discussion of how to interpret positive results, see Mayo, 1988.)

How to Tell the Truth (About Negative Results)
with Metastatistics

Ideally, the policy question of what counts as a substantively important increase in cancer risk is answered at the start, so that the test may be specified in order to have appropriately high probabilities of detecting all and only those increases. Substantively important increases in the formaldehyde case are those increases in cancer risk deemed serious enough to trigger 4(f)—a policy judgment. However, regardless of how the test has been specified—whether based on policy or other values—knowledge of the test's error probabilities, I claim, allows interpreting objectively what the data do and do not indicate about the increased risk—that is, about Δ. The way in which they may be used to this end involves reasoning that is both obvious and familiar.

It can be shown by the analogy with a weighing instrument. The null hypothesis H may be that I have gained no weight since I weighed in last week, say at 125 pounds. To test H, suppose I use this method: I weigh myself on a digital readout scale that expresses weight in whole pound units, and reject H only if a difference is registered. Suppose the result turns out to be the same weight in pounds as last week (125). Am I warranted in concluding I have not gained any weight, even as little say, as one-tenth pound? Because my method had very little chance of detecting such a small increase, even if I had gained it, this negative result would be poor grounds for thinking I had gained no more than one-tenth pound. (My scale is too insensitive.) On the other hand, this negative report is a good indication that I have not gained as much as a full pound. For had I gained a pound, then it is likely that my scale would have registered some gain. Moreover, my negative reading is an even better indication that I have not gained as much as five pounds or more. A simple principle emerges:

> A failure to observe a difference in weight only indicates that my actual weight gain is less than x if it is very probable that the scale would have registered a larger difference in weight, if in fact I have gained as much as x pounds.

This leads to a precisely analogous metastatistical principle for interpreting a negative statistical result, which I label rule (M):

> RULE (M): A failure to observe a statistically significant difference only indicates that the actual increase is less than Δ' if it is very probable that the test would have resulted in a more significant difference than was observed, were the actual increase as large as Δ'.

To apply rule (M), we need a way to calculate the probability that a test T would have resulted in a more significant difference than was observed, were the actual increase as large as Δ'. Error probabilities allow such a calculation. Let us call this probability the *severity of the test* (with a given result) against an alternative Δ'. Focusing on the test T+ in our formaldehyde example, this becomes

> *The severity of (negative) result D_{obs} against an alternative Δ' equals the probability that T+ would have resulted in a difference greater than D_{obs}, were the actual increase as large as Δ'.*[15]

Rule (M) then becomes

> RULE (M): A statistically *in*significant result, D_{obs}, only indicates that the actual increase is less than Δ' if the severity of the result against Δ' is *high*.

It follows from (M) that failing to reject H does not rule out increases as large as Δ' if there is a small probability of getting a more statistically significant result even if the increased risk were as large as Δ'. For in that case, the severity against Δ' is low.

Let us apply rule (M) to one of the negative results that Walker and others cited in defense of the Todhunter interpretation (*Hearing*, pp. 137–38): the Du Pont study. In a mortality study of Du Pont workers, the relative risk of dying from cancer among those in the study exposed to formaldehyde was not statistically significantly greater than among those not so exposed: The null hypothesis H was not rejected.[16] Du Pont concluded that "the data suggested that cancer mortality rates in the company's formaldehyde exposed workers were no higher than the rates among nonexposed workers" (*Hearing*, p. 284). They are inferring, in other words, that the increased risk Δ equals zero. The error in such an interpretation is that the failure to reject the null hypothesis of zero-increased risk is not the same as having positive evidence that the increased risk is zero. For such negative results may be common (i.e., probable) even if the underlying increase in risk is greater than zero. In fact, the Du Pont study had a very small chance of rejecting null hypothesis H even if the actual increase in risk had exceeded zero by substantial amounts—that is, the severity against these alternatives is low. Hence, failing to reject H does not rule out these increases.

For example, the Du Pont study had only a 4 percent chance of rejecting H even if there were a twofold increase in cancer of the pharynx or of the larynx in those exposed to formaldehyde. Thus, failing to reject H does not rule out twofold increases in these types of cancers. The situation was even worse with nasal cancers and not much better with the others. This is indicated in the following chart adapted from a review of the Du Pont study by the National Institute for Occupational Safety and Health (NIOSH) (*Hearing*, p. 548):

	Lung	Pharynx	Larynx
Number of cases	181	7	8
Power to detect odds ratio = 2*	37%	4%	4%
Least significant odds ratio detectable†	2.9	57.5	42.5

*Assumes α = .05 (1 tail)
†Assumes α = .05 (1 tail) and power $(1 - \beta)$ = .80

Although I recommend interpreting tests by considering the severity function rather than the usual NP power function, as used in this chart, this does not alter our current point because high power entails high severity.[17]

Although failure to reject does not indicate that the increase is zero, it does permit an inference about the likely upper bound of the unknown increase Δ. That is, a failure to reject H provides a reason to say that the data provide good grounds for asserting that the increased risk is no greater than such and such (upper bound). To find plausible upper bounds requires determining the increase (i.e., the value of Δ) against which the observed difference does have high severity.[18] The NIOSH chart serves this purpose (converting talk of risk ratios into differences).

In the Du Pont study, the test had a fairly high probability (.8) of rejecting H if the risk of lung cancer were three times higher among exposed than unexposed workers. Hence, a failure to reject H does indicate the increased risk of lung cancer is not as high as threefold. The risk of cancer of the larynx would have to be forty-two times higher among exposed than unexposed workers in order for the test to have a fairly high probability (.8) of obtaining a statistically significant difference (rejecting H). A failure to reject H, then, does indicate that the actual increased risk is not as high as forty-two-fold. For cancer of the pharynx, there would have to be a fifty-seven-fold increase in cancer risk before the test could have a good chance of rejecting H. So, even ignoring some methodological difficulties with the study, its negative statistical results at most indicate that the various cancers are no more than three or forty-two or fifty-seven times as likely among workers exposed to formaldehyde. They clearly do not warrant the conclusion reached by Du Pont and others that the study supports the claim of no increase in (relative) cancer risk among formaldehyde workers. The study does not even support the claim of low increased risk, given what Todhunter himself claimed the EPA counted as low.

Relevance to Assessing the Acceptability of the Evidence

The main purpose of the formaldehyde hearing was not the policy question of how high an increased risk may be before categorizing it as significant enough to trigger 4(f). It was to decide whether the decision not to prioritize formaldehyde represented a shift in the principles being used to reach this categorization. Doing so required understanding whether the evidence was acceptable for the assessments reached. It is necessary to ascertain the implications for protectiveness of the RAP judgments made. Metastatistical reasoning of the sort in rule (M) is the basis for doing so.

Thus (M) may be used to ascertain the approximate bound that the negative result warrants ruling out. This may be used to determine whether the standard of protectiveness being employed is in accordance with what agencies or individuals deem tolerable. (M) also allows one to compare this calculated estimate with the upper-bound risk associated with a different substance. If data on the latter are found to indicate an increased risk no greater than the former, and yet the latter leads to a different risk-management decision, then there are likely to be specific differences in policy values effectively operating in the two cases. These indeed were found. For example, applying the reasoning embodied in (M) showed that the epidemiological data could not rule out there being a risk of cancer as great as or greater than 1 in 100,000. As one scientist (Warren) testified:

> By comparison, the Consumer Product Safety Commission banned urea formaldehyde foam insulation based on an estimated risk of 10^{-5} (one in 100,000). That the EPA would disparage as insufficient merely to justify a closer look at formaldehyde [via 4(f)], risks others consider sufficient to warrant a ban is troubling indeed. (*Hearing,* p. 198)

Because limited sensitivity (or severity) is common in epidemiological studies, positive results are typically not thought to be required to indicate that a substance may pose a significant risk to humans,[19] particularly when, as with formaldehyde, the substance is shown to cause cancer in rats and mice at doses reasonably comparable to those to which people may be exposed. Using the reasoning in rule (M), a number of scientists concluded: "The EPA decision on formaldehyde may constitute an *ad hoc* revision of the principles for assessing carcinogenic risk that have been widely accepted by the scientific community for over a decade" (*Hearing,* p. 179).

Todhunter denies this alleged shift in the agency's policy values for managing carcinogens. For example, Todhunter does not disagree with the general range of risk that agencies have tended to deem of public concern. Nevertheless, he claims formaldehyde does not meet the criteria of section 4(f) for the following reasons: "There is a limited but suggestive epidemiological base which supports the notion that any human problems with formaldehyde carcinogenicity may be of low incidence or undetectable. . . . [The ranges of risks] are of from low priority to no concern" (*Hearing,* p. 253). But even according to Todhunter's own interpretation of this range, to interpret the (negative statistical) data as indicative of a low incidence is, as rule (M) makes clear, to misinterpret it. Granted, the increased hazard may be undetectable with the tests used, but to take that as grounds against a 4(f) designation is to require positive epidemiological results before even recommending that the EPA take a closer look at a substance— which is a shift in the existing policy. Todhunter gives as the second reason for the decision: "There is suggestive evidence that there may be human exposure situations—which may not present carcinogenic risk which is of significance." He is surely correct about this, as it is always true that "there may be human exposure situations which may not present carcinogenic risk" of significance, namely, a zero or extremely low exposure. Were one to take this reason seriously, it seems one could always argue against a 4(f) designation. But this is to play a logical trick on the 4(f) wording. For, as the statute clearly states, a reasonable basis to conclude that exposure presents a significant risk is sufficient to trigger 4(f). Yet Todhunter's second reason is tantamount to construing the 4(f) designation as appropriate only if there is no exposure that does not present a significant risk! Only then would it make sense to say that finding a single exposure that does not present a risk is sufficient *not* to trigger 4(f). It is no wonder that a number of scientists concluded—using considerations identified in rule (M)—that "in order to justify its failure to address formaldehyde under 4(f)...EPA has rewritten both the science and the law" (*Hearing*, p. 195).

Because of the criticisms of the science underlying the EPA risk assessment, in 1985 the EPA finally did place formaldehyde under the 4(f) category, and the NAS–NRC report guidelines emphasized the importance of using power considerations in interpreting negative results. Additional metastatistical rules can be formulated for other types of RAP judgments. The appropriateness of choosing different extrapolation models awaits further biological knowledge, but here too, definite progress has been made, at least for carcinogenic assessments. Despite this progress, it would be erroneous to suppose that we have gotten past the sort of problems that arose in the formaldehyde episode. One thing is certain: Progress in this direction will be hampered if nonseparatists of the metascientific stripe continue to be confused with those of the sociological stripe. For according to the latter view, such metascientific rules cannot help.

CONCLUSION

A major theme running through interdisciplinary discussions on risk is that risk assessment cannot and should not be separated from societal and policy values. However, under the banner of favoring nonseparatism are two radically different views of risk assessment—something that the literature has not explicitly recognized. Typically, all such calls for nonseparation have been taken as evidence for holding, to some degree, what I have termed the sociological view. The untenability of strictly separating the science of assessment from the social and policy values of risk management is typically taken as grounds for denying that scientific methods can provide unbiased risk assessments or can adjudicate objectively between conflicting assessments—that is, for what I call the sociological view. But to lump all nonseparatist positions under the sociological view, I have argued, is a mistake. Other nonseparatists take a different position, which I have called the metascientific view. In contrast with the sociological view, its calls for nonseparation are accompanied by calls for greater scientific and methodological understanding—albeit an understanding that allows for a critical or metascientific scrutiny of the uncertainties involved. By implicitly holding the overly stringent, old image conception of scientific objectivity—one that is precluded by the need to make uncertain judgments without algorithms—adherents to the sociological view are led to deny the attainability or importance of objective assessments of physical risks. As such, what matters most is not whether or not a view espouses separatism, but whether it adheres (implicitly or explicitly) to the old image of science or, alternatively, sets the stage for a new or postpositive image, in which metascientific scrutiny—not algorithms—can appraise objectively the adequacy of risk assessments.

Our case study focused on a metascientific scrutiny of risk-assessment policy (RAP) judgments as they arise in statistical tests of the existence of increases in risk. According to the sociological view, statistical methods in the arena of RAP judgments are tools for manipulation rather than instruments for an unbiased adjudication of conflicting risk assessments. Conflicting risk assessments are seen as largely, if not solely, conflicts over policy values or over different "ways of life." Ironically, this entails an extreme subjectivism or relativism that undercuts the *raison d'être* of the sociological view for many of its adherents: to hold risk assessors and managers accountable to various societal values. As the metascientific view stresses, only by means of a critical understanding of the uncertainties underlying RAP judgments is it possible to distinguish adequately what is warranted by the evidence and what is prejudged by policy values. Thus, to exclude or downplay critical scientific scrutiny from this arena is to forfeit a crucial tool for holding risk policymakers accountable to the degrees of protectiveness deemed acceptable by society.

By concentrating on acceptable evidence—on the entry of values in collecting, interpreting, communicating and evaluating the evidence of risks—this volume attempts to avoid two extremes that have been barriers to progress in this area. The first extreme supposes that issues of evidence of risk can and should be separated from the values that necessarily are part of policies regarding risk (i.e., risk management). The second extreme sees the untenability of such a separation as implying that there is little objective or empirical basis on which to criticize risk assessments—that is, in my terminology, the sociological view. This chapter and, indeed, this volume show that both extremes share the same consequence: Both thwart the goal of determining the acceptability of evidence. We argue instead for a more constructive conclusion, that understanding the interrelations of scientific with value issues enables a critical scrutiny of risk assessments. With further constructive work in this direction, we may begin to make progress in developing criteria for judging the acceptability of evidence on which policy decisions are based.

ACKNOWLEDGMENT

A portion of this research was carried out during the tenure of a National Endowment for the Humanities Fellowship for College Teachers; I gratefully acknowledge that support. For many useful comments on earlier drafts, I thank Marjorie Grene and Harlan Miller. I am grateful to Rachelle Hollander for numerous discussions that helped me clarify the ideas presented here.

NOTES

1. It may also be used more narrowly to include only strictly quantitative risk estimates.

2. For a good account of the practices of the EPA and other agencies during the Reagan administration, see Lash, Gillman, and Sheridan (1984).

3. For an excellent delineation of these positions, see Shapere (1986).

4. They go on to give excellent criticisms of the arguments for social relativism and social reductionism.

5. These are nonscientific, it should be kept in mind, in the traditional old image of science.

6. Many of these fifty-one or so components are straighforwardly statistical. If only in order to ascertain the consequences of adopting one or another answer or choice, one uses statistical considerations.

7. The NAS–NRC report had intended reforms far more amenable to the metascientific view than what actually came to pass. For example, the report stresses:

> The importance of distinguishing between risk assessment and risk management does not imply that they should be isolated from each other; in practice they interact, and communication in both directions is desirable and should not be disrupted. (p. 6)

> Although we conclude that the mixing of science and policy in risk assessment cannot be eliminated, we believe that most of the intrusions of policy can be identified and that a major contribution to the integrity of the risk assessment process would be the development of a procedure to ensure that the judgments made in risk assessments, and the underlying rationale for such judgments, be made explicit. (p. 49)

8. What is more, as Silbergeld notes (Chapter 5, this volume), it has led to such sociological tactics as focusing on differences of opinions about risk rather than the original disagreements, which are not about opinions but evidential uncertainties.

9. Some options, such as choosing to count positive animal studies as indicative of human risks, will always be more protective than will requiring positive human results, say. But the protectiveness of other options will depend on the substance being considered.

10. Among those judging the CIIT assessment valid were the National Toxicology Program, the Mt. Sinai School of Medicine Environmental Cancer Information Unit, and the International Agency for Research on Cancer (IARC). Ashford, Ryan, and Caldart (1983) provide a good discussion in support of the validity of this assessment, in contrast with an assessment by the Consumer Product and Safety Commission.

11. Section 4(f) provides that "Upon the receipt of—

> (1) any test data required . . . or
> (2) any other information available to the Administrator, which indicates to the Administrator that there may be a reasonable basis to conclude that a chemical substance or mixture presents or will present a significant risk of serious or widespread harm to human beings from cancer, gene mutations, or birth defects, the Administrator shall, within the 180-day period beginning on the date of the receipt of such data or information, initiate appropriate action . . . to prevent or reduce to a sufficient extent such risk or publish in the Federal Register a finding that such risk is not unreasonable. (*Hearing*, pp. 476–77)

An excellent discussion of the implications of and requirements under 4(f) occurs in the Primer on 4(f), *Hearing*, pp. 475–523.

12. I have replaced the exponential notation in the original document with equivalent fractions here. For documentation of this and several other of Todhunter's revisions, see *Hearing*, pp. 349–65.

13. Todhunter claimed additional review was not needed because the data had been well reviewed already by the many scientists who concurred with the original PRL-1. The fact that his reassessment of these same data was now being used to support the opposite policy on 4(f) did not seem to faze him.

14. The same test would be used were the null hypothesis to assert that $\Delta \leq 0$.

15. That is, the severity of observed difference D_{obs} against the alternative hypothesis that $\Delta = \Delta'$ equals Prob $(D \geq D_{obs}$, given that $\Delta = \Delta')$, for the given test $T+$.

16. In the Du Pont study, 481 cancer deaths among male employees between 1957

and 1979 constituted the cases. These were matched on relevant factors with controls who did not die of cancer. The statistic observed was the relative odds ratio, the ratio of the odds of having been exposed to formaldehyde among cases and controls. For simplicity, I refer here to the risk rather than the relative risk.

17. The difference between power and severity is that although severity is a function of the particular observed difference D_{obs}, power is a function of the smallest difference judged significant by a given test. Let D^* be the smallest difference test T+ judges significant (i.e., D^* is the *critical boundary* shown in Figure 12.1 beyond which the result is taken to reject H). Then, power is defined as follows:

> The *power* of test T+ against alternative $\Delta = \Delta'$ is equal to the probability of a difference as large as D^*, given that $\Delta = \Delta'$.

Severity, in contrast, substitutes D_{obs} for D^*. The advantage of the severity function, I claim, is that it affords an understanding that reflects the difference that has actually been observed.

18. But how high, it might be asked, must the degree of severity be? We can get a feel for the increase indicated by using benchmarks such as .8, .9, or .95. But by interpreting tests along the lines suggested in (M), the use of statistical tests should no longer be a matter of prespecified error probabilities alone. Instead, we can understand what the actual negative result D_{obs} indicates more or less well by calculating all (or several) of the upper bounds for different degrees of severity. This would yield *severity curves*. Although each upper bound of a given degree of severity is mathematically equivalent to formulating the upper confidence bound at the corresponding level of confidence, the difference is that not all values within the interval are treated equally. It most closely corresponds to forming a series of upper confidence intervals, one for each confidence level. For further discussion of this relationship, see Mayo (1985a). Poole (1987) uses what are essentially severity curves in interpreting statistical results. Such curves are also employed by Kempthorne and Folks (1971).

19. As the formaldehyde options report notes:

> Generally, even the largest and most expertly performed epidemiological studies can seldom detect increases in cancer risk of less that 10% (1 in 10). However, from the public health standpoint, exposures to carcinogens may be considered problems if they increase the risk of cancer by 1 case in 1000 exposed persons or less. (*Hearing*, p. 763)

For a discussion of a study performed to document the problem of insufficiently powerful tests, see Freiman et al. (1978).

REFERENCES

Ashford, N. A., Ryan, C. W., and Caldart, C. C. (1983). "A Hard Look at Federal Regulation of Formaldehyde: A Departure from Reasoned Decisionmaking." *Harvard Environmental Law Review* 7:297–370.

Douglas, M., and Wildavsky, A. (1982). *Risk and Culture*. Berkeley: University of California Press.

Feyerabend, P. K. (1975). *Against Method*. London: New Left Books.

Freiman, J. A., Chalmers, T. C., Smith, Jr., H., and Kuebler, R. R. (1978). "The Importance of Beta, the Type II Error and Sample Size in the Design and Interpretation of the Randomized Control Trial, Survey of 71 'Negative' Trials." *New England Journal of Medicine* 299:690–94.

Funtowicz, S. O., and Ravetz, J. R. (1985). "Three Types of Risk Assessment: A Methodological Analysis." In *Risk Analysis in the Private Sector,* ed. C. Whipple and V. T. Covello, pp. 217–31. New York: Plenum.

Hamlin, C. (1986). "Scientific Method and Expert Witnessing: Victorian Perspectives on a Modern Problem." *Social Studies of Science* 16:485–513.

Kaplan, S., and Garrick, B. J. (1981). "On the Quantitative Definition of Risk." *Risk Analysis* 1:11–27.

Kempthorne, O., and Folks, L. (1971). *Probability, Statistics, and Data Analysis.* Ames: Iowa State University Press.

Kuhn, T. S. (1962). *The Structure of Scientific Revolutions.* Chicago: University of Chicago Press.

Lash, J., Gillman, K., and Sheridan, D. (1984). *A Season of Spoils: The Reagan Administration's Attack on the Environment.* New York: Pantheon Books.

Laudan, L. (1989). "The Sins of the Fathers . . . Positivist Origins of Post-Positivist Relativisms." Unpublished manuscript.

Lynn, F. M. (1986). "The Interplay of Science and Values in Assessing and Regulating Environmental Risks." *Science, Technology, & Human Values* 11:40–50.

Mayo, D. (1985a). "Behavioristic, Evidentialist, and Learning Models of Statistical Testing." *Philosophy of Science* 52:493–516.

Mayo, D. (1985b). "Increasing Public Participation in Controversies Involving Hazards: The Value of Metastatistical Rules." *Science, Technology, & Human Values* 10:55–68.

Mayo, D. (1988). "Toward a More Objective Understanding of the Evidence of Carcinogenic Risk." In *PSA 1988,* vol. 2, ed. A. Fine and J. Leplin, pp. 489–503. East Lansing, Mich.: Philosophy of Science Association.

National Academy of Sciences–National Research Council (NAS–NRC) (1983). *Risk Assessment in the Federal Government: Managing the Process.* Washington, D.C.: National Academy Press.

Poole, C. (1987). "Beyond the Confidence Interval." *American Journal of Public Health* 77:195–99.

Roberts, M., Thomas, S., and Dowling, M. (1984). "Mapping Scientific Disputes That Affect Public Policymaking." *Science, Technology, & Human Values* 9:112–22.

Ruckelshaus, W. (1983). "Risk, Science and Public Policy." *Science* 221:1026–28.

Rushefsky, M. E. (1984). "The Misuse of Science in Governmental Decisionmaking." *Science, Technology, & Human Values* 9:47–59.

Russell, M., and Gruber, M. (1987). "Risk Assessment in Environmental Policymaking." *Science* 236:286–90.

Shapere, D. (1986). "External and Internal Factors in the Development of Science." *Science and Technology Studies* 4:1–9.

U.S. Congress. House of Representatives. Committee on Science and Technology. *Formaldehyde: Review of Scientific Basis of EPA's Carcinogenic Risk Assessment* (1982). Hearing before the Subcommittee on Investigations and Oversight. 97th Cong., 2d sess. May 20.

Weinberg, A. (1972). "Science and Trans-Science." *Minerva* 10:209–22.

Wynne, B. (1982). "Institutional Mythologies and Dual Societies in the Management of Risk." *The Risk Analysis Controversy: An Institutional Perspective,* ed. H. C. Kunreuther and E. V. Ley, pp. 127–43. Berlin: Springer-Verlag.

Contributors

E. William Colglazier, Ph.D
Department of Physics
Director, Energy, Environment,
    and Resources Center
University of Tennessee
Knoxville, Tennessee

Vincent T. Covello, Ph.D
Center for Risk Communication,
    Schoolof Public Health
Columbia University
New York, New York

Ronald N. Giere, Ph.D
Department of Philosophy
Center for Philosophy of Science
University of Minnesota
Minneapolis, Minnesota

Rachelle D. Hollander, Ph.D
Coordinator, Ethics and Value Studies
National Science Foundation
Washington, D.C.

Sheila Jasanoff, J.D., Ph.D
Program on Science, Technology,
    and Society
Cornell University
Ithaca, New York

Jeanne X. Kasperson, M.A., M.L.S.
Alan Shawn Feinstein World
    Hunger Program
Brown University
Providence, Rhode Island

Center for Technology, Environment,
    and Development
Clark University
Worcester, Massachusetts

Roger E. Kasperson, Ph.D
Director, Center for Technology,
    Environment, and Development
Clark University
Worcester, Massachusetts

Deborah G. Mayo, Ph.D
Department of Philosophy
Virginia Polytechnic Institute and
    State University
Blacksburg, Virginia

Valerie Miké, Ph.D
Clinical Professor of Biostatistics
    in Public Health
Cornell University Medical College
New York, New York

Peter M. Sandman, Ph.D
Department of Environmental Journalism
Rutgers University
New Brunswick, New Jersey

Kenneth F. Schaffner, M.D., Ph.D
Department of History and Philosophy
    of Science
University of Pittsburgh
Pittsburgh, Pennsylvania

Kristin Shrader-Frechette, Ph.D
Graduate Research Professor
    of Philosophy
University of South Florida
Tampa, Florida

Ellen K. Silbergeld, Ph.D
Department of Toxicology
University of Maryland
Baltimore, Maryland

Paul Slovic, Ph.D
Research Associate, Decision Research
Department of Psychology
University of Oregon
Eugene, Oregon

Index